Smart and Digital Cities

Fangfen Liu

Smart and Digital Cities

Innovative Design Practices and Future Exploration of New-Type Smart Cities

Springer

Fangfen Liu
Gloryview Technology Co., Ltd.
Guangzhou, Guangdong, China

This work was supported by Key R&D Program of the Department of Science and Technology of Guangdong Province (2019B020214001), Major Science and Technology Project of the Department of Science and Technology of Guangdong Province (2012A08102004) and Science and Technology Plan Project in Industrial High-Tech Fields of Guangdo·y-University-Research (201704020090).

ISBN 978-981-95-2242-2 ISBN 978-981-95-2243-9 (eBook)
https://doi.org/10.1007/978-981-95-2243-9

The original submitted manuscript has been translated into English. The translation was done using artificial intelligence. A subsequent revision was performed by the author(s) to further refine the work and to ensure that the translation is appropriate concerning content and scientific correctness. It may, however, read stylistically different from a conventional translation.

This Springer imprint is published by the registered company Springer Nature Singapore Pte Ltd.
The registered company address is: 152 Beach Road, #21-01/04 Gateway East, Singapore 189721, Singapore

Preface

With improvements in socioeconomic development, residents' requirements for urban development are increasing. The smart city, as a new urban model, has gradually become the new trend for cities in the new era. The construction of smart cities has gone through three stages: concept popularization, pilot promotion, and development innovation. From initial concept innovation to gradual implementation, the construction of smart cities has received considerable attention at the national strategic level. China's smart city construction is unique both domestically and internationally, demonstrating comprehensive development from the indicator system to standard norms and from technology application to demonstration projects.

We, as practitioners and explorers of smart city construction, have written this book together. We hope to promote new smart city development by discussing the development history, core technology, innovative concepts, practical cases, and future scenarios of smart cities for the current era.

How to use intelligent means and new generation information technologies such as artificial intelligence to improve the operation efficiency and management level of cities is a major task. In this era, new smart cities have emerged in new forms—digital smart cities—which not only mean that city management and services have become more effective, precise, and experiential but also represent a new direction in urban development. Only by continuously promoting the construction of smart cities, making full use of technology, and constantly innovating and experimenting can smart cities truly become a new driving force that is able to meet the needs of the people and urban development.

This book was written to respond to this challenge. With advances in technology, new generation artificial intelligence technologies, such as intelligent perception, intelligent recognition, intelligent real-time interaction, intelligent voice interaction, intelligent video interaction, multimodal interaction, and intelligent control, are being widely applied in the construction of smart cities, resulting in broader prospects for city management. However, the development of smart cities still faces a series of problems and several complex challenges, such as privacy protection, information security, and human–machine relationships.

Moreover, in practical applications, the success of smart cities not only depends on the technology itself but also better management of the relationships between people and machines. For example, the interaction mode with the new generation of artificial intelligence and the actual application scenarios must consider factors such as people's emotional reactions and special needs.

Innovation and Design of Smart Buildings, Smart Communities and Smart Cities is the first book in our smart city series, which focuses on introducing the concept, connotation and characteristics of smart cities, top-level design, standard system and application operation. The current book is the second book in this series, which focuses on the efforts made by smart cities in terms of development and the application of the latest technologies (such as GPT, AGI[1]). It introduces the common technologies and innovative concepts of smart city engineering applications; analyzes the current status of smart city development, including the problems faced and their solutions; and explores the future development trends of specific scenarios in smart cities. This book is a continuation of *Innovation and Design of Smart Buildings, Smart Communities and Smart Cities* and continues the practice and exploration of the concept of smart cities.

This book is divided into three parts.

The first part covers the foundation of smart city construction. It first introduces the definition, overall framework, development history, and current status of smart city construction, proposes that the digital city is a new form of the smart city, and analyzes the problems in smart city construction and future development strategies. It also proposes future development trends and construction paths for smart cities.

Then, it describes in detail the eight core technologies applied in smart city construction and discusses how to build a new model of the smart city based on the eight core technologies in the era of digitalization, namely, device interconnections, data interconnections, business interconnections, industry interconnections, and urban–industrial integration.

Finally, it proposes the concept of cloud-edge-end collaborative technology and application innovation for smart cities. Over the years, Gloryview Technology has conducted innovative application research in multiple fields of smart cities, forming a basic technology system for smart cities based on the AIoT. In response to the new demands of smart city construction, it proposes the concept of cloud-edge-end collaborative technology and application innovations for smart cities, namely, "four core technology systems, one innovative application framework", which provides new technical support for the high-quality development of our urban economy and society, on the basis of this system.

The second part concerns the application of smart cities. It first introduces the industry characteristics and product applications of the digital base (Internet of Things platform, data platform, AI platform). Then, based on the digital base, it

[1] **GPT:** Generative Pre-Trained Transformer—a deep learning model trained on large-scale internet data for natural language generation tasks.

AGI: Artificial General Intelligence—a form of artificial intelligence capable of thinking, learning, and performing a wide range of tasks at a human-like level.

discusses the problems, solutions, and practical cases of smart city construction in terms of 6 industries (government affairs, medical care, education, tourism, energy, urban–rural integration) and 3 scales (micro, meso, and macro). This content is designed to help readers understand the current status of smart city construction while highlighting practical experiences and technical achievements for reference.

The third part is the future outlook of the digital city, which explores the future development direction of the digital city in three dimensions: foundation, application, and technology. The basic exploration includes the digital government 3.0 of the advanced model of the service-oriented government, which constitutes the industrial digital transformation driving China's economic development; the application exploration includes the industrial data element service, which promotes data value, building space services and digital operations; and the technical exploration includes the innovations of the metaverse and AGI applications in the future. These application scenarios reflect the development trends of technology and society.

We sincerely hope that this book provides more useful references and inspirations for the construction and development of smart cities. In this era of interfacing with technology, let us jointly explore new ways to develop smart cities and look forward to more cooperation and sharing, jointly promoting the construction process of smart cities and contributing to urban management, industrial development, and better lives for the people.

Guangzhou, China
August 2023

Fangfen Liu

Editorial Committee

Contents

Part II Smart City Applications

About the Author

Fangfen Liu (Chief Editor), is a Chief Engineer of Gloryview Technology, Senior Engineer, expert in the Provincial Science and Technology department, Municipal Industry and Information Bureau, WELL AP, and member of the Advisory Committee of the University of Macau (EAC). She has over 20 years of experience in planning, design, and implementation in the field of information and intelligence. She has reviewed one industry standard, participated in the compilation of three group standards, and has won numerous honors, including "Outstanding Engineering Designer of the Intelligent Building Industry in 2012", "Excellent Award of the Intelligent Building Industry Designer Competition in 2018", "IoT Pioneer of the Year in 2021", and "Outstanding Technical Craftsman in 2023", among others. She has participated in four provincial and municipal scientific research projects, and applied for 11 invention patents (6 authorized) and two utility model patents. She has published 9 papers, edited 1 book, coedited 1 book, and led projects that have won the "Science and Technology Demonstration Project" award, the "Green Building Label" 12 times, platinum WELL, platinum LEED, and other awards and certifications and have achieved nearly zero energy consumption.

Part I
Foundation of the Smart City

Chapter 1
Development History of Smart City Construction

1.1 Definition of the Smart City

The concept of the smart city is the result of continuous innovation and development of the concept of smart buildings, and it continues to evolve over time. At different times, the concept of the smart city has received attention and attracted proposals from various parties, including the Central Political Bureau of the Communist Party of China, national planning outlines, ministry institutions, and standard organizations.

In March 2016, the "13th Five-Year Plan for National Economic and Social Development of the People's Republic of China" (hereinafter referred to as the "13th Five-Year Plan") was released, which proposed for the first time to "build a number of new demonstrative smart cities". On October 9, 2016, during the 36th collective study of the Political Bureau of the Central Committee of the Communist Party of China, General Secretary Xi Jinping noted: "We must deeply understand the role of the Internet in national management and social governance, take the promotion of e-government and the construction of new smart cities as the focus, use data concentration and sharing as the means, build a national integrated big data center, promote the integration of technology, business, and data, and achieve cross-level, cross-regional, cross-system, cross-department, and cross-business collaborative management and services."

The Ministry of Industry and Information Technology of the People's Republic of China (hereinafter referred to as the Ministry of Industry and Information Technology) defined a smart city as follows: "A smart city is the product of industrialization, urbanization, informatization, and intelligence at a specific historical moment. It involves the fusion of new technologies, new thinking, new models, and new applications under the rapid acceleration of the Internet and emerging technologies."

The China Electronics Technology Group Corporation (hereinafter referred to as CETC) defined a smart city as follows: "The new smart city aims to provide full-time, efficient, and orderly urban governance, open and shared data, green and open-source economic development, and clear network space security. Through system

F. Liu, *Smart and Digital Cities*, https://doi.org/10.1007/978-981-95-2243-9_1

planning, information leadership, and reform and innovation, it promotes the deep integration and iterative evolution of the new generation of information technology and urban modernization, realizing a new ecosystem of coordinated national and city development."

On March 13, 2016, a representative of the National People's Congress stated: "The new smart city refers to the deep integration and iterative evolution of system planning, information leadership, and the new generation of information technology with urban modernization, thereby achieving more modern governance, smarter operation, safer development, and happier people. The essence is to enhance the governance system and governance capabilities of the government and society, and to better serve the people." The National People's Congress proposed the construction of 'six ones' to promote the development of new smart cities (as shown in Fig. 1.1), that is, a system architecture, a grid network integrating heaven and earth, a general function platform, a dataset, a city operation center, and a set of standards. The new smart city is based on a platform and a network, and its main function is to enhance governance service system capabilities.

The ITU-T (International Telecommunication Union Telecommunication Standardization Sector) emphasizes sustainable development and defines a smart sustainable city as follows: "An innovative city that uses information and communication technology and other means to improve the quality of life, enhance the efficiency of urban operations and services, and urban competitiveness, while ensuring that it meets the economic, social, environmental, and cultural needs of the present and future generations."

The National Standard of the People's Republic of China "Smart City Terms" (GB/T37043-2018) defines a smart city as follows: "An innovative city that uses information and communication technology to effectively integrate various city management

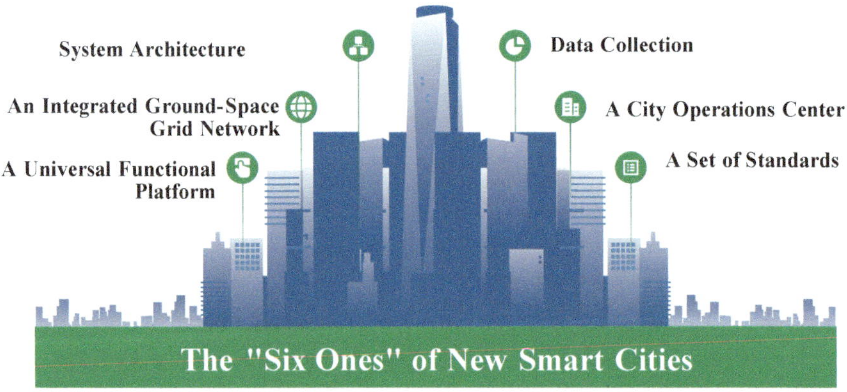

Fig. 1.1 New smart city "Six Ones" construction concept the ISO (International Organization for Standardization) "Smart City Concept Model—Guidance for establishing a model for data interoperability" (ISO/IEC30182-2017) defines a smart city as follows: "Effectively integrating powerless systems, digital systems, and human systems in the built environment to provide citizens with a sustainable, prosperous, and inclusive future."

systems, achieve information resource sharing and business collaboration between various city systems, promote the intelligence of city management and services, enhance the level of city operation management and public services, improve the happiness and satisfaction of city residents, and achieve sustainable development."

On May 15, 2020, the National Development and Reform Commission of the People's Republic of China (hereinafter referred to as the National Development and Reform Commission) defined the new smart city as follows: "The new smart city is a Chinese expression of the concept of a smart city adapted to the actual conditions of our country; the new smart city is a reform and innovation system project [that is being] carried out to enhance the happiness and satisfaction of the people and to enhance the intelligence of urban development methods in response to the real needs of urban economic and social development under the conditions of a modern information society; the new smart city is the implementation of the national new urbanization strategic plan, a smart city with Chinese characteristics, reflecting new policy mechanisms and innovative development models; and the core of the new smart city is people-oriented, and the essence is reform and innovation (the new smart city should be translated into English as an Innovative Smarter City, not New Smart City)".[1]

The proposal of any new concept will have a corresponding era and technological background. We believe that the definition of a new smart city is innovative, originating from the intelligent management capabilities and innovation capabilities of emerging technologies. It is an adaptive innovation model that aims to achieve sustainable development, people's happiness, social stability, and effective government. **The new concept of smart cities emphasizes the following characteristics**:

Data Intelligence: Smart cities need to fully utilize key technologies of cloud-edge collaboration, analyze and intelligently process information through technologies such as edge computing and artificial intelligence to achieve more precise and efficient city operation management.

Rich Application Scenarios: Smart cities need to apply adaptive technologies to more scenarios, such as smart parks, smart health care, and smart energy, to achieve technology adaptation through a wider range of application scenarios.

Self-Organizing Coordination Ability: Smart cities need to introduce technologies such as cloud-edge collaboration and distributed self-organizing networks to achieve self-organization and coordination of the city system, thereby ensuring the stable operation and efficiency of the entire city.

Open Innovation Ecosystem: Smart cities need to establish corresponding open innovation ecosystems through open standards, platforms, and data, thereby attracting more start-ups, developers, and innovators to contribute to the construction of smart cities and achieving cross-border integration of technology and industry.

[1] National Development and Reform Commission of the People's Republic of China. https://www.ndrc.gov.cn/xxgk/jd/wsdwhfz/202005/t20200515_1228150.html

Sustainable Development: Smart cities need to combine adaptive technologies with sustainable development models and achieve the coordinated development of the economy, society, and environment through the efficient use of city resources and environmental protection, thereby promoting the sustainable development of the city.

1.2 Overall Framework of the New Smart City

The overall framework of the new smart city is divided into six levels and two systems, as shown in Fig. 1.2. The six levels include the IoT perception layer, network communication layer, computing storage layer, digital base layer, application layer, and portal layer. The digital base layer includes the IoT platforms, data platforms, and AI platforms. The application layer is divided into industry application classification and scale application classification. Industry applications include smart government affairs, smart health care, smart education, smart tourism, smart energy, and smart urban–rural integration. Scale applications include smart buildings and smart parks. Each application project cooperates with each other in business, their data are interconnected, and from the IoT perception and network communication to computing storage, applications are implemented through the digital base, forming a sustainably developing smart city. The two systems are the standard specification system and the security guarantee technology system.

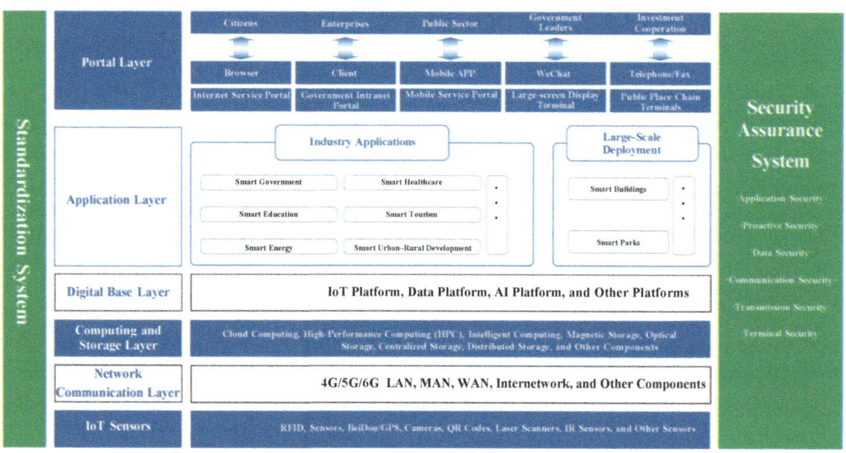

Fig. 1.2 Overall framework of the new smart city

1.3 The Development History of Smart City Construction

Since the American company IBM, proposed the concept of the "Smart Earth" in November 2008, the development of China's "smart city" has gone through three stages: the concept popularization stage (2008–2011), the pilot promotion stage (2012–2015), and the development innovation stage (2016 to present).

1.3.1 Concept Popularization Stage (2008–2011)

Since IBM proposed the concept of the "Smart Earth" in 2008 was the prelude to the development of China's smart city.

November 2008—The American company IBM, first proposed the concept of the "Smart Earth".

November 2010—The Ministry of Science and Technology of the People's Republic of China and other units held the "2010 China Smart City Forum" in Wuhan. The main discussion focused on the innovation and development path selection of smart cities.

December 2010—The "2010 China the Internet of Things and Smart City Construction Summit" was held in Beijing. The conference discussed issues such as the development trend of smart cities and the cultivation of the IoT professionals.

2011—Cities such as Beijing, Shanghai, Ningbo, Wuxi, Shenzhen, Wuhan, and Foshan have successively launched "smart city" strategies, and related plans, projects, and activities have been gradually launched.

During this stage, the concept of the "smart city" was imported from abroad and gradually became understood and accepted. Various forums related to "smart cities" were held throughout China, and the concept of "smart cities" began to be widely popularized. The main feature is that various departments and localities should promote the construction of smart cities according to their understanding, which tends to be relatively scattered and disorganized. From 2008 to 2011, the Chinese government did not clearly define the concept of a smart city.

1.3.2 Pilot Promotion Stage (2012–2015)

After the Office of the Ministry of Housing and Urban–Rural Development of the People's Republic of China (referred to as the Ministry of Housing and Urban–Rural Development) officially issued the "Interim Management Measures for National Smart City Pilots" in 2012, China's smart city pilot projects were successively launched.

December 2012—The Ministry of Housing and Urban–Rural Development issued the "Interim Management Measures for National Smart City Pilot" and the "National Smart City (District, Town) Pilot Indicator System (Trial)", thus officially launching the smart city pilot work. Through the construction of smart cities, the Central Committee of the Communist Party of China (abbreviated as the CPC Central Committee) and the State Council of the People's Republic of China (abbreviated as the State Council) aim to implement innovative development; promote new urbanization; fully build a moderately prosperous society; and explore scientific methods for the construction, operation, management, service, and development of smart cities.

January 2013—The Ministry of Housing and Urban–Rural Development announced the first batch of 90 national smart city pilots. The selection of these pilot cities is based on various factors, such as city size, economic development level, and level of informatization; the construction goal of the pilot cities is to improve the urban management level, optimize public services, and improve urban operational efficiency and people's quality of life.

August 2013—The Ministry of Housing and Urban–Rural Development announced the second batch of 103 national smart city pilots. The goal of the pilot cities is to establish a comprehensive coordination mechanism, implement financial and other guarantee measures, ensure the success of the pilots, and create replicable and promotable experiences.

March 2014—The State Council issued the "National New Urbanization Plan (2014-2020)", in which it "promote[s] the construction of smart cities". The main purpose of this plan is to follow the path of new urbanization with Chinese characteristics, comprehensively improve the quality of urbanization, and promote the coordinated development of cities and rural areas.

August 2014—Eight ministries and commissions, including the National Development and Reform Commission, issued the "Guiding Opinions on Promoting the Healthy Development of Smart Cities". This document proposes that smart cities advance a new concept and model that uses new generation information technologies such as the Internet of Things, cloud computing, big data, and integrated spatial geographic information to promote the intelligent planning, construction, management, and services of cities.

April 2015—The Ministry of Housing and Urban–Rural Development announced the third batch of 97 national smart city pilots. With the support of technological innovation, it focuses on solving real problems that restrict urban development.

During this stage, smart cities move from conceptualization to the introduction of related policies, thus promoting the official implementation of smart city construction in China. Construction is actively carried out nationwide, and smart city pilots are gradually rolled out across the country; however, the project features in various places are not distinctive, resulting in uniformity across cities.

1.3.3 Development and Innovation Stage (2016 to Present)

In 2016, the construction of new smart cities was elevated to a national strategy, national-level support policies were successively introduced, and policies in segmented fields gradually came into play. China's smart city pilots and construction showed a development trend of hierarchical construction, manifesting at multiple points and improving quality and efficiency. Each major city presents different characteristics, thus breaking away from the previous phenomenon of uniformity in smart city construction.

March 2016—The "Thirteenth Five-Year Plan" was released, which proposed the "Construction of Smart Cities" and "New Smart City Construction Actions". The construction of new smart cities was elevated to a national strategy for the first time.

December 2016—The "New Smart City Evaluation Index" (GB/T 33356-2016)[2] national standard was released. It can be used to guide the planning, design, implementation, operation, and continuous improvement of new smart cities.

December 2017—The "New Smart City Development Report 2017"[3] was released, and the evaluation results of the new national smart city were made public for the first time. Overall, there is still considerable room for improvement in the construction of new smart cities in China, which is reflected mainly in the need to improve citizens' sense of gain, unbalanced overall development, and the insufficient role of information resource infrastructure.

June 2018—The "Smart City Top-Level Design Guide" (GB/T 36333-2018) national standard was released. This guide helps standardize the construction of smart cities and improve the quality and level of smart city construction.

January 2019—The "Outline of Smart City Spatio-Temporal Big Data Platform Construction Technology (2019 Edition)" was jointly issued by eight ministries and commissions, including the Ministry of Industry and Information Technology, the National Development and Reform Commission, the Ministry of Science and Technology, the Ministry of Public Security of the People's Republic of China (abbreviated as the Ministry of Public Security), the National Security Department, the Office of the Central Cyberspace Affairs Commission of the People's Republic of China (abbreviated as the Cyberspace Administration of China), the Chinese Academy of Sciences, and the Chinese Academy of Engineering. It aims to standardize the construction of smart city spatiotemporal big data platforms, providing requirements in terms of technical routes, construction ideas, key tasks, and construction principles.

March 2021—The State Council issued the "Outline of the Fourteenth Five-Year Plan for National Economic and Social Development and the Long-Range Objectives Through the Year 2035" (abbreviated as the "Fourteenth Five-Year Plan"), which proposed to "build smart cities and digital villages", "promote the construction of new smart cities in a graded and categorized manner", and "explore the construction of digital twin cities".

January 12, 2022—The State Council issued the "14th Five-Year Plan for Digital Economy Development," which proposed "coordinated promotion of new smart city and digital rural construction" and "deepening the construction of new smart cities."

March 10, 2022—The National Development and Reform Commission released the "Key Tasks for New Urbanization and Urban–Rural Integration Development in 2022," which proposed "improving the level of intelligence," "exploring the construction of the 'urban data brain'," and "promoting urban–rural integrated development."

July 2022—The National Development and Reform Commission issued the "14th Five-Year Plan for New Urbanization Implementation," which proposed "promoting intelligent transformation", "implementing unified management of urban operations, [and] exploring the construction of digital twin cities."

With the gradual increase in the understanding of smart cities and the continuous development of technology, smart cities have been endowed with new connotations

[2] The standard "Evaluation Indicators for New-Type Smart Cities" (GB/T 33356–2016) has been abolished. The current version in effect is GB/T 33356–2022.

[3] The 2017 Report on the Development of New-Type Smart Cities was compiled under the coordination of the Inter-Ministerial Working Group for New-Type Smart City Development, with contributions from 25 participating member agencies.

and requirements. Concepts such as "digital twin cities" and the "urban metaverse" have been proposed in the "14th Five-Year Plan" for new smart city construction initiated by many cities, thus injecting innovative ideas into the construction of smart cities and becoming an important direction for the evolution of smart city development.

Against the backdrop of rapid economic development and continuous advancements in smart city construction, issues such as imbalances in urban and rural development and the environmental impact of greenhouse gas emissions have become increasingly prominent. The Chinese government has successively introduced a series of targeted policy measures for urban–rural integrated development and China's "dual carbon" goals, which have introduced new ideas to the definition of smart cities and the scope of the field and new opportunities for the future development of smart cities.

1.4 Current Status of Smart City Construction in the Context of Digital China and the Digital Economy

In February 2023, the Central Committee of the Communist Party of China and the State Council released the *Overall Layout Plan for Building a Digital China*. The plan adopts the "2522" framework (see Fig. 1.3), which comprises:

- **Two Foundations**—strengthening digital infrastructure and developing a national data resource system;
- **Five Areas of Integration**—embedding digital technologies into the economy, governance, culture, society, and ecological civilization;
- **Two Capabilities**—enhancing the capacity for digital technology innovation and establishing robust digital security safeguards;
- **Two Enabling Environments**—improving both domestic and international conditions for digital development.

he objectives that should be achieved by 2025 include achieving efficient interconnectedness across the digital infrastructure, rapidly increasing the scale and quality of the available data resources, fully unlocking the value of data as a production factor, and substantially improving the quality and efficiency of the growth of the digital economy. Accelerating the development of Digital China serves as a strategic driver for building a modern socialist country and advancing the long-term revitalization of the Chinese nation.

On May 23, 2023, the National Internet Information Office released the "Digital China Development Report (2022),", which noted that the scale of China's digital economy reached 50.2 trillion yuan, accounting for 41.5% of GDP and becoming an important engine for stable growth and transformation. The scale of digital infrastructure has also significantly improved, with the number of 5G base stations reaching 2.312 million, the number of 5G users reaching 561 million, and the number of

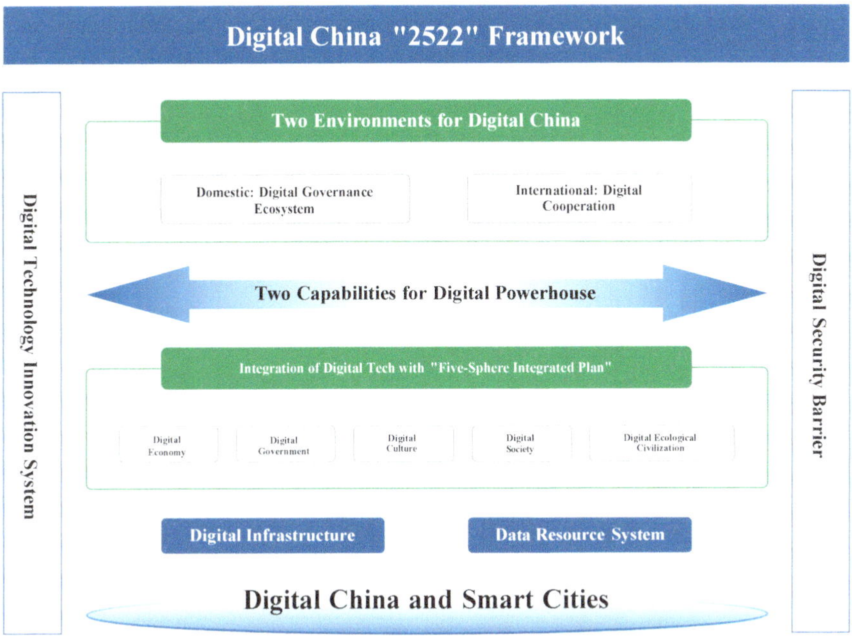

Fig. 1.3 Overall structure of the "2522" digital china construction framework

mobile the IoT terminal users reaching 1.845 billion, making China the first country among the world's major economies to achieve "more things than people." In addition, the construction of China's data resource system is accelerating, with data production reaching 8.1 ZB in 2022, which represents a year-over-year increase of 22.7%, accounting for 10.5% of the global share, ranking second in the world.

In the future, the construction of Digital China will be accelerated; the foundation of Digital China construction will be consolidated; the major arteries of digital infrastructure will be unblocked; a universally accessible and convenient digital society will be built; a confident and prosperous digital culture will be created; a green and intelligent digital ecological civilization will be built; the key capabilities of Digital China will be strengthened; a self-reliant and strong digital technology innovation system will be built; a trustworthy and controllable digital security barrier will be established; the digital development environment will be optimized; a fair and standardized digital governance ecology will be built; and an open and win–win international cooperation in the digital field will be constructed. In this context, smart cities represent improvements in cities in all aspects, all cycles, and all fields. These measures will provide strong support and guarantees the development of China's digital economy.

Smart cities collect and analyze large-scale data through information technology to provide data support for city operations. These in-depth data insights further drive market trends and consumer demand. Moreover, the construction and development of smart cities also encourage innovation in technology and services, thereby providing momentum for the continuous development of the digital economy. More importantly, smart cities need to build and maintain an advanced information and communication technology infrastructure, which is crucial for the healthy development of the digital economy. In this process, by implementing smart city projects, the government can promote the widespread application of digital technology in various industries and fields, thus further creating market space. Finally, by providing smarter and more efficient public services, smart cities improve the quality of life of the public, thereby stimulating digital consumption and injecting momentum into the further development of the digital economy. In general, smart cities provide important guarantees and support platforms for the development of China's digital economy.

1.4.1 The Digital Smart City is the New Form of the Smart City

The construction of smart cities has gone through the stages of concept popularization, pilot promotion, and development innovation and has now entered into a new form in the stage of smart city development innovation—the digital smart city. A digital smart city refers to the use of a new generation of information technology (such as big data, 5G, the Internet of Things, and artificial intelligence) to lead the development of the digital economy, carry out comprehensive and in-depth digital transformation and intelligent upgrading of the city, improve the level of urban management, optimize public services, promote industrial development, and enhance the quality of life of residents, as shown in Fig. 1.4. The development of digital smart cities provides new opportunities for urban innovation, but at the same time, issues such as privacy protection and information security need to be fully considered. Digital smart cities can promote urban economic and social development and its integration with digitalization and thus ensure sustainable cities, happiness among the people, a stable societies, and effective governments. **Digital smart cities constitute an important part of and support for the construction of Digital China. Informatization, digitalization, and the digital economy are inseparable and are important components of the construction of Digital China.**

Digital smart cities represent a new form of smart city development and have unique characteristics compared with those of traditional smart cities. Based on the characteristics of the new type of smart city, digital smart cities are unique in the following aspects:

Reliance on a new generation of information technology: Based on our proposed concept of "data intelligence", digital smart cities rely on a new generation of information technology to achieve comprehensive, accurate, and in-depth digital transformation and intelligent upgrading of urban data.

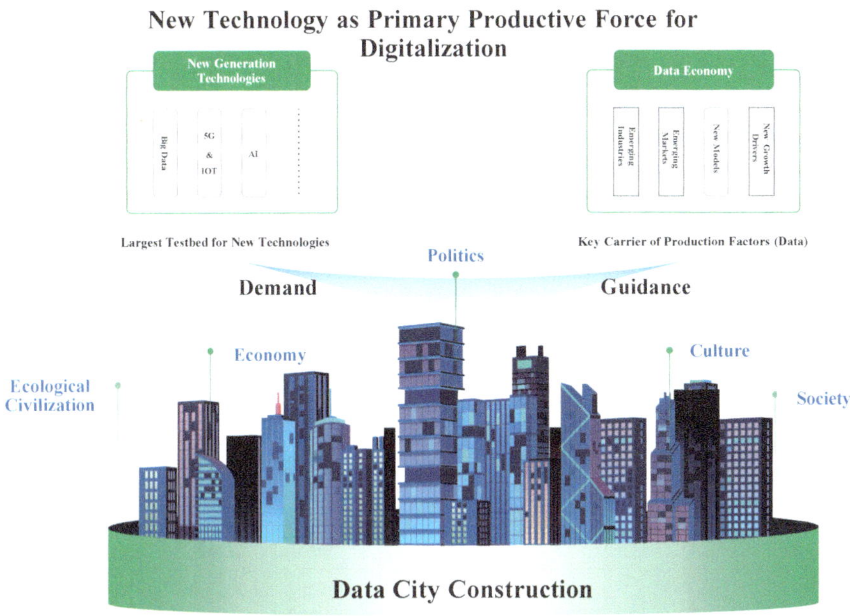

Fig. 1.4 Productivity and key production elements of digital smart city construction

Data-driven: Digital smart cities are based on data. On the basis of the "rich application scenarios" of the new type of smart city, through the collection, integration, and analysis of data from various facilities and equipment in the city, intelligent, refined, and efficient management of the city is realized.

Comprehensive intelligence: The "self-organizing coordination ability" of a new type of smart city, in the form of the digital smart city, reflects the comprehensive data intelligence of various facets of city operations, including urban planning, transportation, environmental protection, public safety, and social services, to achieve the automation of various city affairs.

Multidimensional interconnection: Based on the "open innovation ecology" of the new type of smart city, digital smart cities interconnect various resources inside and outside the city in multiple dimensions, enabling comprehensive interconnection and data sharing between various fields within the city, between cities, and between cities and industries.

Future-oriented: Digital smart cities not only focus on current city management and services but also give more attention to the sustainable development and intelligent upgrading of future cities, given the orientation toward "sustainable development" of the new type of smart city. Thus, they meet the increasing demands of people with respect to city quality, quality of life, and environmental protection.

1.4.2 Problems and Development Strategies in the Construction of Smart Cities in China

Although China has made certain achievements in the construction of smart cities, as this process has intensified, some problems have gradually emerged. These problems can be analyzed from two perspectives: the government perspective (as shown in Fig. 1.5) and the service system (as shown in Fig. 1.6):

Given the analysis of the ten common problems in the construction of smart cities in China and the mechanism and efficiency problems of the smart city service system, several aspects urgently need to be resolved:

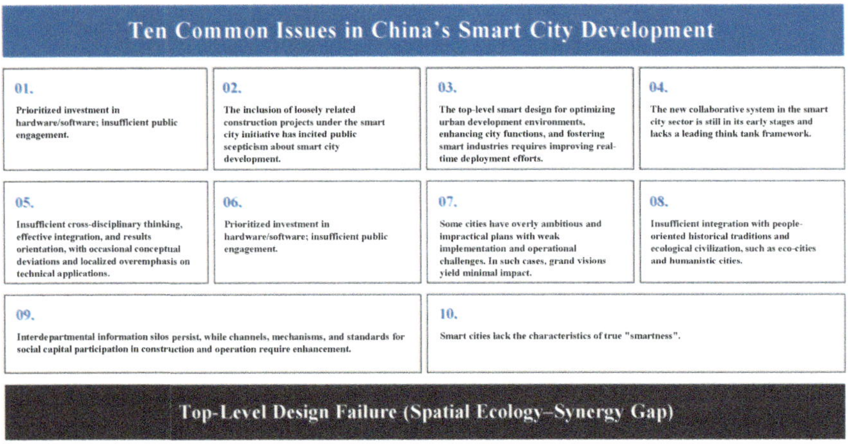

Fig. 1.5 Common problems in the construction of smart cities in China from the government perspective

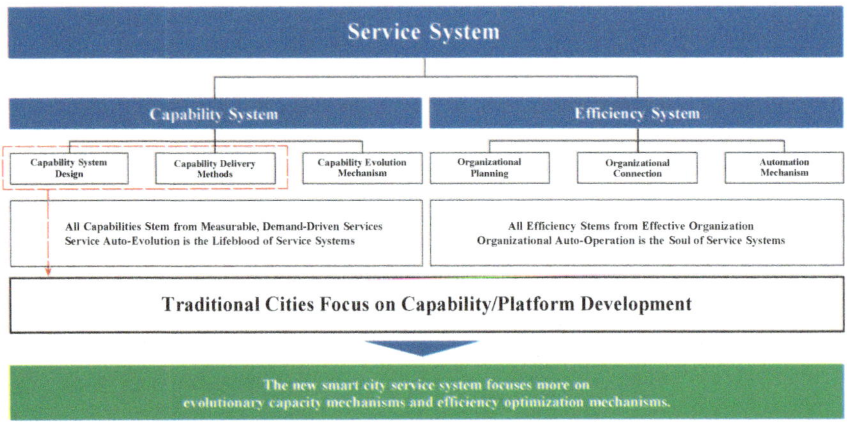

Fig. 1.6 Problems in the smart city service system

Insufficient Top-Level Design

The construction of smart cities in China is a complex and important topic involving top-level design in technology, systems, and humanities. However, the top-level design is currently insufficient. The top-level design of smart city construction is a complex and long-term task that requires the government to strengthen efforts and promotions in various aspects, such as experience sharing, industry innovation, technology research and development, planning and design, and citizen education, to effectively promote the healthy and orderly development of smart cities.

The Struggle of Management Systems and Institutions in Achieving Synergy with Technology

Since each smart city is built at different levels, there is a large difference in the technologies applied by different cities, making unified management and supervision difficult. Second, the speed of technology updates is rapid, and management systems and institutions often need more time to update. As a result, some management systems and institutions are unable to keep pace with technology applications.

In the process of building smart cities, management systems and regulations play crucial roles in regulating and standardizing the construction of smart cities, such as via urban construction management standards and data privacy protection. However, management systems and technical systems lack coordination with each other; thus, the application effect has not reached its optimum.

Difficulties in Sharing Smart City Data

Data sharing is a core issue in the construction of smart cities. Despite increasing attention, there are still difficulties. The main reasons for these difficulties are twofold. First, management systems and regulations have not been reformed according to the new governance model of smart cities. The data management of each department is relatively independent and lacks coordination and linkage. There are differences in the awareness, willingness, and interests of each department regarding data openness. Second, the information systems of different departments are not interconnected. Owing to the different data formats and standards of different systems and manufacturers, data sharing is difficult. As the construction of smart cities involves a wide range of industries and varying levels of technological maturity, including mature, developing, and emerging technologies, the construction goals and applications are diverse. The lack of overall planning and top-level design makes it difficult to unify standards.

Lack of Deep Integration Applications

Currently, most smart city applications are still focused on single vertical domains. Thus, it difficult to integrate and analyze data from various industry application fields, and there is a lack of macroscopic and deeper predictive and decision-making applications. The formation of consensus-based, industry specific, categorized effective data tagging, and universal industry specific, categorized tag parsing applications is therefore an important direction for future work.

Severe Data Security Situation

For a long time, the underlying standards, architecture, products, and ecosystems of the information and communication technology (ICT) industry have been formulated and controlled by foreign companies or institutions, exposing China's ICT industry and even a large number of users to risks such as choking, data leakage, and information security. Therefore, it is particularly important to reduce foreign dependence by gradually achieving independent innovation supply for computer information technology and other hardware and software aspects.

Moreover, due to the severe data security situation, which prominently manifests as difficulty in guaranteeing intrinsic security, severe process security situations, and unclear responsibilities for system security, the marketization of data elements faces many challenges. These challenges manifest prominently as insufficient effective data supply, an absence of the data element market, an imperfect technical system, an urgent need for improving the legal system, and a lagging institutional management system. Therefore, to enhance the role of element allocation, promote the marketization of data elements, and release the value of market elements, further research on the basic concept of data elements is needed. Accelerating innovation of the new generation technology of data elements, speeding up the establishment of a data asset management methodology for China's digital transformation, and balancing data application and data security are also necessary.

Irreconcilable Contradiction Between Economic Development and Environmental Protection

Rapid economic development causes environmental harm, leading to a reduction in the quality of the natural environment. In contrast, emphasizing environmental protection hinders economic development and prevents rapid economic growth. Therefore, economic development and environmental protection are diametrically opposed to each other. Since the reform and opening up, China's economy has developed rapidly, and its overall economic strength has substantially improved. However, owing to extensive past economic development, ecological destruction and environmental pollution have become prominent issues. Moreover, in addressing global climate change, China has demonstrated the responsibility of a major country and has played an important leading role. Balancing the relationship between economic development and environmental protection is a problem that urgently needs to be solved. We can consider achieving a win–win model of economic development and environmental protection through the development of new technologies and applications.

Regional Development Imbalance

Under the strong leadership of the Central Committee of the Communist Party of China with Comrade Xi Jinping at its core, China has achieved historic achievements and undergone historic changes in the new era, especially in winning the battle against poverty and building a moderately prosperous society in all aspects. However, owing to rapid economic development and the continuous development of smart city construction, the gap in regional development has gradually widened. This phenomenon is reflected mainly in the imbalance of development between the eastern and western regions and between urban and rural areas, among which the development gap between cities and rural areas is more prominent. The uneven development of cities and rural areas, the unsmooth flow of urban and rural elements, and the imbalanced allocation of public resources have induced deeper social contradictions. The imbalance in regional development has become one of the main obstacles to achieving the common prosperity of the people. Therefore, strengthening integrated urban–rural development through technology advances presents an opportunity for the future development of smart cities.

Insufficient Citizen Participation

Engaging the public, businesses, and the market is a major challenge in our smart city construction and management. In the construction of smart cities, the lack of citizen participation is reflected mainly in the insufficient resonance of demand perception and the lack of comprehensive application services. The construction and implementation of smart cities have not fully adapted to the diverse needs of citizens. In actual operation, some city government departments and enterprises have not fully researched and analyzed the needs of citizens and therefore have not truly achieved a people-oriented service model. Some services of smart city construction applications are not perfect, and the citizen experience is poor. For example, inaccurate parking location information and inconvenient operation methods affect the effectiveness of smart parking applications.

Difficulties and Limited Benefits of Smart City Operation

The construction of smart cities is a highly complex and long-term task, and difficulties in implementation and low benefits severely restrict the realization and development of smart cities. The main reasons are as follows:

First, there is a lack of unified standards and norms. Owing to the different technical standards and norms adopted by different manufacturers, there are multiple systems in smart cities that cannot cooperate and coordinate with each other, which severely affects the sharing of information.

Second, it is difficult to obtain data, the quality of the data is poor, and data islands are severe. Smart cities need a large amount of diverse data support. However, in reality, different departments and institutions have different datasets; these datasets cannot be integrated and shared, making it difficult to achieve fine-tuned management and operation of smart cities.

Finally, the investment risk of smart city construction is high, and the return period is long. Owing to the large investment scale and long construction period of smart city construction, it is not easy to evaluate the investment results in real time, which introduces uncertainty and risk to investors and inhibits the investment of governments and enterprises.

In summary, the construction of smart cities is a systematic project that requires support from various resources and continuous investment. Only by overcoming difficulties in implementation and the limited benefits can the sustainable development of smart city construction be truly achieved.

In response to the current problems, the following strategies can be adopted for the future development of smart cities:

Strengthen Top-Level Design

More attention should be given to the evolution mechanism of capabilities and the mechanism of efficiency, integrating government services through a unified government platform, promoting cooperation in terms of the background departments, and sharing data and rights. This design can also create more value for the entire society.

Enhance the Citizen Experience Design

We should specifically carry out citizen experience design improvements to address the usage needs of citizens and improve the convenience and aesthetics of citizens' use through personalized and intelligent design.

Increase Publicity and Popularization Efforts

To improve citizens' understanding and awareness of smart city construction, enterprises and governments need to educate and inspire them through various means, such as publicity and promotion and community visits.

Accelerate Improvements in Application Services

It is necessary to consider citizen experiences and needs in application services and adjust, update and optimize the services in a timely manner, thus attracting more citizens to participate in improving them.

Provide More Humanized Technical Support

Citizen feedback should be considered, technical support and assistance should be provided through various channels, and problems should be acknowledged and solved in a timely manner.

Strengthen Data Security Protection

Key issues such as privacy protection and information security need to be considered in the process of data disclosure, adoption, and processing. This need is also one of the important factors for citizens' willingness to participate in the construction of smart cities.

Promote the Integration of Industry and Cities

The participation of all parties in society should be promoted through digital government and drive industrial innovation.

Give Attention to Smart City Operations

Smart city construction is a systematic project that requires support from various resources and continuous investment. Only by overcoming the limited benefits can the sustainable development of smart cities be truly achieved.

1.5 Development Trends of Smart Cities

The development trends of smart city construction can be divided into the following aspects:

Policy Guidance and Planning: This trend involves strengthening decision-making and engineering top-level design and coordinated development and accelerating the construction of the digital economy.

Sustainable Development and Environmental Protection: This trend includes promoting the dual-carbon trend and green low-carbon development, as well as the construction of a digital energy ecosystem.

Data Asset Trading and Data Security: This trend includes promoting industry development through data asset flow and new models of digital city security.

Emerging Technology Applications and Development: This trend includes the important applications and development prospects of information technologies such as the new generation of artificial intelligence and blockchain in smart cities.

Urban–Industrial Integration and Smart City Operations: This trend entails accelerating the connection of smart industries to urban development and innovating the operation and management models of intelligent and digital cities.

Infrastructure Construction and Operations: This trend involves accelerating the construction of intelligent platforms and computing centers to enhance the capabilities of urban infrastructure.

1.5.1 Policy and Top-Level Planning

Decision-Making and Engineering Top-Level Design: Guiding the Decision-Making and Implementation of Urban Development

Top-level design includes decision-making top-level design and engineering top-level design. As shown in Fig. 1.7, the decision-making top-level design should be prioritized, as the scope of smart city construction is very broad and involves multiple aspects, such as policy, technology, society, the economy, and the environment. During the decision-making top-level design phase, these aspects need to be comprehensively analyzed and considered, and comprehensive, feasible, and forward-looking implementation plans should be formulated to lay the foundation for the smart city.

Specifically, decision-making top-level design needs to focus on concepts, visions, directions, logic, paths, and systems. Among them, concepts and visions are the core ideas and goals of smart city development, which need to be considered from multiple aspects, such as urban development, resident needs, and industry changes, to clarify the characteristics and functions of smart cities; direction and logic refer to the overall planning and implementation of various tasks necessary for achieving the overall and synergistic construction of smart cities; paths refer to the specific ways and measures

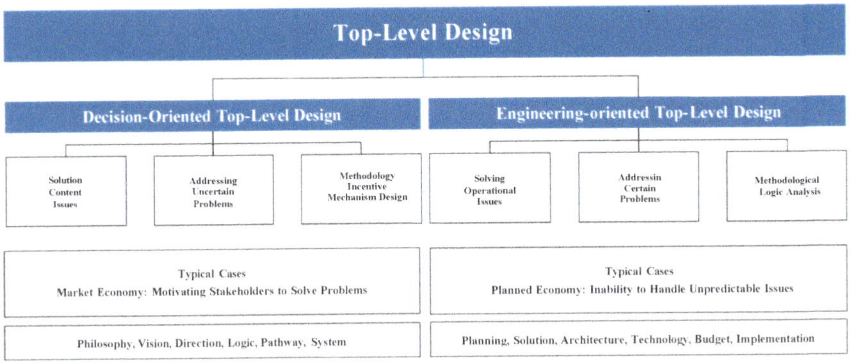

Fig. 1.7 Top-level design system diagram

to implement smart cities, including the layout and arrangement of technology, data, talent, funds, etc.; and systems refer to the design and improvement of organization, operation, management, and other tasks to ensure the smooth progress and sustainable development of smart city construction.

Based on the decision-making top-level design, it is necessary to further compile the engineering top-level design; clarify the construction goals, timetable, budget, and division of responsibilities for each specific project; and ensure the coordinated advancement and optimized implementation of each project. Moreover, attention should be given to promoting factors such as technological innovation, entrepreneurial innovation, and management innovation to enhance the leading edge and competitiveness of smart city development.

In summary, decision-making top-level design and engineering top-level design constitute the foundation and key to smart city construction and need to be comprehensively considered and meticulously implemented to promote sustainable urban development and improve residents' quality of life.

In the construction of smart cities, decision-making top-level design provides the basis and support for the effective implementation of engineering top-level design. Only by fully considering all the aspects and details of smart city construction and establishing a comprehensive, feasible, and forward-looking implementation plan can the construction of smart cities be smoothly promoted, thereby contributing to the sustainable development of cities.

The main differences between decision-making top-level design and engineering top-level design are as follows:

Decision-Making Top-Level Design: Decision-making top-level design is the first stage of the smart city construction process. Its main purpose is to formulate the development direction and overall strategy of the smart city to meet the needs of socioeconomic development and sustainable urban development. During the decision-making top-level design phase, comprehensive research and analysis are needed to determine the city's development goals, policies, and systems and to plan the city's infrastructure and functional modules. Decision-making top-level design should be forward-looking, comprehensive, and feasible, providing directional guidance for subsequent engineering top-level design.

Engineering Top-Level Design: Engineering top-level design is an important stage in the implementation process based on the guidance of decision-making top-level design. Its main purpose is to translate the direction and strategy determined in the decision-making top-level design into specific technical and practical solutions. Therefore, it needs to be feasible and operable.

Decision-making top-level design and engineering top-level design in smart city construction have different characteristics and functions. Decision-making top-level design starts from a macro perspective and enables comprehensive strategic planning for the future development of a city, whereas engineering top-level design focuses more on details, providing technical solutions and planning for specific technical schemes. The two are complementary and closely integrated important stages in the construction of smart cities.

The top-level design process for smart cities can be divided into the following steps: demand research and analysis, overall vision planning, supporting system design, implementation strategy formulation, specific planning and design, and project implementation acceptance and handover (as shown in Fig. 1.8). In the planning and design stage, the preliminary design draft needs to be reviewed by experts. A panel of city experts and relevant professionals is formed to review and discuss the preliminary design. Based on the opinions and suggestions of the expert review, the preliminary design is modified and improved, with the goal of improving and perfecting the plan. The final top-level design should be scientific, operable, sustainable, and feasible, fully meet the city's needs and match the future development direction.

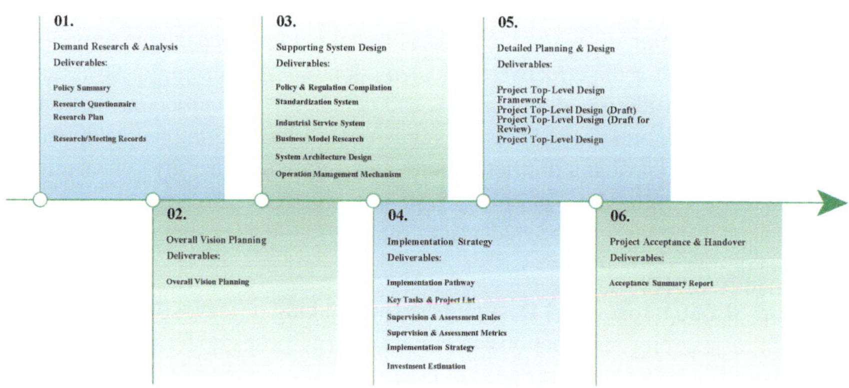

Fig. 1.8 Top-level design planning process diagram

To address the issue of top-level design failure in smart cities, the country has actively introduced relevant policies and measures. Among them, the latest top-level design policy and regulation for smart cities is the "Smart City Top-Level Design Guide" (**GB/T 36333–2018**). This standard was released on June 7, 2018, and was implemented on January 1, 2019. The standard mainly includes the overall requirements, basic process and demand analysis, overall design, architecture design, implementation path design, and other facets of the top-level design of smart cities. It is applicable to the top-level design of smart cities and can guide top-level design in the field of informatization.

Accelerating the Construction of the Digital Economy: Driving the Sustainable Development Engine of Smart Cities

With the continuous development and popularization of information technology, the digital economy has gradually become a new force in global economic development. As one of the core carriers of the digital economy, smart cities are becoming an important strategic means for governments to accelerate the construction of the digital economy and promote sustainable development. For this reason, the Chinese government is making unremitting efforts to guide and promote the sustainable development of smart cities. There are many government policies to guide the sustainable development of smart cities, including the following:

On January 12, 2022, the State Council issued the "14th Five-Year Plan for Digital Economy Development", which noted that, in conjunction with the construction of new smart cities, the plan will accelerate the integration of urban data and the cultivation of industrial ecosystems and improve the level of urban data operation and development; promote the integrated development of digital urban and rural areas; promote the integrated development of cloud and network collaboration, calculations and networks; and provide systematic artificial intelligence services for smart cities. Moreover, this process should be standardized, promote a healthy lifestyle and sustainable.

On March 10, 2022, the National Development and Reform Commission issued the "Key Tasks for New Urbanization and Urban–Rural Integration Development in 2022": Promote the construction of new smart cities, improve the level of intelligence, and promote the integrated development of urban and rural areas.

On June 23, 2022, the State Council issued the "Guiding Opinions on Strengthening the Construction of Digital Government": Promote the construction of digital government, accelerate digital transformation, improve government governance capabilities and service levels, and provide strong support for the comprehensive construction of a modern socialist country.

The government is continuously strengthening and deepening its efforts to guide and promote the sustainable development of smart cities, with the aim of creating a new form and development model of the digital economy and promoting the sustainable development of the national economy.

1.5.2 Sustainable Development and Environmental Protection Category

Building a Digital Energy Ecosystem: Promoting Green and Low-Carbon Energy in Smart Cities

Promoting the greening and low-carbon development of economic and social development is a key step in achieving high-quality development. Against the background of the "dual carbon" policy, the development of high-level urban capabilities has deepened the high-level, multidimensional coordination of digital cities. The opportunity for the comprehensive pilot of digital and green transformation development (dual transformation coordination), jointly carried out by the Office of the Central Cyberspace Affairs Commission of the Communist Party of China, the National Development and Reform Commission, the Ministry of Industry and Information Technology, the Ministry of Ecology and Environment of the People's Republic of China (hereinafter referred to as the Ministry of Ecology and Environment), and the National Energy Administration, will entail the coordinated transformation of digitalization, greening, and low carbon, creating momentum for the high-quality development of smart cities.

In this context, building a digital energy ecosystem has become a core task in promoting green low-carbon development. First, the construction of a smart city energy internet is the core component of the digital energy ecosystem. The smart city energy internet realizes the informatization and intelligent management of all the links of energy production, transmission, distribution, and consumption through digital technology, the Internet of Things, etc., thereby achieving the goals of low energy consumption, high efficiency, and low emissions. Moreover, the smart city energy internet can organize scattered energy resources to provide stable, reliable, and high-quality energy services for urban residents.

Second, a digital energy ecosystem needs to establish an ecological environment for green and low-carbon development. To achieve this development, the digital energy ecosystem needs to vigorously promote clean energy and low-carbon technologies; accelerate the conversion of coal to gas, coal to electricity, and coal to heat; construct clean energy power generation, grid parity, and energy storage facilities; and achieve the greening and decarbonization of energy production and consumption. Moreover, the digital energy ecosystem also needs to optimize the energy utilization structure, reduce energy waste, improve energy efficiency, and ultimately achieve the goal of sustainable urban development.

Finally, the digital energy ecosystem needs to collaborate with all parties to build a global energy internet. The construction of a digital energy ecosystem requires the joint participation of all parties, such as governments, energy companies, and research institutions. In particular, building a global energy internet and achieving transnational transmission of clean energy, such as solar and wind energy, will help promote structural adjustments in energy production and consumption methods, promote green and low-carbon development, and facilitate global climate governance.

1.5.3 Data Asset Trading and Data Security Classes

Data Empowers Industry, Digital Boosts Interconnection: Industry Data Asset Trading Trend

The Chinese government has already issued a series of policies related to the development of the digital economy. For example, in 2021, the "Fourteenth Five-Year Plan" further clarified that it is necessary to establish basic systems and standards for data resource property rights, trading circulation, cross-border transmission, and security protection and to promote the development and utilization of data resources. In addition, the 2022 government work report clearly stated that it is necessary to promote the development of the digital economy, strengthen the overall layout of digital China, build digital information infrastructure, gradually build a national integrated big data center system, promote the large-scale application of 5G, promote the digital transformation of industries, and develop smart cities and digital villages. In addition, the Pudong New Area People's Government has issued the "Fourteenth Five-Year Plan for Industrial Development in Pudong New Area", which mentions the need to fully grasp the new scenarios and opportunities in economic development, urban transportation, people's lives, etc., during digital transformation, exploit the technical advantages of technologies such as 5G, big data, and artificial intelligence, further advance leading technologies and address key technologies, and promote the digitalization of industries and the industrialization of digital technology.

In addition to policy support, with the increasing development of big data and artificial intelligence technology, data have become indispensable resources in today's society. In this context, the trend of data trading is gradually emerging. The flow of data assets promotes the vigorous development of the smart city industry. Through data trading, information is transferred from data producers to data users, continuously promoting the innovative development of the industry and forming a new pattern of digitalization and intelligence in smart cities.

New Mode of Urban Security Guarantees: Promoting Data Security Guarantees in Smart Cities

The construction of smart cities involves the collection and use of a large amount of data in various aspects of urban transportation, the environment, public safety, and other contexts. Hence, problems such as data leakage, data hijacking, and data misuse can arise, which pose potential security risks to the city. Second, the issue of personal privacy protection, such as ID numbers, phone numbers, and license plate numbers, is receiving increasing attention. If this information is used improperly, it will infringe on citizens' privacy and may even trigger negative social public opinion and legal responsibility.

Therefore, research on and response to the issues of data security and privacy protection in smart cities have become new modes of guaranteeing urban security.

The Chinese government has already issued a series of policies concerning digital security. For example, on June 10, 2021, the 29th meeting of the Standing Committee of the 13th National People's Congress voted to pass the Data Security Law, elevating data security to the level of national security, and proposed corresponding requirements for the safe export management of important data. In addition, the State Council issued the "Guiding Opinions on Strengthening the Construction of Digital Government", which specifies the need to strengthen the construction of the network security guarantee system and promote digital government and digital transformation. Data circulation and privacy computing in smart cities are also important issues. Some experts have noted the need to protect personal privacy in smart cities while ensuring data circulation to prevent personal privacy leaks.

1.5.4 Emerging Technology Applications and Development Classes

New Generation Information Technology for Smart City Applications: Driving Smart City New Technology Trends

With the advent of big data and high computing power, new technologies are constantly emerging and being applied to various fields, such as city management, services, and life. Among them, the continuous development and integration of new generation technologies, such as artificial intelligence, blockchain, 5G, the Internet of Things, geographic information systems, new materials, and autonomous driving, provide new ideas and means for the construction of smart cities. For example, the application of technologies such as artificial intelligence-assisted autonomous driving, intelligent power inspection, and industrial robots can increase production efficiency and quality, improve the working environment, and inject new vitality into urban economic development. Blockchain technology, as a new idea and means to ensure city data security and privacy protection, ensures that city data are not tampered with and leaked through actions such as decentralization and distribution, facilitates safe sharing and management, and ensures the compliant use of city data. In addition, 5G and the industrial internet promote the high-quality development of intelligent manufacturing, ChatGPT.[4] Emerging as a dark horse among internet technologies, intelligent service robots are widely used in various scenarios. Cities use such technologies for purposes such as vehicle–road collaboration, intelligent traffic control, smart parking, and cloud computing to achieve real-time monitoring of traffic flow and congestion indices, improve traffic efficiency, and reduce carbon emissions, which are highly important for the sustainable development of cities. In summary,

[4] ChatGPT is a large language model based on the GPT-3.5 architecture and was developed by OpenAI. GPT stands for "Generative Pre-trained Transformer," a type of pre-trained model built on the Transformer architecture. ChatGPT has been trained on vast amounts of text data, giving it extensive linguistic information and contextual understanding.

the continuous emergence and application of new technologies will provide greater development opportunities for the construction of smart cities, creating more favorable conditions for the continuous improvement of urban governance and service levels.

1.5.5 Integration and Operation of Production and Cities

The Smart Industry Connects City Development: Accelerating the Integration and Collaboration of Industry and City Development

The integration of production and cities is a current trend in urban development. In addition, the integration and collaboration of the smart city industry and urban development has become an important means for cities to achieve sustainable development. The smart industry connects and integrates various city domains through digital and intelligent means, thereby promoting the upgrade and transformation of urban industries.

The development of the smart city industry is a process of not only digital and intelligent transformation but also industrial integration, urban development, and innovation. The smart industry connects and builds various domains of the city, thus achieving the transformation, upgrade, and optimization of traditional industries. This trend of the integration of production and cities will continue to develop, providing continuous energy and momentum for urban development.

Intelligent and Digital City Management and Operations: Innovative Digital Modes

With gradual improvements in the digital infrastructure of smart cities, the acceleration of the application of digital technology in various contexts in the city's operations and the upgrade of diverse application scenario demands, services spanning the entire process and all sectors of city physical space and digital space planning, design, management, operation, maintenance, and security have become the focus. In the next stage, the operational value of smart cities, covering data resourceization, data assetization, data scenarization, etc., will be highlighted. Data collection, storage, processing, circulation, analysis, mining, application, governance, and opening will attract considerable attention. Accelerating the deep integration of digital technology and application scenarios, strengthening data governance and classification aggregation, promoting the scenario matching of urban big data, and facilitating the orderly capture and sharing of public data will become the focus.

1.5.6 Infrastructure Construction and Operations

Accelerate the Construction of Intelligent Platforms and Computing Power Centers: Promoting the Core Infrastructure of Smart City Construction

The construction of intelligent platforms such as urban brains, urban intelligent platforms, and urban operation management centers is accelerating. Various places are striving to build city data aggregation centers and exchange centers, gradually expanding intelligent applications from city management and services to economic development and other fields. With the emergence of smart city scenarios, the attributes of computing power resources are becoming increasingly obvious. With the support of a new generation of artificial intelligence, big data, network technology, etc., the construction of a perfect computing power scheduling mechanism will help build a flexible and intelligent city computing power center that can cross domains and be scheduled, allocated, traded, and regulated. Accelerating the construction of intelligent platforms and computing power centers can meet growing data processing needs, solve complex problems, promote innovation and development, improve efficiency and sustainability, and enhance the security and emergency response capabilities of cities. These aspects are crucial to the success and sustainable development of smart city construction.

1.6 Analysis of the Construction Paths of Smart Cities

The construction paths of smart cities can be divided into innovation-driven, industry-driven, management service-driven, sustainable development-driven, multiobjective development-driven, and "six-in-one"smart city construction paths. These paths have different emphases and applicable scenarios when promoting urban development. There are various types of investment models, such as government self-construction and self-operation, government bonds, government-enterprise co-construction and co-operation (PPP), concession period operation (BOT), concession construction operation (BOO), the EPC model, the F-EPC model, the green low-interest loan finance + N model, and the government industry guidance fund model. These investment models can provide financial support and risk control for smart city construction while adapting to different development needs and resource situations. In terms of operation models, various models exist, such as self-construction and self-operation, franchise operation, service outsourcing, and enterprise operation. These models can be chosen by cities of different scales and development levels to better achieve the management and development goals of smart cities (as shown in Table 1.1). In summary, the development of smart cities requires the coordinated action of various factors, including different construction paths, diversified investment models, and flexible and diverse operation models, all of which in turn require the joint efforts of the government, enterprises, and all sectors of society.

Table 1.1 Comparative analysis of the different construction paths of smart cities

<table>
<tr><th>Type</th><th>Financing model</th><th>Operating model</th><th>Management object</th><th>Construction scale</th><th>Construction cycle</th><th>Representative cities</th></tr>
<tr><td>Innovation-driven</td><td rowspan="6">Government self-construction and operation (traditional model)
Government bonds
Joint construction and operation by the government and enterprises (PPP model)
Build-Operate-Transfer (BOT model)
Build-Own-Operate (BOO model)
EPC model and F-EPC
"Green low-interest loan finance + N" model
Government industry guidance fund model</td><td>Government self-construction and operation</td><td>Enterprise</td><td>Small</td><td>Moderate</td><td>Masdar</td></tr>
<tr><td>Industry-driven</td><td>Enterprise operation</td><td>Enterprise</td><td>Average</td><td>Longer</td><td>Nanjing/Foshan</td></tr>
<tr><td>Management service-driven</td><td>Service outsourcing</td><td>Public sector</td><td>Larger</td><td>Moderate</td><td>Hong Kong/ Seoul</td></tr>
<tr><td>Sustainable development-driven</td><td>Franchise operation</td><td>Public sector</td><td>Larger</td><td>Long</td><td>Amsterdam</td></tr>
<tr><td>Multiobjective development-driven</td><td>Self-construction and operation, franchise operation, service outsourcing</td><td>Enterprise, public sector</td><td>Large</td><td>Long</td><td>Shenzhen/ Kitakyushu</td></tr>
<tr><td>"Six-in-one" type</td><td>Self-construction and operation, franchise operation, service outsourcing</td><td>Enterprise, public sector</td><td>Medium</td><td>Moderate</td><td>Nanjing/ Hangzhou/ Suzhou/ Chengdu</td></tr>
</table>

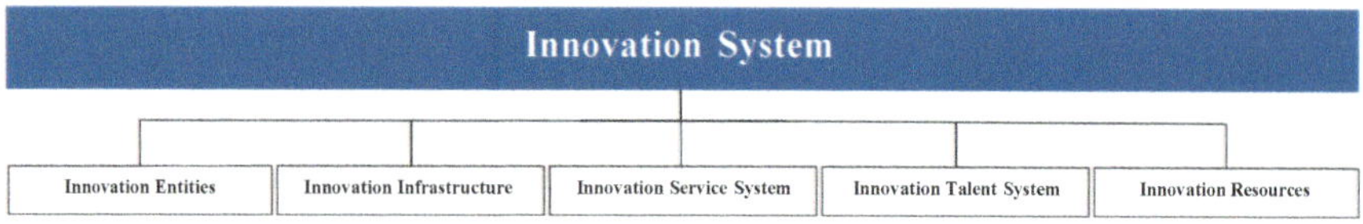

Fig. 1.9 Innovation-driven smart city path diagram[5]

1.6.1 Innovation-Driven Smart City

The innovation-driven path refers to the establishment of a comprehensive innovation system based on the application of emerging information and network technologies. This innovation system includes the main body of the smart city, the construction of innovative infrastructure, the innovative management service system, the innovative talent system, and the innovative resource environment, as shown in Fig. 1.9.

The goal of the innovation-driven path is to promote the development and overall improvement of a city through the use of innovative technologies. In this path, innovation is not only aimed at technology but also considers the comprehensive development of social, economic, and environmental aspects. By establishing and improving the infrastructure and management service system related to smart cities, talent cultivation and the effective use of innovative resources can be promoted, and the enthusiasm of all market entities can be fully stimulated.

1.6.2 Industry-Driven Smart City

The industry-driven path is oriented toward high-tech information industries, establishing a city development path centered on smart industry chains or industry clusters (as shown in Fig. 1.10). Here, smart industries include emerging industries and traditional industries that have undergone intelligent transformation through information and communication technology (ICT). The representative cities of this path include Foshan and Nanjing in China. Foshan, located in the heartland of the Pearl River Delta, boasts unique geographical resources and strives to build a smart city development model by forming industry linkages and functional complementarity closely connected with surrounding industries. The aims of this model are to develop high-tech industries, lead with smart industry chains and industry clusters, integrate intelligently transformed traditional industries, promote the application of digital, intelligent, and artificial intelligence technologies, stimulate urban economic development and enhance industrial competitiveness. This path involves aspects such as industrial upgrading, urban planning, and talent cultivation and requires the joint promotion of government and market entities to create a superior development environment that supports sustainable urban development.

[5] 2016–2021 China Smart City Development Leadership Survey and Investment Strategy Analysis Report [R]. Shenzhen Zhongzi Linghang Investment Consulting Co., Ltd.

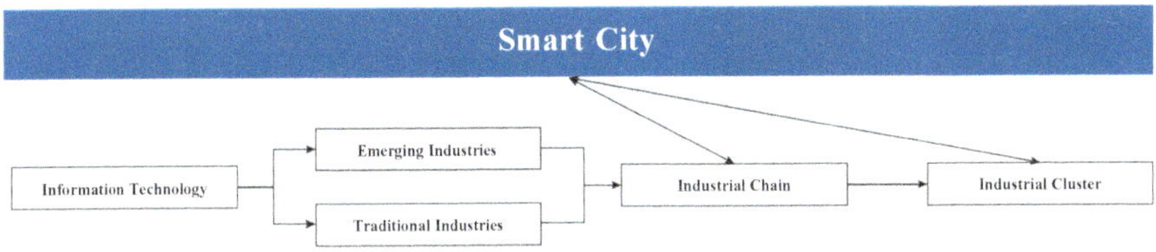

Fig. 1.10 Smart city industry-driven path diagram[6]

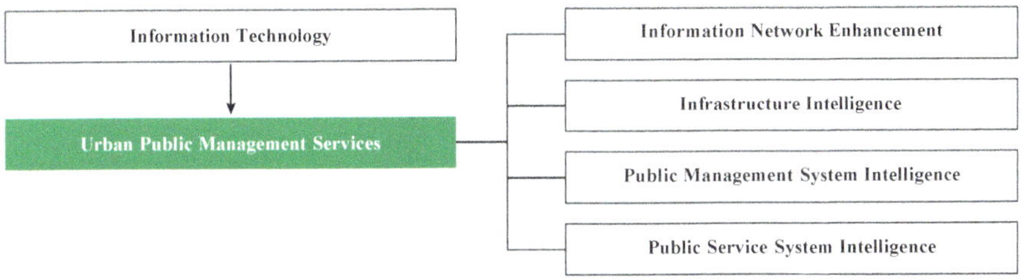

Fig. 1.11 Management service-driven path diagram[7]

1.6.3 Management Service-Driven Smart Cities

The management service-driven path optimizes and enhances a city's public management service capabilities through technology, making public management and services more precise, efficient, intelligent, and convenient (as shown in Fig. 1.11). This path includes improving information networks, intelligently transforming infrastructure construction, and comprehensively enhancing the intelligence of public management and service systems. The path aims to improve the quality and efficiency of public services and promote the modern transformation of urban management services. A representative city is Hong Kong. Since 2007, Hong Kong has implemented a smart city construction model of "government investment, service purchase, enterprise participation, construction and operations", with the main goal of providing information services to all citizens. Through government investment in infrastructure construction, the optimal allocation of resources is achieved, further improving the city's management and service levels. This path requires government guidance and market participation, focusing on the unity of innovation and benefits and creating a more convenient and beneficial urban service environment.

[6] 2016–2021 China Smart City Development Leadership Survey and Investment Strategy Analysis Report [R]. Shenzhen Zhongzi Linghang Investment Consulting Co., Ltd.

[7] 2016–2021 China Smart City Development Leadership Survey and Investment Strategy Analysis Report [R]. Shenzhen Zhongzi Linghang Investment Consulting Co., Ltd.

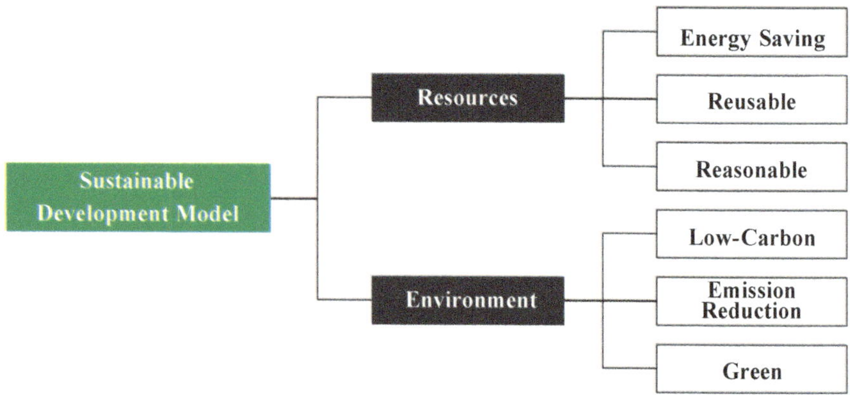

Fig. 1.12 Sustainable development smart city path diagram[8]

1.6.4 Sustainable Development of Smart Cities

The sustainable development path is a city development path that starts with environmental protection and sustainable resource development. Its core goal is to manage environmental resources intelligently, achieve rational, efficient, and reusable use of resources, and establish an environmental resource system that can achieve sustainable development (as shown in Fig. 1.12). This path emphasizes environmental protection and resource conservation; advocates the conservation and rational use of energy, water resources, etc.; achieves efficient reuse and recycling of resources; and reduces the pressure of resource waste and environmental pollution, thereby protecting the ecological environment and human health. This path represents a sustainable urban development model, which emphasizes the coordinated development of the environment, economy, and society.

1.6.5 Multi-objective Development Smart City

The multiobjective development path refers to a comprehensive city development path that considers multiple aspects, such as intelligent upgrading of industries, improvements in public management services, improvements in residents' lives, and the sustainable use of the resource environment during smart city construction (as shown in Fig. 1.13). This path requires the development of a new generation of information technology and is guided by intelligent applications, focusing on the development of smart industries and smart innovation and promoting the construction of and improvements in smart application systems. A representative city on this path is Shenzhen, China, which aims to form a comprehensive model of life, industry,

[8] 2016–2021 China Smart City Development Leadership Survey and Investment Strategy Analysis Report [R]. Shenzhen Zhongzi Linghang Investment Consulting Co., Ltd.

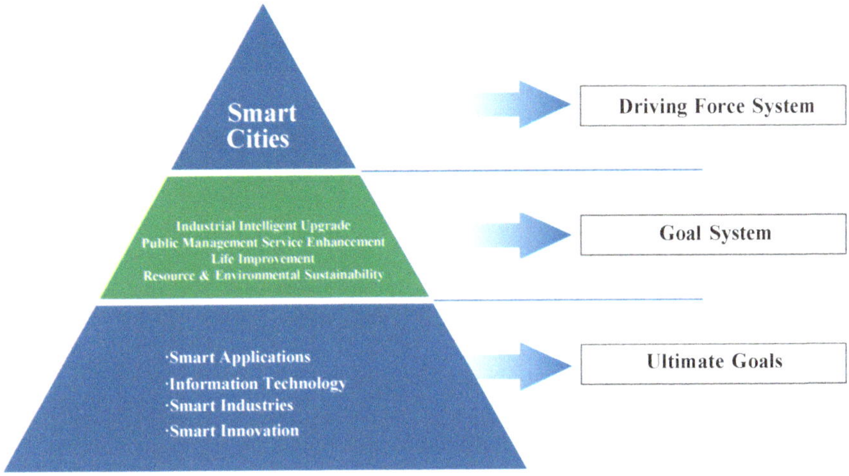

Fig. 1.13 MultiObjective development smart city path diagram[9]

and social management based on the integration of key city information. The core of the multiobjective development path is to balance the interests and needs of various aspects of development and achieve comprehensive, balanced, and sustainable urban development. This path requires the participation of the government, enterprises, and all sectors of society, jointly promoting the development of the city, guided by intelligence and sustainable development, to achieve high-quality, efficient, and sustainable urban development.

1.6.6 "Six-In-One" Smart City

The construction path of a "six-in-one" smart city involves selecting a path from the six major areas of smart city construction, including the main tasks of construction, key projects, choosing a model suitable for local smart city construction, defining each implementation stage, and specifying related safeguard measures[10] (as shown in Fig. 1.14). The areas of smart city construction can be summarized into six aspects: smart livelihoods, smart government/governance, smart applications, the smart population, the smart environment, and infrastructure. Among them, smart livelihoods include culture, education, health, transportation, streamlined services, and integrated online services; smart government/governance includes data governance and process reengineering; smart applications include smart industries; the smart population includes citizens, enterprises, and institutions, along with activities such as

[9] 2016–2021 China Smart City Development Leadership Survey and Investment Strategy Analysis Report [R]. Shenzhen Zhongzi Linghang Investment Consulting Co., Ltd.

[10] Compiled from publicly available sources by the Guomai Research Institute.

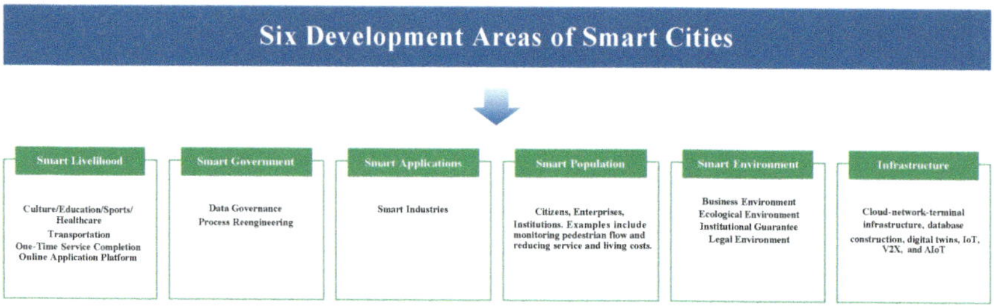

Fig. 1.14 Six major areas of smart city construction

monitoring population flow and reducing service and living costs; the smart environment includes the business environment, the ecological environment, institutional safeguards, and the rule of law environment; and finally, infrastructure includes the cloud, network, database construction, digital twins, the Internet of Things, vehicle networking, and intelligent networking. This path has been applied in many cities, such as Nanjing, Hangzhou, Suzhou, and Chengdu.

The construction path of a smart city can be divided into innovation-driven, industry-driven, management service-driven, sustainable development, multiobjective development, and "six-in-one" smart city construction paths. These construction paths are formulated according to the characteristics and needs of different cities. For example, the innovation-driven type is suitable for cities with strong scientific and technological innovation capabilities, the industry-driven type is suitable for cities with a good industrial base, the management service-driven type is suitable for cities with a high level of city management, the sustainable development type is suitable for cities that focus on environmental protection and sustainable development, the multiobjective development type is suitable for cities with high demands in multiple aspects, and the "six-in-one" type is the most inclusive construction method, which is tailored to local conditions.

Chapter 2
Next-Generation Core Technologies for Smart Cities

2.1 Technology Overview

Enhancing the operational efficiency of cities through digitization and continuously solving practical problems in urban governance constitute an important means to improve the level of urban governance. In July 2022, the Ministry of Housing and Urban–Rural Development and the National Development and Reform Commission issued the "14th Five-Year Plan for National Urban Infrastructure Construction", which proposed accelerating the construction of new urban infrastructure and promoting the intelligent transformation and development of cities. The "14th Five-Year Plan for Digital Economy Development" issued by the State Council in January 2022 defined the digital economy as "the main economic form following the agricultural economy and industrial economy. It is a new economic form that promotes more unified fairness and efficiency with data resources as the key element, modern information networks as the main carrier, and the integration of information and communication technology applications and the digital transformation of all elements as the key driving force."

For both the development of innovative models of smart cities and the demand for digital economy services, it is necessary to build a technology ecosystem that contributes to the development of smart cities with the help of the new generation of core technology systems of smart cities. With respect to the development needs of urban intelligence, using the smart city technology system to compensate for the shortcomings of urban public services, promote high-quality industrial development, and advance precise urban governance is an important means to achieve the new type of smart city entailing device interconnections, data interconnections, business interconnections, industry interconnections, and industry–city integration (as shown in Fig. 2.1). With respect to digital industrial transformation, the development of technology to replace labor is an important driving force in promoting a significant increase in human social productivity. Establishing an

F. Liu, *Smart and Digital Cities*, https://doi.org/10.1007/978-981-95-2243-9_2

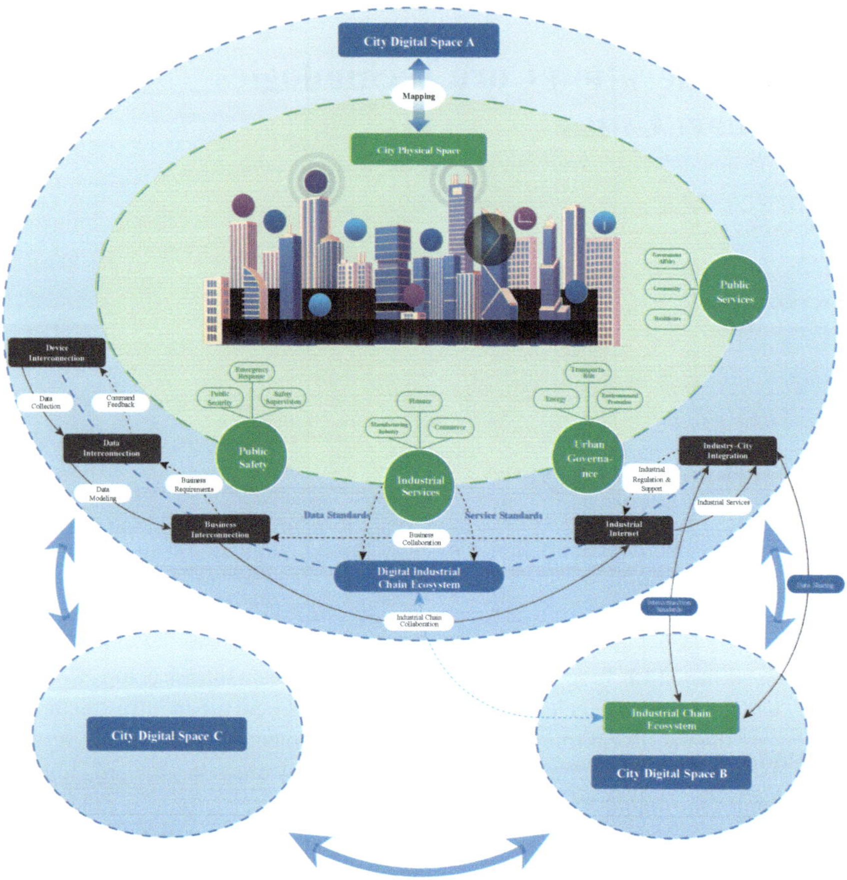

Fig. 2.1 New model of the smart city with device interconnections, data interconnections, business interconnections, industry interconnections, and industry–city integration

industrial intelligence support technology system with the help of the new generation of core technology systems, which efficiently collects, transmits, processes, and shares massive amounts of information, is important to make the information acquisition–calculation–decision–feedback process more efficient.

Smart cities and industries will have three major information processing capabilities[1]:

Comprehensive perception of information: A smart city will contain a large number of sensing terminals, which capture various metadata of people's lives, production, and the urban environment in city operations and services through the sensor network.

[1] Kexia Zhang, Liming Lin. Analysis of Key Technologies in Smart City Construction [J]. Science & Technology Information, 2014(04).

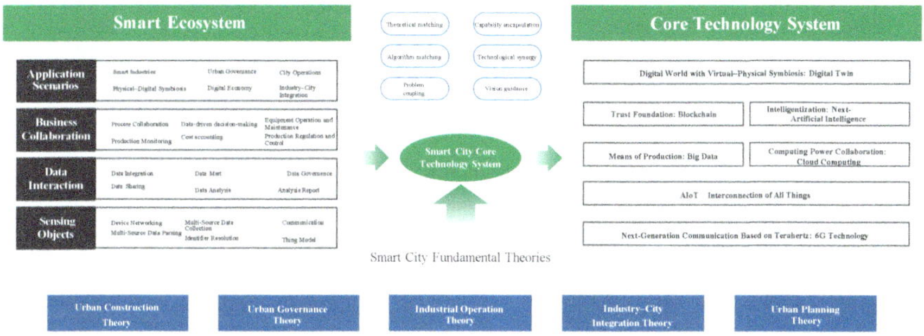

Fig. 2.2 Core technology elements of smart cities

Massive data processing capability: Cities and industries will have the ability to store massive cross-departmental and cross-industry heterogeneous data and the ability to analyze, calculate, and process massive amounts of heterogeneous data efficiently. In addition, they can develop intelligent capabilities based on data analysis and knowledge management.

Intelligent management service capability: To support the industry applications of smart cities, establish a unified public management platform for comprehensive applications of smart cities oriented toward services. In addition, provide universal, intelligent applications and services for residents' lives, comprehensive city management, and production in various industries.

This chapter delves into the eight core technologies applied in the construction of smart cities—next-generation artificial intelligence, big data, cloud computing, blockchain, digital twins, the Internet of Things, 6G, and terahertz. It considers in detail how these eight core technologies in the digital age construct a new model of smart cities that integrates device interconnections, data interconnections, business interconnections, industry interconnections, and industry–city integration. Moreover, this chapter summarizes the development trends of core technologies in smart cities, revealing the development direction for smart cities and industrial digitization.

The core technology elements of smart cities are shown in Fig. 2.2:

2.2 Next-Generation Artificial Intelligence: The Core Technology of Smart Cities

2.2.1 The Digital Age and Next-Generation Artificial Intelligence

Smart cities represent the transformation of traditional industries, urban operations, and economic activities through digital technology. As an important information application method in the digital age, digital technology is key to strengthening the information acquisition–decision–feedback process in industrial operations. Effectively coordinating digital technology with the process of industrial development is an important means of shifting toward a new stage of organizational development. Historically, the development of technology to replace labor has been a significant driver of substantial increases in societal productivity. The results derived from the application of new technologies include improvements in societal, industrial, and organizational productivity; product diversification and enrichment; and increased organizational complexity. The complexity of industrial organizations demands higher and more precise levels of information acquisition, computation, and application.

As an information processing technology within the digital technology system, next-generation artificial intelligence is another key technology for further enhancing productivity development and has become the infrastructure for achieving device interconnections, data interconnections, business interconnections, and industry chain interconnections. In the long run, the role of next-generation artificial intelligence technology in industrial digitization will be the foundation of the new generation of computational systems. A computational system built on next-generation artificial intelligence technology can efficiently complete the information acquisition–calculation–decision–feedback cycle; build artificial intelligence model systems to facilitate the intelligent networking of devices; conduct intelligent analysis of data; enable intelligent interconnections between business systems and intelligent collaborative decision-making between industry chains; and handle the massive amount of information produced by complex systems of devices, products, businesses, organizations, and information applications. Modern smart cities and smart industries, supported by information technology, have extended their networks in all aspects, thereby capturing massive amounts of multimodal data. Using next-generation artificial intelligence to establish model systems to process massive amounts of information and use this information to establish a decision-making basis is an important means to make the information acquisition–calculation–decision–feedback process more efficient.

2.2.2 The Core of Next-Generation Artificial Intelligence Technology

In July 2017, the "New Generation Artificial Intelligence Development Plan" issued by the State Council noted that China's new generation of key common artificial intelligence technologies should be algorithm-centric; be based on data and hardware; be focused on enhancing perception recognition, knowledge computation, cognitive reasoning, motion execution, and human–machine interaction capabilities; and form an open, compatible, stable, and mature technology system. The technical core of the new generation of artificial intelligence is a new type of information processing technology system comprising various technical tools that synergistically integrate big data, algorithms, and computing power.

From the perspective of algorithm design, the design of next-generation artificial intelligence models greatly enhances the accuracy of the model through a large number of parameters. The main advantages of next-generation artificial intelligence technology over previous artificial intelligence technology are as follows:

Solves the problem of the fragmentation and diversification of traditional artificial intelligence models, which greatly improves the universality of the model: When dealing with different scenarios, artificial intelligence models often need to be specifically developed, tuned, optimized, and iterated, which requires high labor costs, leading to the workshopization of artificial intelligence models. The design of next-generation artificial intelligence models adopts the "pretraining + downstream task fine-tuning" method: it first captures information from a large amount of labeled or unlabeled data, storing that information in a large number of parameters, and then performs fine-tuning, greatly improving the universality of the model.

Possess self-supervised learning capabilities and reduces training and development costs: The development of self-supervised learning capabilities can be seen as reducing the dependence on data labeling, and a large amount of unlabeled data can be directly applied. On the one hand, it reduces labor costs, and on the other hand, it makes small sample training possible.

Breaks free from the shackles of structural change and overcomes the upper limit of model accuracy: In the past, improvements in model accuracy were achieved mainly via changes in the network structure. As model design technology gradually matures and begins to converge, it is becoming difficult to overcome the accuracy limit by optimizing the neural network structure. However, research has shown that larger data scales do indeed increase the upper limit of model accuracy.

From the perspective of model development paradigms, the development of the new generation of artificial intelligence models eliminates the issue that domain models are limited to domain knowledge. These new models possess better generalization capabilities, require lower training and learning costs, and can quickly complete transfers between various domains. Over the past decade, building training models through "deep learning + high computational power" has been the mainstream technical approach to realizing artificial intelligence. As the three elements of deep learning, data, and computational power are all available, the world has exhibited fervor for "mass model training," creating a large number of artificial intelligence companies. However, in the nearly 10 years since the rise of deep learning technology, artificial intelligence models have been trained for specific application scenario needs, i.e., small models, which are examples of traditional, customized, workshop-style model development methods. From the research and development to the application of traditional artificial intelligence models, the entire process includes defining the problem, determining the requirements, collecting data, designing the model algorithm, training and tuning, deploying the application, and maintaining operations. Thus, in addition to requiring excellent product managers to accurately determine the model requirements, AI R&D personnel need solid professional knowledge and collaborative capabilities to complete a large amount of complex work. In

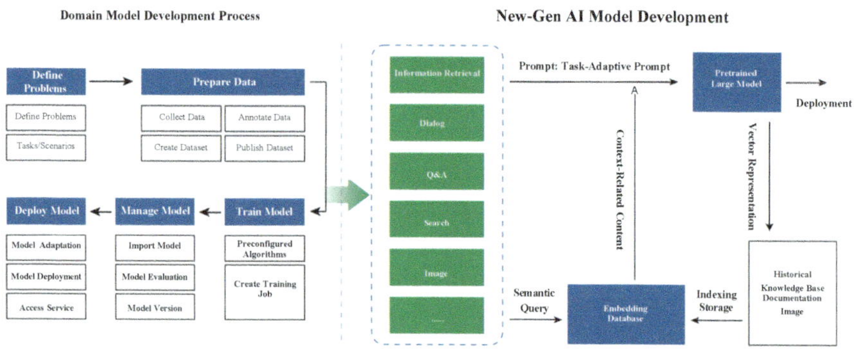

Fig. 2.3 Development process of the new generation of artificial intelligence models

the implementation phase, the models obtained from the "one scenario, one model" approach to domain model development are not universally applicable to many vertical industry scenario tasks.

The new generation of artificial intelligence model development learns from massive, multitype scenario data and summarizes the general capabilities under different scenarios and different businesses, learning a set of features and rules to become a model base with generalization capabilities. When developing applications based on the new generation of artificial intelligence models or facing new business scenarios, the new generation of artificial intelligence models can be fine-tuned, such as secondary training on small-scale labeled data for specific downstream tasks, or not be fine-tuned, in which case they can achieve general intelligent capabilities, allowing multiple application scenario tasks to still be completed. Therefore, the general capabilities of the new generation of artificial intelligence models can effectively cope with diversified and fragmented artificial intelligence application requirements, providing the possibility for the large-scale promotion of artificial intelligence applications. The development process of the new generation of artificial intelligence models is shown in Fig. 2.3:

With respect to the development direction of the new generation of artificial intelligence technology systems, these systems (as shown in Fig. 2.4) will integrate big data intelligence, swarm intelligence, cross-media intelligence, human–machine hybrid enhanced intelligence, and autonomous unmanned systems into a "data, model, computational power collaboration, tool platform, application ecology" technological infrastructure. This infrastructure can help make the shift from artificial knowledge expression technology to big data-driven knowledge learning, from single-type data processing to cross-media cognition, learning and reasoning, from "machine intelligence" to human–machine hybrid "enhanced intelligence", from a focus on "individual intelligence" to that on "swarm intelligence", and from robots to autonomous unmanned systems.

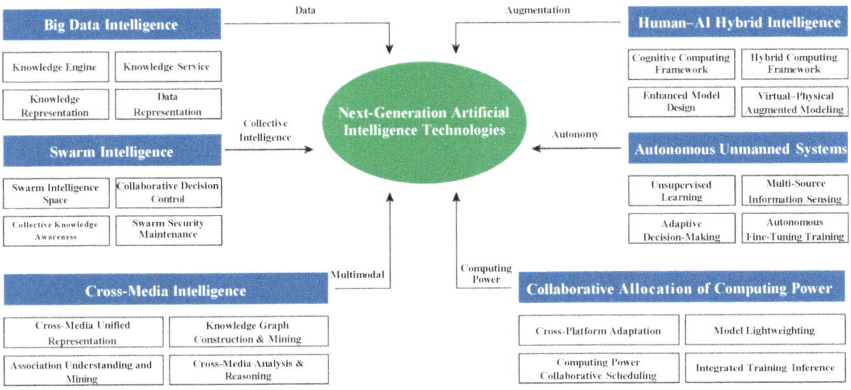

Fig. 2.4 New generation of artificial intelligence technology systems

2.2.3 Generative Artificial Intelligence: A New Development Stage of General Artificial Intelligence

The year 2022 was a watershed year for artificial intelligence. It included the release of several consumer-oriented applications such as ChatGPT, DALL·E[2] and Lensa,[3] the common theme of which is the use of generative artificial intelligence (AI-generated content, abbreviated as AIGC), marking a paradigm shift in the field of artificial intelligence. Current artificial intelligence uses pattern detection or follows rules to help analyze data and make predictions, whereas the emergence of the Transformer[4] architecture has opened a new field: generative artificial intelligence can mimic the human creative process by creating novel data similar to the data it was trained on, thereby elevating artificial intelligence from an "enabler" to a (potential) "collaborator".

Generative artificial intelligence constitutes a major change in the way content is produced, i.e., "professionally generated content" and "user-generated content". Traditional AI can only perform analytical tasks; it is not good at creating and perceiving. The generative artificial intelligence revolution stems from the fact that machines are beginning to possess their own creative abilities, meaning that, in addition to mechanical work, machines are learning to perceive and understand the world and generate new creative content. The AIGC technology scenario is shown in Fig. 2.5:

General artificial intelligence (AGI), also known as general AI, is a type of artificial intelligence that can understand or learn from humans and perform intellectual

[2] DALL·E is an AI tool capable of generating images based on textual descriptions.

[3] Lensa is an AI-powered photo enhancement and editing application.

[4] Transformer is a deep neural network feature extractor. It is widely used in natural language processing tasks such as machine translation and text generation. It also serves as the foundation for pretrained language models, enabling knowledge transfer across a variety of tasks.

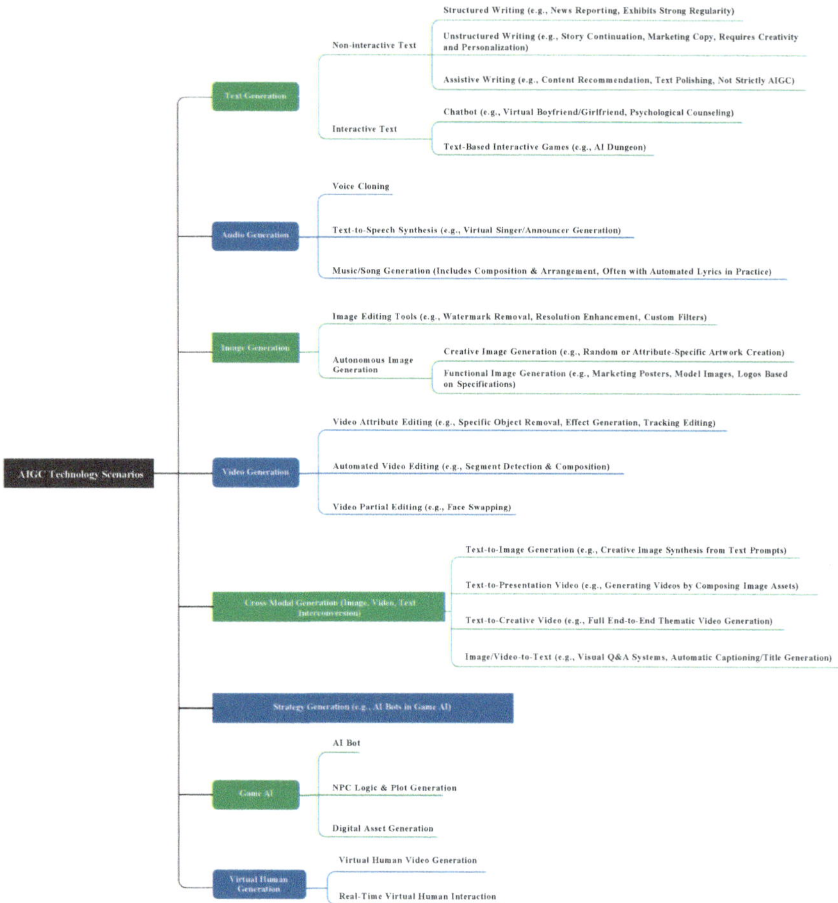

Fig. 2.5 AIGC technology scenarios

tasks. General AI has always been one of the important goals of AI technology development. Its main feature is the ability to transfer the neural network model trained for a single task to adapt to new tasks. The same model can process and integrate multimodal data such as text, images, and voice. Moreover, it only needs to use the relevant parts of the model to solve specific tasks without fully activating the model. Generative AI is a technology that creates new content from existing text, audio files, or images. In November 2022, OpenAI released the generative AI product ChatGPT. Since its debut, ChatGPT has caused a sensation worldwide owing to outstanding "human-like" performance. As a typical generative AI, ChatGPT is supported by a large language model (LLM) created in the field of natural language processing. Its powerful data, computing power, and language model give it excellent text content generation capabilities, ushering in a new mode of human–computer interaction. ChatGPT can be applied to many scenarios and products in real life, including text production, web page translation, and other daily language processing tasks, as well as site production, code writing, script drawing, and other professional technical fields, and can be a "good assistant". GPT-4, which was launched in March 2023, is a multimodal AI model that, while retaining the algorithmic foundation of reinforcement learning from human feedback (RLHF), can better handle conversion and "understanding" between text and images, audio and video, marking another breakthrough in the field of generative AI.

The basic capabilities of the large models of generative AI can already solve the difficult problem of actively creating content according to the goal of general AI. Thus, the development of generative AI will greatly accelerate the pace of achieving general AI. Generative agents based on large models can be used to automate any process. When given a goal, they can create tasks for themselves, complete tasks, create new tasks, rearrange the priority of task lists, complete new priority tasks, and cycle through this process until the goal is reached.

2.3 Big Data: The Key to Eliminating Data Silos

2.3.1 Big Data Technology in the Digital Age

The digital transformation of traditional industries, city operations, and economic activities can be seen as a process of processing and making feedback decisions from information that is collected and generated in various ways. The storage, backup, transmission, mining, and integration of information are the basis for ensuring that digital technology play a transformative role. In brief, data constitute the new basic production material in the digital age. With the gradual interconnection of devices, users, industries, organizations, and society, a large amount of data is generated every moment, which is profoundly changing the way information is processed. This situation has led to the production of large and complex data and corresponding processing technology, namely, big data and big data technology, as these data cannot be handled (stored and processed) by conventional software tools within a certain time period.

From the perspective of industrial operations, city operations, economic activities, and decision-making processes, big data constitute the basic material for achieving a successful information acquisition–calculation–decision–feedback loop. The effective use of big data to circulate, intersect, associate, and integrate analysis within various devices, businesses, organizations, and social structures is the key to ensuring the success of digital processes. Against the backdrop of the rapid development of the internet, cloud computing, mobile, and the Internet of Things, big data present characteristics such as multimodality, massiveness, complex associations, and redundancy. Using big data technology to process massive amounts of data more efficiently, effectively, and accurately is an important means to realizing the new smart city entailing device interconnections, data interconnections, business interconnections, industry interconnections, and city–industry integration.

2.3.2 *The Core of Big Data Technology*

The term "big data" refers to massive, high-growth, diverse information assets that require new processing models to provide stronger decision-making power, insight discovery, and process optimization capabilities. To some extent, big data constitute the technological frontier of data analysis. In simple terms, big data technology entails the ability to quickly obtain valuable information from various types of data. The strategic significance of big data technology lies not in capturing massive amounts of data but in professionally processing these data in a meaningful way.

The meaning of big data technology evolves with the development of traditional information technology and data applications, but at the core, the big data technology system is the basic technology for storing, computing, and processing massive amounts of data. In the development process of big data technology, big data present characteristics of large volume, multiple dimensions, complex processing, and low value density (as shown in Fig. 2.6).

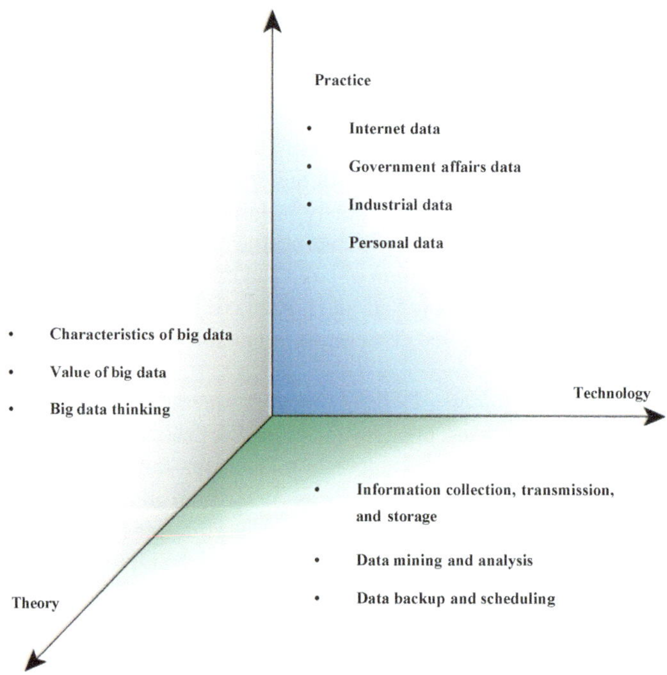

Fig. 2.6 Characteristics of big data

Given the novel characteristics of big data, existing technical architectures and routes are no longer capable of efficiently processing such massive amounts of data. For relevant organizations, if the vast amount of information collected cannot be processed in a timely manner to provide useful feedback, it would result in losses. Considering the current characteristics of big data, the essential function of big data technology is to summarize or discover general patterns from massive, low-value density, redundant data. Compared with previous data processing technologies (such as statistical sampling), big data technology embraces the heterogeneity and redundancy of data. Through correlation analysis of multisource and multimodal data, big data technology can increase fault tolerance. With respect to the development of big data technology, the focus has also evolved from a sole concern with efficiency improvement to the consideration of efficiency improvement, business empowerment, strengthening security, and promoting circulation as equally important.

Efficiency Improvement: Using cloud-native thinking to improve information processing efficiency. The integration of cloud-native technology and big data technology realizes the integration of business data storage and calculation. With the further decoupling of scheduling, security, parsing, and other modules, each module adapts to the underlying resource units such as containers, achieving elastic scaling and thereby significantly improving resource utilization.

Business Empowerment: Using open platforms to improve data utilization efficiency. Open platforms gradually componentize big data processing technologies such as data integration, data scheduling, data security, data backup, data storage, and data governance, further reducing the difficulty of data development and utilization and improving the data scheduling coordination ability of different departments. Data development work is gradually extending from the technical department to various business departments, thus accelerating the integration of data with various business operations.

Strengthening Data Security: Zero trust supplements built-in security. The concept of zero trust aims to eliminate network boundaries for fine-grained access control and currently represents an advanced exploration of data security. As an improvement over traditional network boundary protection methods, this concept does not aim to establish secure areas or trusted users inside or outside the company network; rather, it treats all operations inside and outside the enterprise as untrustworthy. In relation to the concept, design, and implementation of zero trust, various solutions have been proposed, such as lightweight zero-trust network access models and the adherence of all network access to the principle of resource minimization.

Promoting Circulation: Using privacy computing[5] to promote the circulation of data elements. Owing to considerations of data security and privacy protection, data holders have led to the erection of "data walls". Currently, the rise of privacy

[5] Privacy computing refers to a set of technologies that enable data analysis and computation without exposing the underlying data, thereby achieving the goal of making data usable but not visible.

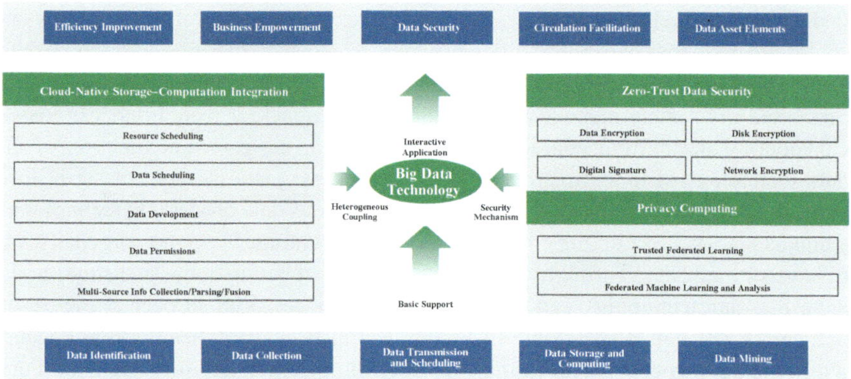

Fig. 2.7 Big data technology system

computing is changing this situation, and trusted federated learning[6] has emerged. Privacy computing is a technology and system for two or more parties to perform joint calculations. Participants can collaborate to perform joint machine learning and analysis on data without revealing their own data. Privacy computing can achieve the conditions where "data are available but not visible" and "data are static, whereas the model is dynamic"; it can eliminate data islands, strengthen privacy protection, strengthen data security compliance; and is expected to become the optimal solution for data security and privacy protection.

The big data technology system is shown in Fig. 2.7:

2.3.3 Integrated Big Data Technology System: A New Stage in the Value Mining of Data Elements

With the convergence of information technology and human production and life, global data are showing explosive growth and massive aggregation, which has a significant impact on economic development, social governance, national management, and people's lives. In recent years, China's government, enterprises, and research institutions have invested considerable resources in conducting research related to big data. Big data have led to substantial progress in the areas of policy, technology, industry, applications, etc. In addition, the big data technology environment has undergone some changes. Some new technologies have emerged, which mainly showing the following main trends:

[6] Trusted federated learning is an enhanced form of federated learning that not only ensures the privacy of raw data and the provable security of models but also guarantees a highly efficient learning process, model usability, the interpretability of decision-making mechanisms, and model traceability and auditability, which are important for regulatory compliance.

Basic Technology: Responding to New Trends in Big Data

With the continuous and explosive growth of data volume and the continuous expansion and deepening of various applications, data further present trends such as exponential growth in the data scale, dynamic skew of data, sparse associations, complex applications, multisource heterogeneous data, and unstructured data becoming the main data. These trends increase the requirements for the development of big data basic technologies such as data collection, transmission, storage, backup, and mining. In the future, basic technology will develop toward the fast replacement of streaming architecture, technology cloudification, heterogeneous computing, open compatibility with various businesses, etc. The traditional "scalability-first" big data processing system design will be replaced by the "performance-first" system design, such as Spark[7] and Flink.[8] The significant market share of systems in the big data processing ecosystem reflects this trend. The rise of specialized big data processing systems in fields such as graph computing (graph accelerators, graph computing frameworks, etc.) and deep learning frameworks (TensorFlow, PyTorch, etc.) also demonstrates the application of this system design philosophy in the technical ecosystem.

Fusion Technology: Guiding Data Value Mining

With the continuous explosive growth of data and the constant expansion and deepening of various applications, mainstream methods based on deep learning, which focus on single-source, single-modal data and have uninterpretable models, can no longer meet development needs. The major challenges currently facing big data analysis include how to overcome the barriers caused by multisource heterogeneous data, integrate the knowledge contained in multidomain or even all-domain data, achieve interpretable analysis results, and thus enhance their usability.

With the further development of advanced applicable technologies such as the new generation of artificial intelligence, cloud computing, the Internet of Things, graph theory, knowledge graphs, etc., big data technology will further integrate these technologies to form a big data value mining system. On the one hand, this system will promote the further development of big data analysis, namely, from focusing on associations to exploring causality, thereby achieving interpretability of analysis results. On the other hand, as the basic production material for the new generation of artificial intelligence technology and industrial digitization, the integrated system will further promote the new generation of artificial intelligence technology toward general artificial intelligence, further strengthen the information acquisition-calculation-decision-feedback loop in industrial digitization, and gradually realize

---

[7] Spark is a fast and general-purpose computing engine designed for large-scale data processing.

[8] Flink is a framework and distributed processing engine used to perform stateful computations on both unbounded and bounded data streams.

the new smart city model of device interconnections, data interconnections, business interconnections, industry interconnections, and industry–city integration.

Cross-Domain Data Storage and Calculation: Promoting Data Element Circulation

Since the 18th National Congress of the Communist Party of China, great importance has been attached to the high-quality development of data elements at the national level, including the proposal to build a digital economy with data as the key element. China's data field laws and regulations have continuously improved the basic data system of "four beams and eight pillars" that was initially formed, government data sharing and opening has continued to deepen, the vitality of the data circulation industry, and data security compliance protection capabilities have gradually improved. The core goal of big data technology is to maximize the value of data, with the key being to eliminate the "data island" and achieve efficient sharing and collaboration of data elements. Traditional data management is limited to single enterprises, businesses, data centers, etc. In the future, big data management will develop from the traditional single-domain mode to the cross-domain mode, spanning spatial, jurisdictional, and trust domains. However, crossing spatial domains causes high and unstable network latency; crossing jurisdictional domains causes data and application heterogeneity, greatly increasing data management complexity; and crossing trust domains requires the ability to tolerate various malicious errors. The changes brought about by the cross-domain mode will bring new opportunities and challenges to big data technology.

2.4 Cloud Computing: The Core Technology Supporting Multinetwork Integration

2.4.1 Digital Economy and Cloud Computing

In January 2022, the State Council issued the "14th Five-Year Plan for Digital Economy Development", which noted that the digital economy is a new economic form that promotes more unified fairness and efficiency, with data resources as the key element, modern information networks as the main carrier, and the integration of information and communication technology and the digital transformation of all elements as important driving forces. The essence of the digital economy is an economic model driven by digital technology, and the only way for all digital technologies to create value is by providing a service. The development of the digital economy has two typical characteristics. One characteristic is decentralization. The concept of a "unified large market" proposed by the state aims to meet the demand for decentralization in the new form of the digital economy, overcoming the circulation barriers of various production elements and forming a freely flowing market system. In this system, personnel, organizations, products, and economic forms flow freely. The other characteristic is digital service, that is, reliance on emerging digital technologies to serve in a decentralized manner. The first dimension of digital service is that digital representatives need to use many new digital technologies, combined with innovative application scenarios, to digitize physical resources and virtual resources, consequently forming digital twins or digital assets.

In the development process of the digital economy, individuals, organizations, industries, cities, etc. in the economic society will share a service-centric digital space. This situation will constitute a new paradigm linking services, and cloud computing will serve as the infrastructure to effectively and deeply combine various services. Various service objects share and communicate with various services within the digital space through cloud computing technology. From the perspective of another core industry in the digital economy, namely, digital service, cloud computing technology will further promote the essence of industrial digitization through decentralization, service sharing, and end-to-end capabilities. That is, this technology will promote the efficient and effective operation of data, models, decisions, and other information in the information acquisition–calculation–decision–feedback loop so that various service objects can obtain what they need and share services in the new smart city model of device interconnections, data interconnections, business interconnections, industry interconnections, and industry–city integration.

2.4.2 *The Core of Cloud Computing Technology*

Cloud computing is a form of distributed computing, which refers to the process of breaking down large data computing tasks into countless small programs through the network "cloud". These small programs are then processed and analyzed by a system composed of multiple servers to produce results that are returned to the user. In the early days of cloud computing, it was simply a form of distributed computing that handled task distribution and the merging of computational results. Therefore, cloud computing is also known as grid computing. With this technology, it is possible to process tens of thousands of data points in a very short time (a few seconds), thus achieving powerful network services.[9]

From the perspective of the effective collaboration of data, algorithms, and computing power required by the digital economy and new smart cities, the core function of cloud computing technology is to achieve the efficient collaboration of computing resources, the convenience of algorithm services, the sharing of various services, and the openness of digital space. It is a necessary and key technology for effective collaboration and sharing among various services in the digital space.

[9] Ziming Xu, Yangfeng Tian. The Development History and Applications of Cloud Computing [J]. Information Recording Materials, 2018, 19(8): 66–67.

Typically, cloud computing services are divided into three types: Infrastructure as a Service (IaaS), Platform as a Service (PaaS), and Software as a Service (SaaS). Infrastructure as a Service is one of the main service categories, providing virtualized computing resources, such as virtual machines, storage, networks, and operating systems, to individuals or organizations that are cloud computing service providers. Platform as a Service is a service category that provides a platform for developers to build applications and services through the global internet, providing on-demand development environments for developing, testing, and managing software applications. Software as a Service is a type of service that provides on-demand, pay-per-use software applications via the internet. Cloud computing providers host and manage software applications, allowing their users to connect to and access applications via the global internet. A complete cloud computing technology system includes virtualization technology, resource service and scheduling technology, middleware service technology, and cloud platform management, as shown in Fig. 2.8:

Virtualization technology is one of the core technologies of cloud computing. Cloud computing provides users with virtual resources, not physical resources. Virtualization is achieved by running multiple virtual machines simultaneously on a physical host. Each virtual machine can run a separate operating system. On the surface, these virtual machines appear to be independent servers, but in reality, they share the CPU, memory, hardware, network cards, and other resources of the physical host. Moreover, the virtual machines on the same physical machine are isolated from each

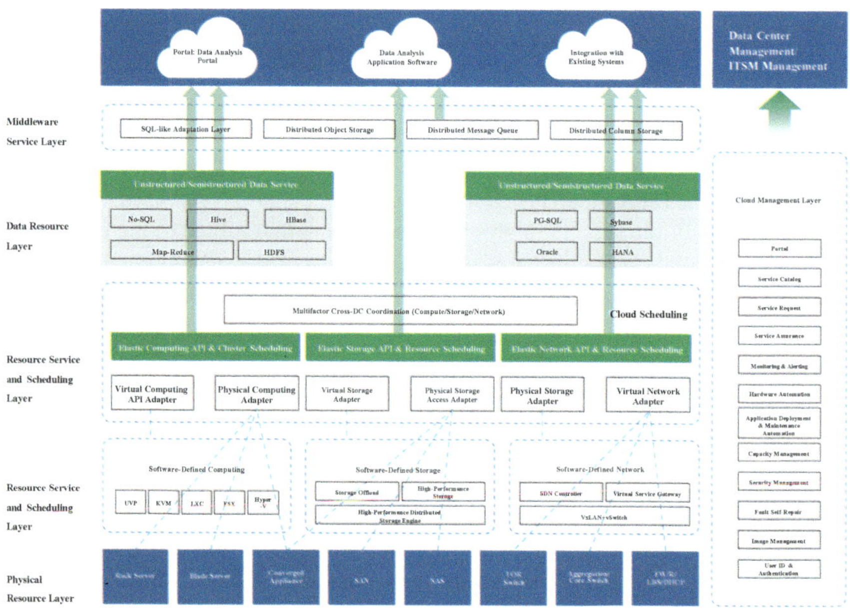

Fig. 2.8 Cloud computing technology system

other, so a fault in one virtual machine will not affect the other virtual machines on the same physical machine.

Resource service and scheduling technology is a key technology for coordinating and scheduling data and computing resources and for realizing the service-oriented nature of computing resources. It includes resource scheduling, service orchestration, and applications. Resource scheduling is based on the SDN[10] bearer network, which enables network connections to access computing power resources from the cloud, edge, and end. For the central cloud, traditional OpenStack[11] is used for resource scheduling and management. For the edge and end side, Kubernetes[12] is used for resource scheduling. For the network side, SDN, DCL[13] and other technologies are used to interconnect data centers. Service orchestration facilitates the service-oriented orchestration of underlying resources and the unified scheduling of computing network resources. Through the capability library, the PaaS platform is further sunk, promoting the further opening of computing network capabilities and providing different business access paths for upper-layer users and developers, thereby realizing self-service capabilities. The application layer connects to the operation support system, business system, and the company's R&D system, providing full-stack computing network resource service capabilities.

Middleware service technology connects different applications, systems, or platforms. Applications can be automatically deployed through the middleware platform. During the deployment process, the middleware platform can dynamically create virtual machines through the service interface of the IaaS layer and distribute deployment packages to complete program deployment. After the deployment is completed, the virtual resource pool resources in the background can be dynamically scheduled according to the concurrent access of the business system, hence realizing the true elasticity and scalability of computing and storage capabilities.

Cloud platform management includes the allocation, monitoring, maintenance, and updating of cloud computing resources. The most critical issue that cloud platform management needs to solve is how to make numerous servers work together to deploy various businesses in the most convenient way, discover and recover system faults at the fastest speed, and then use intelligent and automated means to enable large-scale systems to achieve the most reliable and stable operation.

[10] SDN (software-defined networking) is an innovative network architecture that enables network virtualization by decoupling the control plane from the data plane.

[11] OpenStack is an open-source cloud computing management platform project, designed primarily to provide Infrastructure as a Service (IaaS) to users.

[12] Kubernetes is an open-source platform for managing containerized applications across multiple hosts in a cloud environment.

[13] DCL (data control language) is a category of SQL commands used to control access permissions for data.

2.4.3 Cloud-Edge Collaboration: The Trend of Collaborative Computing Networks

Cloud Computing Services: A Multi-cloud Hybrid Will Be the Mainstream

With the continuous development of cloud computing technology, organizations need flexible and diverse solutions, including joint computing environments between private and public clouds and between public clouds. This design allows businesses to move applications and data between different cloud platforms. It will become mainstream in cloud computing in the next few years because it can provide greater flexibility and scalability while protecting sensitive business data and applications. The trend toward multi-cloud computing will lead to better performance, greater scalability, and improved security, thus providing better solutions for organizations. From the perspective of market demand, various integrated innovative platforms, cloud-native applications, and private enterprise clouds need to work together to encourage businesses to choose multicloud/hybrid cloud architectures. From the perspective of business development, the normalization of business security, data sovereignty, privacy supervision, cost optimization, and the need for innovation in cloud services and new technologies will promote businesses to adopt hybrid cloud methods to advance the digitization of devices, data, business, and supply chains.

Cloud-Native Sinking, Cloud-Edge Collaboration Integration

In recent years, the popularity of cloud-native technologies and concepts, including containers, microservices, and DevOps, has continued to increase. These technologies and concepts emphasize loosely coupled architectures and simple and convenient scalability, aiming to achieve a consistent cloud computing experience on different infrastructures through unified standards. Compared with virtual hosts, cloud-native technology is more suitable for edge cloud computing scenarios, providing integrated application distribution and collaborative management for the cloud-edge and solving the problems of large-scale application delivery, operation, and control on the edge side.

On the other hand, with the development of industrial digitization and smart cities, especially the gradual application of the new generation of artificial intelligence technology in business system computing, the optimization of computing resources is becoming increasingly important. Cloud-edge collaboration integration will become an important part of cloud computing technology research. By optimizing the structure of computing resources, allocating high-frequency calls and low-latency business needs to establish edge data centers, and promote the construction of edge networking of 5G bearer networks, the conditions for sinking computing power and networks to the edge can be created. Moreover, the industrial internet, vehicle networking, remote medical care, and other industry policies explicitly mention edge

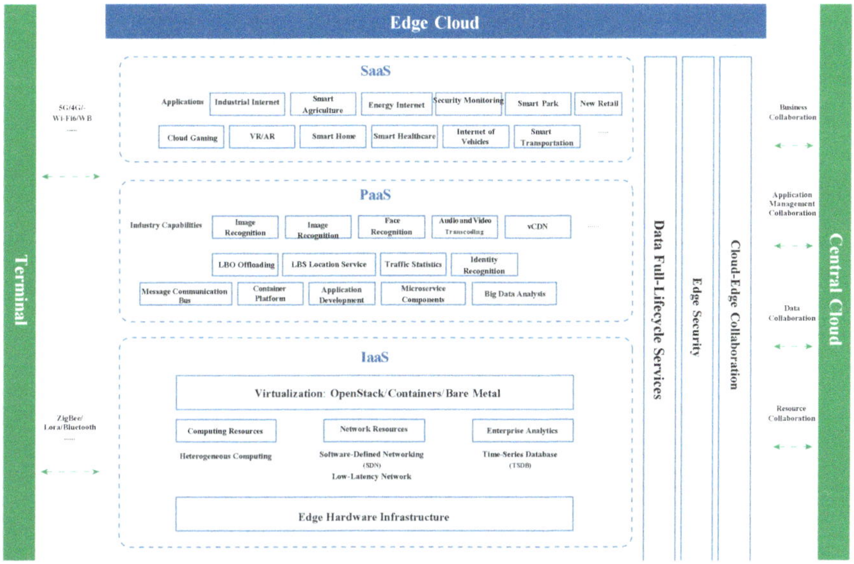

Fig. 2.9 Cloud-edge-end collaboration technology architecture

computing; promote key technology research, standard system construction, and software and hardware product development; and promote the integration of edge clouds in typical industries. The cloud-edge-end collaboration technology architecture is shown in Fig. 2.9:

Computing Power Collaboration, Supercomputing Toward Intelligent Computing

With the new generation of artificial intelligence moving from "single point breakthroughs" to "ubiquitous intelligence" in the process of industrial digitization, a smart society characterized by digitization, networking, and intelligence is accelerating. As the cornerstone of artificial intelligence, intelligent computing power constitutes a key means of constructing a new type of information infrastructure with multielement integration in the computing network. It has become the core engine for the high-quality development of the digital economy, and the construction of an intelligent computing power infrastructure has reached a climax.

The current intelligent computing power services exhibit ubiquitous diversity. Specifically, for developers, cross-architecture application optimization deployment has high development costs. For computing power service providers, it is impossible to achieve reasonable planning of heterogeneous computing power and dynamic migration of applications, resulting in low resource utilization. For cloud-edge computing power collaborative scheduling, the construction of a collaborative system of cloud network computing power, edge cloud computing power, edge node computing power, and computing tasks to maximize the reasonable use of cloud computing resources still needs continuous optimization in key technologies, such as computing power virtualization and heterogeneous computing power unified programming, computing power native interfaces and heterogeneous computing power compilation optimization, hardware native stacks and runtime support mechanisms, computing power network orchestration and resource collaborative scheduling, and multinetwork cloud-edge computing power collaborative scheduling, to provide high-throughput, high-quality, high-security, high-availability, cost-effective cloud computing services.

2.5 Blockchain: The Underlying Technology of the Innovative Model of Smart Cities

2.5.1 Innovation in Smart City Models and Blockchain

The effective and efficient operation of modern society is based largely on trust among people, organizations, people, and organizations. A major feature of future smart cities and even future societies is a "trust society". The establishment of a trust society will be based largely on contracts signed between various entities. Traditional and current contracts often require the intervention of legal institutions to foster trust. This method of ensuring social interaction through external forces is known as a centralized trust architecture. People feel safe to interact, mainly because they believe that the manager will enforce the agreement, punish breaches, and combat fraud and predatory behavior. This model is also the mainstream social governance model in the world today. In the process of social governance, the central institution becomes the storage node for various interaction information. However, the problem with a single information storage node is that information is easily destroyed, tampered with, or even destroyed by humans. Although interaction records can be maintained through firewalls and continuous backup mechanisms, it is unclear whether this layered design can withstand information damage.

With respect to the development of industrial digitization, the communication protocols between devices, the boundaries of data sharing, information sharing between businesses, information transmission between the upstream and downstream portions of the industrial chain, and the information-sharing process between industry and city management all rely on a central institution to set standards and on downstream adaptation to the upstream. The centralized operation mechanism often leads to increased challenges in industrial digitization, and information sharing between devices, data, businesses, and industrial chain formats and between industry and cities becomes extremely costly.

By using blockchain technology to solve the problem of trust centralization at the base level, we can abandon traditional governance models and build a new underlying trust architecture in society and industry. Blockchain technology is a key technology for the operation, management, and operation of innovative smart cities. Blockchain technology changes the centralized method of information storage, transmission, and sharing. Interactions between users form a "block" after verification by a consensus algorithm, and the "block" recording information is stamped with a timestamp and automatically generated on all nodes on the chain in order. There is no central information storage node on the blockchain, and users can view all the information records on the node. The records on the blockchain are difficult to tamper with; a user can tamper with the records on their own node but cannot tamper with the records of all other nodes. If a node on the blockchain is damaged, as long as there is one node that keeps the record, when the link is re-established, all the information on all the nodes can be automatically restored.

2.5.2 The Characteristics of Blockchain

Blockchain is essentially a technical system composed of multiple technologies. The cooperation between these technologies enables the trustworthy circulation of value, which is the core value of blockchain. To achieve trustworthy value circulation, we need to rely on "decentralized (or weakly centralized) management, hard-to-alter data storage, trustworthy peer-to-peer transactions" or a combination of these technologies. We call the value realized by these technologies the local value of blockchain. Some businesses can meet their needs through the local value of blockchain.

In a broad sense, blockchain technology is a new distributed infrastructure and computing paradigm that uses a blockchain data structure to verify and store data, a distributed node consensus algorithm to generate and update data, cryptography to ensure the security of data transmission and access, and smart contracts composed of automated script code to program and operate data. In a narrow sense, blockchain technology is a chain data structure that combines data blocks in chronological order and ensures the immutability and nonforgery of distributed ledgers through cryptography. From the perspective of technology application, the thinking applied to solve problems with blockchain can be defined as a way of thinking that analyzes and solves problems with blockchain technology. Solving business problems with blockchain thinking can initially proceed in the following four dimensions.

Weakened central management. Unlike traditional businesses, blockchain can weaken centralized management to varying degrees, which can manifest as partial decentralization and complete decentralization.

Blockchain storage. Compared with traditional databases, the distributed storage of blockchain can prevent tampering with and the traceability of on-chain data. The prevention of data tampering increases trust in the data, greatly reducing the possibility of data fraud.

Value circulation. Blockchain can complete the anchoring between actual value and data assets. The trustworthy network built by blockchain can support the sharing and circulation of digital assets in multiple scenarios.

Building trust. The tamper-proof nature of data builds trust in data. Smart contracts can automatically execute once triggered through programmable code, thus building trust in peer-to-peer transactions.

The workflow of blockchain technology is shown in Fig. 2.10:

2.5.3 *Blockchain Under Multi-technology Integration: A New Stage of Technology, Application, and Model Innovation*

The Deep Integration of Multiple Technologies Expands Blockchain Technology Development

Blockchain technology will be deeply integrated with artificial intelligence, cloud computing, big data, the Internet of Things, and other technologies to increase the number of application scenarios and increase the development of the real economy. For example, in the supply chain finance industry, the use of the Internet of Things technology alone to connect warehousing logistics, financial institutions, and supply chain enterprises cannot ensure that all connected devices can trust each other and thus transmit transaction data to each other. However, using the consensus mechanism provided by blockchain technology can ensure that all connected devices in the Internet of Things trust each other. Through information exchange and communication, the data generated by various connected devices can be uploaded to the blockchain network, realizing the deep integration of supplier flow, logistics capital flow, and other information and deepening sharing. Blockchain's smart contract technology can enable each smart device in the Internet of Things to execute functions such as exchanging information or verifying identities with other nodes based on established or implanted rules, thus ensuring that the data provided by each data provider are not misused.

Blockchain Will Innovate Industry Applications to Address Credit Demand

Currently, blockchain technology has been applied in various industries, such as education and health care. Leveraging the technical features and principles of blockchain have positioned this technology to become one of the foundational technologies for industry business systems and scenario innovation, addressing the credit needs of various sectors. Under the new smart city model, which requires device interconnections, data interconnections, business interconnections, industry chain

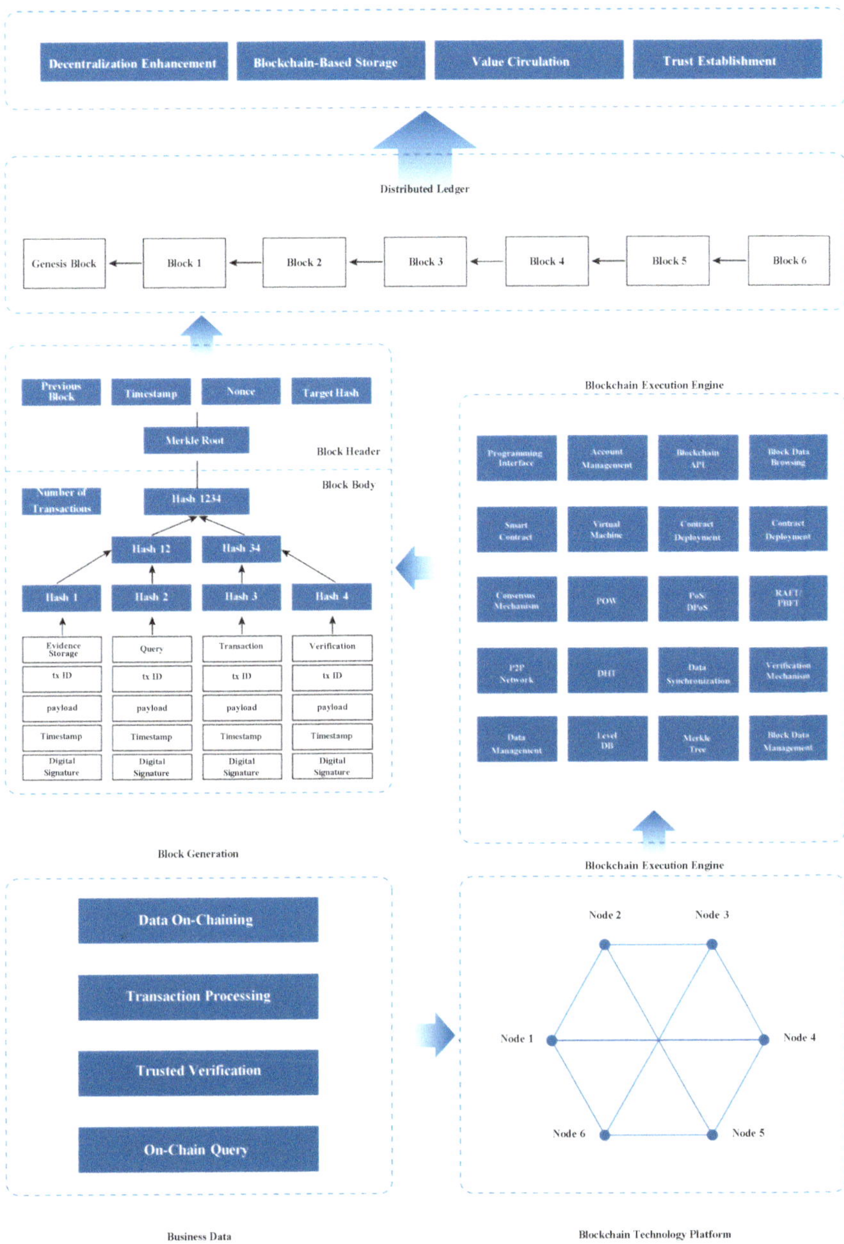

Fig. 2.10 Workflow of blockchain technology

interconnections, and industry–city integration, blockchain technology, as a key technology, will innovate the basic business systems under the smart model. This innovation will establish a new system of devices, data, businesses, industries, and industry–city collaboration. Under this system, the construction of a distributed ledger based on device information will innovate the Internet of Things products and applications, the construction of a data sharing service system based on blockchain will innovate data interconnection models, the construction of industry chain interconnections based on smart contracts will innovate upstream and downstream collaboration, the construction of a business ecosystem based on consensus mechanisms will innovate business couplings, and the construction of a credit system based on blockchain will innovate an operation model for industry development and regulation. The innovative application scenarios of blockchain are shown in Fig. 2.11.

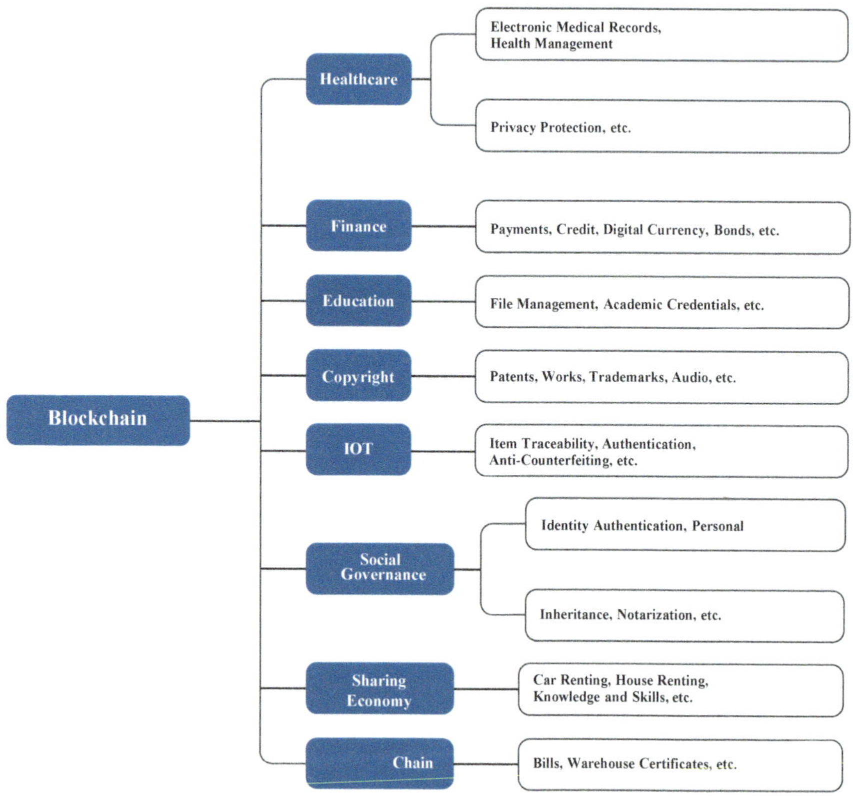

Fig. 2.11 Innovative blockchain application scenarios

Integrating Blockchain Technology into the Metaverse and Innovating Interaction Modes

Advanced artificial intelligence, the Internet of Things, augmented reality and virtual reality, cloud computing, and blockchain technology will play a role in creating the virtual reality space of the metaverse, where users will interact with computer-generated environments and other users in the course of real-life experiences.

A centralized metaverse requires more intensive user participation, more extensive use of internet services, and greater disclosure of personal user data. All of these factors strongly suggest a higher risk of cybercrime. Giving power to centralized institutions to manage, control, and distribute user data is not a sustainable development model for the future of the metaverse. Therefore, developing decentralized metaverse platforms will provide users with autonomy. Decentraland[14], Axie Infinity[15] and Starl[16] are examples of decentralized metaverses driven by blockchain.

In addition, the advanced solutions of Blockchain 4.0[17] can help metaverse users regulate their security and trust needs. For example, on a metaverse gaming platform, users can buy, own, and trade game items with great potential value. To prevent these assets from being counterfeited, ownership needs to be proven through immutable and scarce items such as NFTs.[18]

2.6 Digital Twin: The Core Technology Constructing a Symbiotic World of Virtuality and Reality

2.6.1 Symbiosis of Virtuality and Reality and Digital Twins

As mentioned earlier, a major characteristic of the digital economy is its service orientation. The ability of people, equipment, organizations, environments, industries, or abstract objects to conveniently obtain various services within the digital economy is an important indicator of development toward a service orientation. At the microscale, the various objects, business forms, and subsystems of the digital

[14] Decentraland is a distributed shared virtual platform designed to address the issue of intermediary profit-taking by platform operators in centralized systems. It supports a decentralized, open-source model that enables low-cost transactions between content creators and players.

[15] Axie Infinity is a decentralized turn-based strategy game built on the Ethereum blockchain.

[16] Starl is the sole governance token of the StarLink metaverse ecosystem. StarLink is a blockchain-based virtual environment that simulates a realistic space station located within the galaxy.

[17] **Blockchain 4.0** refers to the next stage beyond Blockchain 1.0–3.0. It emphasizes faster speed, greater ease of use, and enhanced user experience, with primary applications in Web 3.0 and the metaverse. Blockchain 1.0 denotes the cryptocurrency era typified by Bitcoin; Blockchain 2.0 combines smart contracts with digital currency, a model exemplified by Ethereum; Blockchain 3.0 extends decentralized applications beyond finance.

[18] An **NFT (non-fungible token)** is a blockchain-recorded token that represents a multidimensional, uniquely identifiable data object with complex attributes.

economy are composed of numerous entities with diverse structures and highly uncertain behaviors. Moreover, the relationships between local and overall, short-term and long-term, disorder and order, competition and development, and service and regulation among these entities are becoming increasingly complex, thus requiring a large amount of information to describe, with very high complexity. At the macroscale, based on data, analytical models, and a model system of virtual–real symbiosis, we can identify some simple and elegant rules to describe the digital economy and extract its common evolutionary laws. Therefore, to reveal the inherent structure of the digital economy, it is important to construct a parallel system modeling framework that combines virtuality and reality and uses dynamic real data to support the analysis of system operation rules, which will provide an effective, efficient, and economical means to achieve a service orientation in the digital economy.

Currently, digital twin technology, as a key technology for achieving the integration of information and physical systems in intelligent manufacturing and implementing the concept of intelligent manufacturing, has been widely applied in fields such as intelligent research and development, intelligent production, intelligent management, intelligent services, and industrial chain collaboration. It plays a significant role in promoting the digitalization, networking, and intelligent development of industries. From the perspective of industrial digitalization needs, the cyclical system of industrial operation—i.e., information acquisition–calculation–decision–feedback—requires not only information transmission and processing technology at each stage but also the revelation of the inherent laws of industrial operation and the construction of an efficient model system to guide industrial operations. On the industrial side, building a digital system that coexists with reality and virtualizes processes, equipment, materials, workshops, production lines, factories, and various forms of industrial chains, even to the consumer end, is an important method to help reveal the inherent laws of industrial operations. By constructing the relationship between the physical model and the virtual model of each industry subject, obtaining the digital representation of the existing subject, and combining the current and historical condition data describing the subject or process, a virtual representation of a real system and its state can be created. Through the iterative process of information acquisition–calculation–decision–feedback, the inherent laws of industrial operations will gradually become clear on the virtual side.

Given the new demands of digital economy services and industrial digitalization, digital twin technology is a key and important technology for promoting "service-oriented, intelligent" development. The use of digital twin technology will greatly reveal the inherent laws of complex systems, thereby promoting the transformation of digital economy services and industrial digitalization.

2.6.2 *Evolution of Digital Twin Technology*

The core of digital twin technology is to fully utilize data such as physical models, sensor updates, and operation history; integrate multidisciplinary, multiphysical

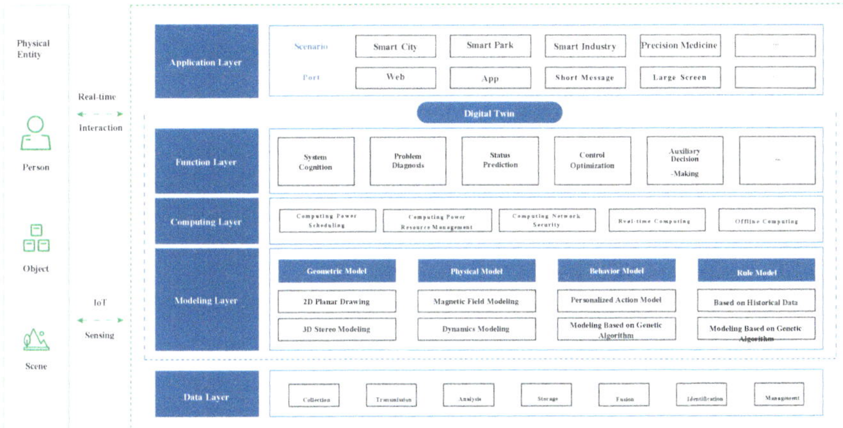

Fig. 2.12 Digital twin system architecture

quantity, and multiscale, and multiprobability simulation processes; and complete mapping in virtual space, thereby reflecting the entire lifecycle process of the corresponding entity equipment. The fusion of multisource heterogeneous data, data-driven precise mapping, intelligent analysis to assist decision-making, and the collaboration of disciplinary theory and modeling simulation are key to the successful operation of the digital twin technology system.

The complete digital twin system architecture is shown in Fig. 2.12. It includes a perception layer, a data layer, a calculation layer, a function layer, and an application layer.

Perception Layer: The perception layer includes various new types of infrastructure equipped with advanced the Internet of Things technology for physical entities.

Data Layer: The data layer includes high-precision data collection to ensure calculation accuracy, high-speed data transmission to ensure real-time interaction, and full lifecycle data management to ensure reliable storage and retrieval.

Calculation Layer: The calculation layer is the core of the digital twin body, which fully utilizes various advanced key technologies to utilize the underlying data and support the upper function.

Function Layer: The function layer is the direct value manifestation of the digital twin body, enabling system cognition, problem diagnosis, state prediction, and decision-making assistance functions. System cognition refers to the abilities of the digital twin body to accurately describe and present the state of the physical entity and to autonomously analyze and make decisions in addition to perception and calculation, with analysis and decision-making being higher-level functions and the goal and trend of intelligent system development. System diagnosis refers to the digital twin body's real-time monitoring of the system, which includes the ability to determine an impending unstable state, that is, "presensing". State prediction refers to the ability of the digital twin body to predict the future state of the physical entity based on system operations data, that is, "foreknowledge". Decision-making assistance refers to providing references for various decisions in the system operation process based on the results presented, diagnosed, and predicted by the digital twin body.

Application Layer: The application layer is the ultimate value manifestation of the digital twin body for various scenarios. It manifests as various products in different industries, which can significantly promote the digital transformation of all walks of life. Currently, digital twins have been applied to smart cities, smart industries, smart health care, vehicle networking, and other fields, with digital twin cities and digital twin manufacturing being the most established applications.

2.6.3 Digital Twins in the Future Metaverse with Multi-technology Integration

The New Generation of Artificial Intelligence Promotes the Precision of the Digital Twin System

The advantages of digital twins can be exploited based on the fundamental characteristics of real-world load systems, such as "from reality to virtuality, controlling reality with virtuality, interactions between virtuality and reality, and coevolution". Following the chain of understanding, computation, simulation, and empowerment, the inherent laws of complex systems can be revealed using digital twin systems. Within a functioning digital twin system, which is based on the theory of complex adaptive systems, the inherent complex laws of digital governance objects in various fields, such as the economy, politics, culture, society, and ecological civilization construction, are explored, aligning the top-down digital model with the development logic and laws of the governance objects. The new generation of artificial intelligence, with its constructed "data, model, computational power collaboration, tool platform, application ecology" technological system, will play a significant role in establishing a model system coupled with theoretical knowledge and digital models. In the process of virtual–real mapping and dual-loop driving, the new generation of artificial intelligence will transform the logical relationships among multisource heterogeneous information such as data, phenomena, and events fed back from physical space into understandable and applicable knowledge, representing it as an "information–model computation–feedback virtual system, which greatly reflects the inherent laws of the system and enhances the precision of the digital twin system.

Digital Twin Technology is the New Stage of Development Toward the Metaverse

As a carrier for one-to-one mapping, replication, and simulation of the real world in digital space, digital twins also constitute a digital infrastructure that symmetrically exists with the physical world in the future metaverse. Thus, they constitute an important part of the metaverse, greatly enhancing the construction efficiency and realistic experience of the metaverse. For the development of the metaverse, constructing a virtual world that matches the operating laws of the real world will accelerate the creation of the metaverse world. With the advent of the Internet of Things, the convenience of data acquisition and the need for effective data processing, the new generation artificial intelligence, the intelligent Internet of Things, digital twins, and other advanced applicable technologies are forming a new type of information processing technology system, which is being used to construct intelligent digital twin bodies that can be used to map the physical world and event processes accurately. A large part of the metaverse will be based on the use of these digital twin bodies to construct large virtual worlds of people, organizations, and societies.

2.7 Intelligent Internet of Things: AIoT Building the Internet of Things

2.7.1 Internet of Things and Intelligent Internet of Things

The development of smart cities and the digital economy and the interconnection and intercommunication of people, things, organizations, industries, business forms, service systems, and social governance in a digital space are notable features. In an interconnected digital space, the convenience of various objects in obtaining services is an important indicator of the level of intelligence of a city or economic structure. To continuously improve the service level of the digital space, the first necessary achievement is the Internet of Things, that is, creating intelligent connections between the four key elements of people, processes, data, and things. By connecting the information generated by objects, organizations, things, processes, economic forms, society, etc., with the Internet of Things technology, a key element network is formed. The new generation of artificial intelligence, cloud computing, blockchain, and other advanced technologies are used to intelligently process the information within the element network. With the help of the virtual–real coexisting world constructed by digital twin technology, various service objects can effectively, efficiently, and freely obtain the required services in the coexisting world. As a subset of the Internet of Things technology, the intelligent Internet of Things technology is an important component of achieving the Internet of Things. Compared with the intelligent Internet of Things, which facilitates intelligent connections between the four key elements of people, processes, data, and things, the Internet of Things interconnects physical devices/objects to form an ecosystem connecting objects/physical devices or to create an ecosystem from things to things.

As an important support for the digital economy, industrial digitalization aims to achieve device interconnections, data interconnections, business interconnections, and industry interconnections, which is an important goal for enhancing the integration between industrial forms and high-quality industrial development. The Internet of Things, using the new generation of information and communication technology as the main means, ubiquitously connects everything and provides services such as information perception, transmission, and processing, which can promote the tripartite integration of "human–machine–thing"; provide key support for the digital, networked, and intelligent transformation of various industries; and injects strong momentum into the development of the digital economy. On the one hand, the Internet of Things constructs a new type of network with low latency, high reliability, and wide coverage, effectively connecting "human–machine–thing"; penetrating different information systems; meeting the differentiated network requirements of digital transformation in various industries; and achieving interconnection, intercommunication, and interoperability across enterprises, fields, and industries. On the other hand, it builds a computing processing platform to facilitate the digital transformation of various industries; optimizes business processes by developing and deploying intelligent applications and services for different scenarios on the platform; and accelerates high-quality industrial development.

2.7.2 Intelligent Internet of Things Technology System

The essential goal of intelligent Internet of Things technology is to build a network of intelligently interconnected physical devices that can collect, transmit, and exchange data without human intervention. Its goal is to form an intelligent ecosystem connecting objects/physical devices, that is, to create an ecosystem from things to things. The national standard Internet of Things general layered model (as shown in Fig. 2.13) is divided into four layers: a perception layer, a network layer, a platform layer, and an application layer:

Perception Layer: This layer is responsible for the collection, automatic recognition, and intelligent control of information from the physical world. Various objects in the physical world do not inherently possess communication capabilities. Intelligent nodes such as sensors, actuators, smart devices, and RFID readers collect information from the physical world and interact with the network layer through communication modules. In the diagram above, the main components of the perception layer include IoT terminals, IoT end nodes, perception extension networks, and IoT access gateways. IoT terminals are terminals that achieve direct access without passing through an access gateway and can interact with relevant functional entities in the IoT network and platform layer. IoT end nodes can rely on the perception extension network or connect to the network or platform layer through the access gateway without relying on the perception extension network. The perception extension network is located

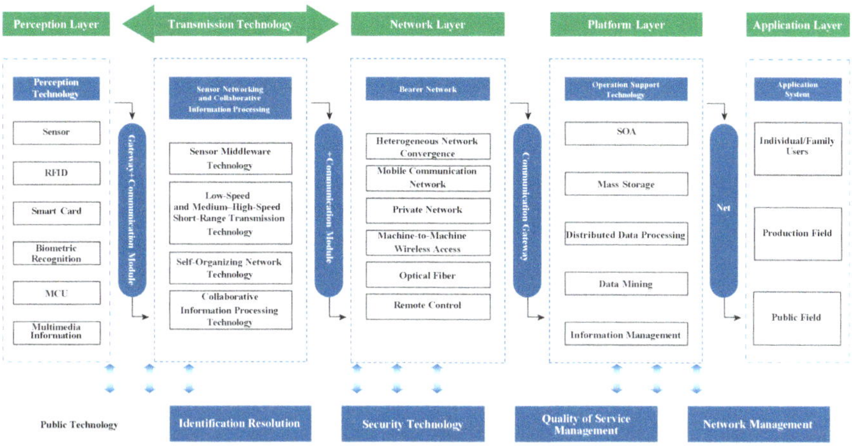

Fig. 2.13 General layered model of the internet of things

between the IoT end nodes and the IoT gateway, which extends communication on the existing network basis and expands the coverage area and service field of communication. Some IoT perception extension layer nodes may only have short-range communication capabilities. To connect to a wide area network, they need to gain access through the perception extension network and the IoT access gateway.

Network Layer: This layer is responsible for supporting the transmission, routing, and control of information from the perception layer and for providing support for communication between people and things and between things and things in the IoT. The network layer includes the core network. The core network of the IoT can take various forms, including existing public networks, such as mobile networks, fixed networks, the internet, or various dedicated networks of governments and enterprises. At minimum, the IoT core network should provide network connection capabilities to support the bidirectional transmission and control of IoT information. It can directly support the development and deployment of upper-layer IoT applications. Accordingly, depending on the type of IoT network, it may also provide network control functions, business control functions, interconnection functions, mobility management functions, etc. It may be further divided into multiple levels, such as network access and transmission, network core control, etc.

Platform Layer: The IoT application support platform provides secure and reliable device connection communication capabilities, supports device data collection to the cloud, governs engine data flow, and ensures cloud data delivery to the device end. In addition, it provides convenient and quick device management capabilities and supports physical model definition; structured data storage; and remote debugging, monitoring, operation and maintenance. To simplify and reduce the complexity of the development and deployment of upper-layer IoT applications, the IoT application support management platform provides some common capabilities and support to IoT applications and provides open interfaces so that applications can access and use network resources and capabilities. These functions allow upper-layer IoT applications to shield the underlying device and network connection methods.

Application Layer: The application layer includes various specific applications corresponding to both public services and industry services. Industry services can be industry public services for the public or industry-specific services to meet specific application needs within the industry. Public services are basic services provided to meet the general needs of the public, such as smart homes and mobile payments. Industry-specific services are services provided to meet the unique needs of the industry, such as smart grids, intelligent transportation, and intelligent environments. Some of these industry services can also be provided to the public, such as intelligent transportation, which can be referred to as industry public services.

2.7.3 Development of Intelligent IoT Under Multi-information Fusion

Multisensor Fusion to Solve Data Heterogeneity Problems

Multisensor fusion has a dual meaning. The superficial meaning refers to the physical integration of two into one, in this case, integrating multiple sensors into a compact sensor device, a typical example being the IMU inertial unit.[19] The deeper meaning is the fusion of data from multiple sensors. Multisensor data fusion can be likened to the human brain's comprehensive processing of information detected by various functional organs, thereby making judgments about the environment and situation. In scenarios such as consumer electronics, autonomous driving, and robotics, through the configuration and management of a large number of various types of sensor nodes, the redundancy and complementarity of multisource data are used to compensate for the errors and defects of single-sensor signals. Data heterogeneity and data conflicts are resolved through data models and fusion algorithms, ultimately providing consistent conclusions or effective decision support. This feature is a technical barrier that manufacturers need to establish. It requires repeated debugging and balancing between sensor combination schemes, costs, computing power, and communication resource allocation. Taking autonomous driving as an example, its signal sources include radar, infrared, images, etc. By mining the inherent connections between redundant and complementary data, high-precision environmental perception images and positioning results are constructed, which then guide the car to perform automatic obstacle avoidance, cruise control, and other driving tasks.

[19] An IMU (inertial measurement unit) is a device used to measure an object's three-axis orientation angles (or angular velocity) and acceleration. A typical IMU consists of three single-axis accelerometers and three single-axis gyroscopes.

Intelligent Sensing Devices: Edge Computing Overcomes Computing Power Limitations

Under the dual drive of 5G communication and the development of the Internet of Things, the number of terminals and the volume of data continue to accumulate, causing bottlenecks in centralized processing architectures. The advantages of distributed local processing in relieving communication and storage burdens, reducing latency, and ensuring data security are becoming apparent. Intelligent sensors integrate traditional sensing units with computing units and AI algorithms, enabling sensors to have information processing capabilities beyond measurement. Through the decentralization of computational power and algorithms from the center to the edge, intelligent sensors autonomously complete the inspection, diagnosis, and calibration of real-time metadata; optimize data quality; autonomously complete data analysis; and execute decision feedback. In emerging application scenarios such as industrial control, medical services, and facial recognition, there are extreme demands for real-time response, minimal error, and citizen privacy protection. Intelligent sensors, empowered by computational power and algorithms, can overcome the limitations of current sensing technology, meet the needs of efficient and secure processing of large amounts of real-time data in the above scenarios, and reduce the cost and resource consumption related to cloud analysis.

Integration of Privacy Computing with Blockchain to Ensure the Security of IoT Data Value

The number of IoT device connections and the magnitude of data generated are growing explosively, and the market demand for data value mining and secure data circulation is increasingly urgent. The integration of privacy computing and blockchain technology can provide security guarantees in the circulation of data across entities, thus offering an important solution to balance data security and the release of data element value. Privacy computing, which is based on cryptography and machine learning technologies, can obtain computational results from invisible ciphertexts, achieving data exchange and open sharing while protecting entity information security. As an important supplement, blockchain technology, with its distributed storage, nonforgery, and traceability features, ensures the authenticity and reliability of the information source. The year 2021 marked the first year of privacy computing. With the coordinated development of the industrial chain and the continuous traction of data trading and open sharing needs, the technology integration of IoT + privacy computing + blockchain will accelerate penetration into various industries.

2.8 Terahertz: Revolutionizing the Underlying Technology of Perception Networks

2.8.1 Next-Generation Network Communication and Terahertz Technology

Terahertz waves are electromagnetic waves with frequencies in the range of 0.1 to 10 terahertz and are used in the analysis, imaging, security inspection, and communication of materials. This emerging technology, which bridges photonics and electronics, has unique advantages. However, in the past, it was referred to as the "terahertz gap (Thz gap)" owing to the lack of effective generation and detection sources and thus was not widely applied. With the significant advancements in ultrafast optoelectronics and low-scale semiconductor technologies since the 1980s, terahertz waves have yielded effective light sources and testing methods, promoting the rapid development of terahertz science and technology. Terahertz wave technology has a wide range of applications in materials science, chemistry, life sciences, security detection, wireless communication, and other fields. For example, terahertz waves can be used for material composition analysis, nondestructive testing, drug detection, cell biology research, food safety inspection, wireless communication, and other applications. Therefore, terahertz technology is a comprehensive technology with broad application prospects and challenges. The position of terahertz waves on the electromagnetic spectrum is shown in Fig. 2.14:

The characteristics of terahertz waves include low quantum energy, which does not cause harmful ionization reactions, and good penetration for most dielectric materials and nonpolar liquids. Terahertz waves can penetrate paper, knitted fabrics, plastics, wood, gypsum, ceramics, and other materials but exhibit poor penetration of metals and water.

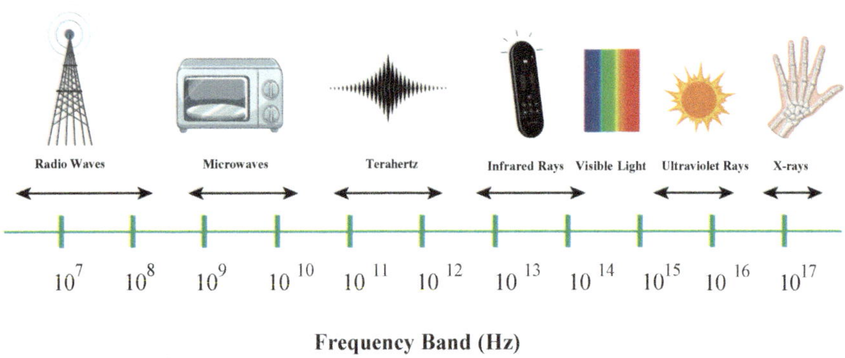

Fig. 2.14 Position of terahertz waves on the electromagnetic spectrum

Mobile communication is moving toward higher frequency bands for in-depth research, and terahertz waves are considered a hot technology for 6G communication that has received much attention. The academic community has widely conducted research and discussions on this topic. As demand, technology, and industry mature, these technologies will gradually attract the attention of the industrial community. The terahertz frequency band inherits the propagation characteristics of microwaves and optics, making it very suitable for long-distance wireless communication, intersatellite communication, and ground mobile communication. However, in the application of ground mobile communication, terahertz technology faces greater challenges and requires more research. For example, the measurement and modeling of terahertz wireless channels, the maturity of device materials suitable for civilian communication, high-speed high-precision capture and tracking technology, high-rate MIMO[20] transmission technology, the definition of application scenarios, network architecture, and networking methods are still in the early stages of research, and many gaps remain in these areas.

2.8.2 *Terahertz Technology and Application System*

To meet the growing demands in communication networks, the integration of microwave communication and optical communication, which is considered a key wireless technology for addressing real-time traffic demand in mobile heterogeneous network systems, can solve problems such as spectrum scarcity and capacity limitations in current wireless systems. Terahertz communication technology, with its advantages of Tbit/s-level data transmission rate, low impact from weather conditions, high security, and the ability to achieve multipoint communication, can provide global application services ranging from nanocommunication to satellite communication.[21]

Nanoscale Applications

Two typical applications in nanoscale scenarios are nanonetworks and on-chip and interchip communication. Nanonetworks connect many nanorobots into a network, which is used in fields such as environmental sensing and human medicine. Moreover, adding an increasing number of computing cores to each chip to achieve vertical expansion of computing power is currently the main method for improving computational capacity. To achieve shared data among cores and synchronize their activities,

[20] MIMO (multiple-input multiple-output) is an antenna system that uses multiple antennas at both the transmitter and receiver ends to form multiple channels, thereby significantly increasing channel capacity.

[21] Sha Xie, Haoran Li, et al. A Review of Terahertz Communication Technologies for 6G Networks [J]. Mobile Communications, 2020, 44(6): 36–41.

terahertz technology will provide fast and reliable services to facilitate this type of interchip communication.

Wireless Access Services

Terahertz communication can further improve the quality of multimedia application services, such as the Super Hi-Vision video format with a resolution of 7680 × 4320, which requires a data rate exceeding 24 Gbit/s. In addition, terahertz communication can support emerging services with massive data and high latency requirements, such as extended reality and holographic communication. Terahertz communication can also meet the needs of places with massive data traffic, such as entrances to public buildings, shopping mall lobbies, and event venues. For users seeking high-speed mobility, terahertz technology can also support high-speed mobile communication, such as autonomous driving.

Ultralarge Capacity Backhaul

In the 6G era, communication between cellular base stations (backhaul) or between cellular base stations and remote radios (front end) at several bits/s will increase the backhaul capacity. Terahertz technology, with its extremely high data transmission rate (peak transmission data rate reaching 1 Tbit/s), will become one of the effective solutions for achieving the standard capacity of 100 Gbit/s backhaul links.

Secure Communication

Solving eavesdropping and interference attacks is key to ensuring user communication privacy and security. The presence of highly directional narrow beams and extremely wide communication bandwidths in terahertz technology makes eavesdropping and interference extremely difficult. Currently, a typical application of terahertz secure communication in military scenarios is that different combat units can form a self-organizing terahertz network within a short distance, thereby preventing the leakage of combat information and ensuring communication security.

Terahertz Space Communication

There is no terahertz absorption loss in space, so satellite communication under the terahertz frequency band has a fast transmission speed, long transmission distance, and the advantage of miniaturization of terahertz communication terminals, making terahertz space communication, such as real-time space viewing, very promising.

The application scenarios of terahertz communication are shown in Fig. 2.15:

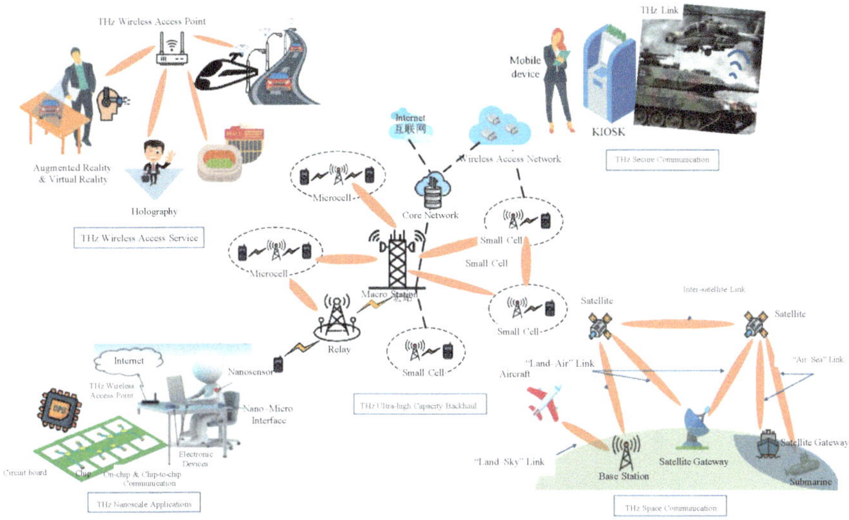

Fig. 2.15 Terahertz communication application scenarios[22]

2.9 Sixth-Generation Mobile Communication Technology (6G): the Key to Building a Triple Network Convergence Ecosystem

2.9.1 Virtual and Real Coexistence and 6G Technology

To meet the development needs of the new smart city model, which is oriented toward device interconnections, data interconnections, business interconnections, industry interconnections, and industry–city integration, it is necessary to build a freely connected digital fusion world. Moreover, to meet industrial digitalization needs aimed at strengthening the industrial operation cycle system—i.e., information acquisition–calculation–decision–feedback"—it is necessary to build a digital world that can adapt to various types of industries. From a technical demand perspective, to build a ubiquitous digital world, freely connect the physical world and the digital world, and achieve interaction and high integration between them, it is necessary to achieve precise real-time digital collection of the physical world through ubiquitous perception and information capture capabilities and build a digital world through ubiquitous connection and fusion computing (including computing power, storage, intelligence) capabilities while supporting digital applications in various industries through strong communication capabilities. Therefore, communication, information, and computing are the three most important basic capabilities for building a freely

[22] Sha Xie, Haoran Li, et al. A Review of Terahertz Communication Technologies for 6G Networks [J]. Mobile Communications, 2020, 44(6): 36–41.

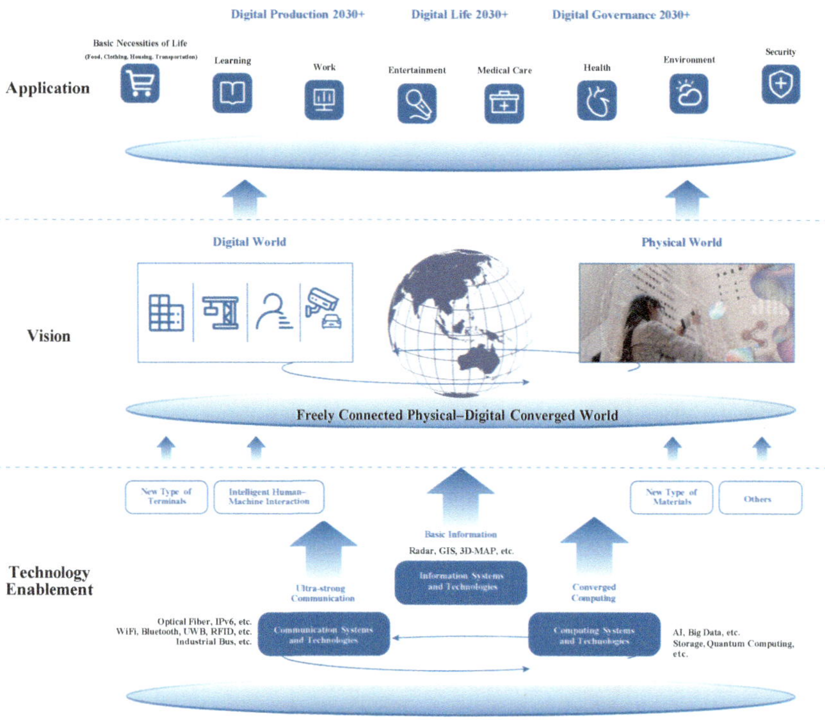

Fig. 2.16 The use of 6G to construct a freely connected physical–digital fusion world

connected physical and digital world. 6G, which is characterized by superstrong communication, basic information services, and fusion computing, will natively support communication, information, and computing services, becoming the network information base for supporting the efficient and sustainable development of future society. The construction of a freely connected physical–digital fusion world by 6G is shown in Fig. 2.16:

2.9.2 Potential Enabling Technologies of 6G

According to the "6G Overall Vision and Potential Key Technologies" white paper released by the China Academy of Information and Communications Technology, the potential key technologies based on 6G (as shown in Fig. 2.17) include new air interfaces and network architectures with inherent intelligence, enhanced wireless air interface technologies, new physical dimension wireless transmission technologies, new spectrum usage technologies, communication perception integration technologies, and other new wireless technologies. They also include

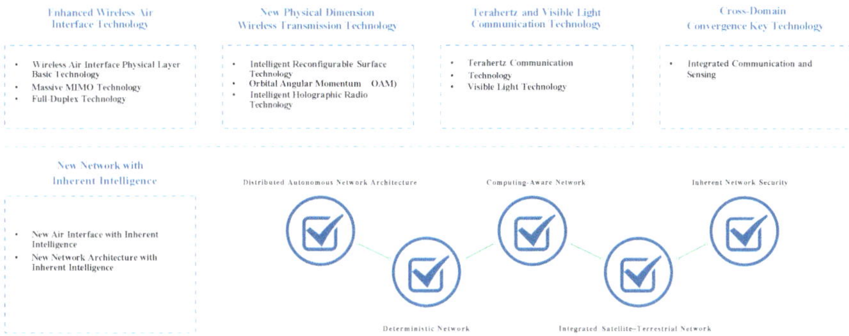

Fig. 2.17 Development direction of potential key 6G technologies

distributed network architectures, computational power-aware networks, deterministic networks, integrated satellite and terrestrial networking, and inherent network security technologies.

The 6G network is a key component in achieving deep integration of the real physical world and the virtual digital world, building a new world of ubiquitous intelligence and digital twins. From the perspective of the basic service system for building the digital world, the basic services that the 6G network can provide include ultrastrong communication services, basic information services, and integrated computing services.

In terms of ultrastrong communication services, cellular mobile communication has aimed to provide seamless communication and connection services in wireless mobile scenarios since its inception. 6G will continue this historical mission and provide ultrastrong communication services.[23] Moreover, 6G will further expand the coverage of geographical space, lower the access threshold for terminals, improve the availability of connections, expand the number of customers, and achieve free connection and information transmission between the physical world and the digital world.

In terms of basic information services, the terminal or base station equipment of the 6G system can measure the received signal while transmitting radio waves for communication, thereby wirelessly perceiving a radio wave propagation environment and the target objects in it and obtaining information such as location, speed, direction, material, imaging, etc. Hence, this equipment can support a rich array of perception applications and scenarios. In addition, as a ubiquitous connection system, 6G will generate a large amount of valuable basic data information in the process of supporting the connection between the physical and digital worlds.

[23] 6G Services, Capabilities, and Enabling Technologies [R]. vivo Communication Research Institute, 2022.

In terms of integrated computing services, computational power, which includes network, computing, and storage resources in multiple dimensions, captures the new productivity in the digital economy era. 6G will inherently support integrated computing functions, provide integrated computing services, including artificial intelligence, through the integration of computing and networking technologies, and better enable a high level of integration of the physical and digital worlds.

2.9.3 6G: A Technological Trend of Virtual and Real Coexistence

Perception Internet

The Perception Internet refers to real-time interactive media interconnection services that involve full holographic collaboration of vision, hearing, touch, taste, smell, emotion, and thought. A typical use case of the Perception Internet, "shadowing real-time shared perception," means that within a predetermined duration, through permission and trust control, a person can truly experience another person's senses or even life through their own vision or other senses. For example, a mother can truly experience whether her child's feet are being rubbed by a new pair of shoes.

AI Service Internet

The AI Service Internet refers to collaborative intelligent interconnection services that anyone, any machine, any organization, or any behavior can enjoy in the future. A typical use case of the AI Service Internet is "highway autonomous driving," in which autonomous cars or fleets are able to design the best route according to real-time navigation and positioning robots, smartly avoid collisions with people or objects outside the car, and reach the destination in the shortest time and with the least energy consumption.

Industry Service Internet

The Industry Service Internet refers to collaborative or virtual twin sensing and execution interconnection services required for any field or platform, cyber-physical system (CPS), or digital twin service. A typical use case of the Industry Service Internet is "haptic feedback robot surgery," in which minimally invasive surgeries such as coronary artery and laparoscopic surgeries are completed remotely through human–machine collaboration and with the help of multichannel auxiliary videos (including augmented reality videos) and haptic feedback.

Business Service System Based on the Integration of Three Networks with 6G Technology

The Perception Internet focuses on the real-time sharing of full holographic perception, the AI Service Internet focuses on ubiquitous intelligence, and the Industry Service Internet focuses on collaborative automation between humans and machines or between machines. To achieve a business service system integrated with the three networks, new challenges are posed to the bandwidth, latency, connection density, spatial capacity, and reliability of communication networks. Based on a general outlook on the characteristics of 6G network technology and the deconstruction of the requirements for building a new technology system for smart cities centered on the business service system integrated with the three networks, with the development of 6G network technology, the transformation of the business service system integrated with the three networks will present the following characteristics:

On the Perception Internet side, it is expected to achieve ubiquitous connectivity, and the acquisition, transmission, and analysis of data by data, sensing devices, information systems, etc., will achieve higher rates, lower latency, and wider connection density. The networking of sensing devices will be more convenient, denser, and broader.

The following three transformations will occur on the AI Service Internet side: the AI Service Internet will be integrated into the 6G network and provide services to the outside world, thereby creating new market value, namely, the AI Service Internet engine. Using the intelligent capabilities of the AI Service Internet engine, it will provide services such as intelligent control; the AI Service Internet will cooperate with the end-edge-cloud, achieving intelligent scheduling of various types and dimensions of resources, including communication capabilities, computing, and storage, and improving the overall performance of the network; the AI Service Internet can achieve wide-area data measurement and monitoring in 6G, realizing rapid automated operation and maintenance of the network, rapid detection, and rapid self-repair, that is, AI-native network maintenance.

On the Industry Service Internet side, the advantages of the high throughput and ultralow latency of 6G network technology, coupled with the basic service system formed by high-density information transmission and collection, large-scale networking on the edge, and the interconnection of cloud AI models via 6G network technology, will make the development and application of industry innovation applications more convenient. The promising prospects of real-time digital twins and metaverse planning, such as ultrahigh bandwidth and ultralow latency, are expected to be realized and solve the problems of insufficient clarity and fatigue caused by immersive experiences and real-time interaction in the industry and dizziness caused by high latency.

Fig. 2.18 Business service system of three-network integration under 6G network technology

The future application value of 6G network technology is reflected mainly in the extensive and scalable collection of data, the real-time nature of interaction, the convenience of obtaining data, etc., thereby leading to more accurate digital models, higher value calculation results across regions and time experience, and more intelligent feedback and services, thus realizing the wisdom of the entire society. The business service system of three-network integration under 6G network technology is shown in Fig. 2.18:

Chapter 3
Cloud-Edge-End Collaboration Technology and Innovative Application Concepts for Smart Cities

As a new driver to enhance urban development, smart cities create a new ecological model conducive to innovation through information technology, integrate urban systems and services to improve the efficiency of resource operations, enhance urban management and service levels, and optimize industrial layouts and operations. The construction of smart cities has proceeded through three stages, namely, the concept popularization stage, the pilot promotion stage, and finally, the development and innovation stage, where it has now entered the new form of "digital smart city".

Given the different stages of smart city construction, different smart city information technology systems and innovative applications are needed to adapt to the needs of urban intelligence. On the one hand, the technology system needs to adapt to the changes in the services provided by the city, and on the other hand, the development of a new generation of information technology also promotes the development of more smart forms of the city. In the "pilot promotion" stage of smart city construction, the integration of the Internet of Things, cloud computing, and the internet to achieve the interconnection and interoperability of various city services is the main form, emphasizing the application of existing technologies, models, and systems. Given the current technology status and smart city construction needs at this stage, we have developed an AIoT-based smart city basic technology system to achieve the integration and interconnection of different intelligent terminal devices, different system platforms, and different application scenarios, thus ensuring everything is integrated. Based on this technology system, for important areas in the city—public buildings, agricultural production facilities, industrial parks, urban governance, energy control, etc.—we have built an innovative application framework for various city services and developed a series of suitable and innovative products. Some research results corresponding to the process of city construction and operation have been well demonstrated. In the new type of smart city construction stage of building a digital space in the "digital smart city", device interconnections, data interconnections, business interconnections, industry interconnections, and city–industry integration will be the main modes of new smart city construction, which will manifest as reliance on

F. Liu, *Smart and Digital Cities*, https://doi.org/10.1007/978-981-95-2243-9_3

a new generation of information technology, data-driven, fully intelligent, multi-dimensional interconnections and future-oriented construction. Given the needs of digital smart city construction, we propose a cloud-edge-end collaborative technology system and innovative application framework for smart cities to solve the real dilemmas faced by smart city construction, provide technical support for smart city construction through an innovative cloud-edge-end collaborative technology system, build an integration and application framework that reflects both generality and industry innovation concepts for smart cities, smart medical care, smart energy, smart agriculture and other fields, and provide a good development environment for smart city construction through technology, management, and model innovation.

3.1 AIoT-Based Smart City Basic Technology System

With the continuous advancement of urbanization and the rapid development of information technology, building a new type of smart city that encompasses device interconnections, data interconnections, business interconnections, industry interconnections, and urban–industrial integration is one of the important trends in urban development. From the perspective of urban development, the foundation of building a new type of smart city is to achieve the interconnection of everything, including people, objects, data, business formats, industries, and organizations. From the perspective of industrial development, on the one hand, competition between industries is becoming increasingly fierce, and a higher degree of collaborative integration between industrial formats is being demanded, which also places greater requirements on the level of interconnection and interoperability between people, equipment, processes, and business formats in the industrial sector to reduce costs and improve efficiency. From the perspective of technological development, Wi-Fi, Bluetooth, cellular network technology, and low-power wide-area network technology are all gradually maturing. Artificial intelligence technology is also being applied in various scenarios. Applications for the currently rapidly developing AI algorithms include machine vision, speech recognition, natural language processing, machine learning, etc.

Given the new demands of urban, industrial, and technological development, traditional IoT technology can no longer meet the interconnections of people, objects, data, business formats, industries, and organizations in the era of ubiquitous smart connectivity. The AIoT technology system, formed by the integration of AI technology and IoT technology, innovates the interconnection mode between various objects and has become the focus of the current development of interconnection technology. Through the Internet of Things, massive amounts of data from different dimensions are generated and collected, stored in the cloud and at the edge, and then analyzed through big data and higher forms of artificial intelligence to achieve the datafication and smart connectivity of everything. Ultimately, the integration of Internet of Things technology and artificial intelligence aims to form an intelligent ecosystem, within which different intelligent terminal devices, different system platforms, and different application scenarios are interconnected and interoperable, thereby achieving ubiquitous integration.

In response to the new demands for AIoT technology integration in cities, industries, business development, and industry, we are constantly exploring the integration of cutting-edge technologies such as the Internet of Things, big data, and artificial intelligence, as well as application scenarios, to form a smart city basic technology system based on the AIoT (as shown in Fig. 3.1). This system is based on the AIoT technology system formed by the integration of artificial intelligence and Internet of Things technologies. The Internet of Things collects underlying data, and artificial intelligence technology processes and analyzes the data. By integrating GIS (geographic information system), BIM (building information modeling), and human–computer interaction technologies, we build advanced and applicable innovative application scenarios. Moreover, we propose a layered distributed integration concept for smart cities, building a smart city basic technology system based on the AIoT with advanced and applicable technologies according to application scenario requirements. The system includes intelligent sensor node and terminal development technology, IoT access platform technology, GIS-based data visualization and analysis technology, BIM visualization technology, video surveillance image recognition technology, intelligent human–computer interaction technology, indoor high-precision location service technology, general basic office platform technology, layered distributed integration technology, etc. It can meet the needs of a massive amount of IoT device access, spatial management and analysis, a large amount of intelligent information data processing, data analysis and prediction.

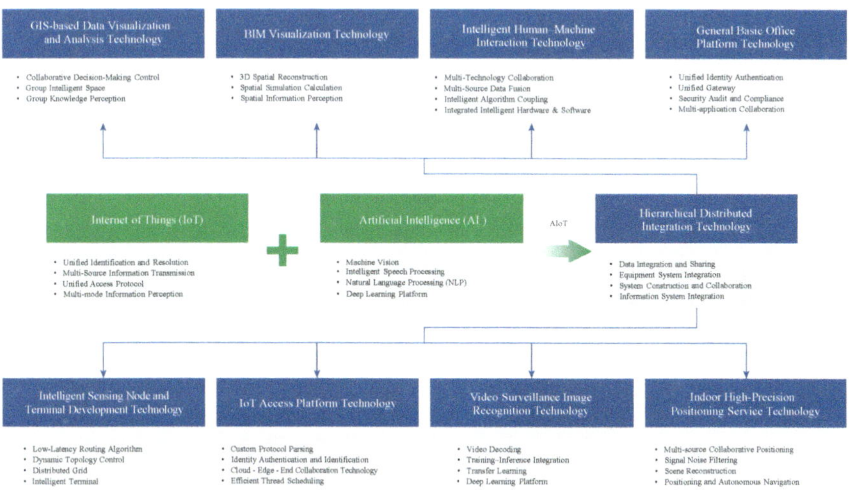

Fig. 3.1 Basic technology system of a smart city based on the AIoT

Intelligent Sensor Nodes and Terminal Development Technology

To address the needs of multisource information collection, transmission, and processing in the era of ubiquitous interconnection, we fully leverage the frontier research theory of IoT technology and innovative application system design experience. In response to the difficulties of information collection equipment docking under multiple sensors, nonstandard transmission protocols, and the low efficiency of multisource information processing, we have built an intelligent sensor node and terminal development technology based on the intelligent IoT. This technology features dynamic topology control strategies, multilevel energy consumption control, low-latency, low-control complexity routing algorithms for large-scale networks, and distributed networked system design, which are key technologies for building IoT intelligent sensor nodes and intelligent terminals. In engineering applications, owing to the adaptability of sensor nodes to sudden environmental changes and the stability of transmission networks, short-term local device failures can lead to data loss, which is the main reason for the failure of the "sensor node–model control–strategy control–feedback control" process. For this reason, we have built an abnormal data estimation completion mechanism and related data collection, transmission, and parsing key technologies. First, all devices are connected to a unified IoT system, and devices are networked within the platform to form a device network. By constructing a virtualized spatial structure, on the one hand, faulty devices can be quickly located through the IoT platform, and on the other hand, to address the problem of local data loss caused by local device failures, the interpolation method is used to estimate the data in the fault area through the data collected by nearby devices, which to some extent reduces the risk of control decision failure caused by data loss.

Based on this technology, a complete intelligent sensor and terminal development system has been formed, which can quickly develop the required sensor nodes and terminal products in a short time according to business needs. This technology integrates IoT communication and intelligent sensing technology to achieve data collection and transmission in various scenarios.

Dynamic Topology Control Technology: With the large-scale and wide-area deployment of sensing devices brought about by industrial upgrades, sensor networks composed of sensors, gateways, and communication facilities will become increasingly large and dense. To ensure data reception over a large range, it is necessary to further increase the number of gateways and communication facilities. Traditional routing methods are more applicable to smaller-scale networks, but when the number of nodes in the network increases dramatically, the complex connection relationships between many nodes exponentially increase the difficulty of path selection. Hence, it is difficult to select appropriate relay nodes for data forwarding under such a complex network topology; even if a routing metric standard that meets communication quality requirements is used, the method will consumes a large number of redundant resources when deciding the routing path, which is unacceptable for a resource-limited sensor network. For this reason, we have developed dynamic topology control technology, which dynamically covers a wider geographical area, includes more sensors or terminal devices, and can adapt to the selection of network layer routing paths in high-density environments.

Low-Latency Routing Algorithm: The application of wireless sensor networks is widespread. In applications where real-time data collection is needed, low-latency routing protocols play an important role. By analyzing the communication process of sensor nodes, the expected energy consumption of the nodes is estimated, allowing the nodes to choose neighboring nodes with lower energy consumption as forwarding candidates. When the final forwarding node is determined, the protocol makes decisions by combining the duty cycle information of the next-hop neighbor nodes of the candidate nodes, allowing the sending node to choose the candidate node that can forward the data more quickly to reduce latency, thereby achieving a balance between energy consumption and latency performance.

Distributed Grid Technology: In a data collection network composed of sensing devices, we can often obtain information about the deployment location of the sensing devices only, which cannot meet the needs of large-scale spatial digital characterization and precise control. For this reason, we have developed distributed grid technology. By analyzing the spatial form of digital characterization, optimizing the deployment of sensor nodes, forming a device network with sensing devices, splitting the sensor network into multiple grids with the help of distributed grid technology, and estimating the information distribution of the grid based on the information collected by the sensing devices in the grid using estimation methods such as interpolation and Bayesian probability, we can obtain the digital characterization of the entire network.

Intelligent Terminal Development Technology: In response to the need for multisource heterogeneous information processing collected by multiple sensor nodes, we have developed development technology for intelligent terminals; built a multiprotocol unified access module for sensor nodes; realized the rapid access of IoT devices; built a multisource heterogeneous information processing algorithm development platform; performed the fusion analysis of image, text, voice, structured and other multimodal data; and built edge computing technology suitable for terminal devices, thus achieving the collaborative scheduling of cloud-edge computing resources.

IoT Access Platform Technology

Based on the needs of high-concurrency access strategies, massive data connections, and distributed computing of massive data for IoT applications, we use custom protocol parsing computing, terminal identity authentication and identification computing, edge-cloud collaboration technology, etc., to create an IoT access platform. The aim is to adapt to common sensor access protocols and communication protocols and provide diversified application programming interfaces (APIs) for upper-layer applications. The connection platform can connect standard or nonstandard protocols of various hardware layers, as well as third-party subsystems. In addition, the platform provides efficient thread scheduling functions, generating the optimal multithread concurrent communication pipeline according to the actual network connection and ensuring a millisecond-level rapid response to the data.

Custom Protocol Parsing Computing: In response to the development of IoT wireless access technology, wireless access technology systems, including short-distance communication technologies such as ZigBee, WiFi, and Bluetooth and long-distance wireless communication technologies such as LoRa, SigFox, eMTC, and NB-IoT, are becoming increasingly complex. Faced with the problem that the access protocols produced by complex and diverse access technologies cannot be quickly matched, we have developed custom protocol parsing computing technology to quickly access various protocols via the IoT platform and uniformly output a unified standard data forwarding protocol to the business application system, thereby reducing the cost of access and output.

Identity Authentication and Identification: Identity authentication technology is an important component of security technology. Its main function is to identify user identities and prevent illegal users from invading the network to tamper with and misuse resources. The purpose of identity authentication is to establish a trust relationship between two parties so that both parties can trust each other's identity without concern. Generally, identity authentication information refers to unique information or secret information of the authenticator, which a third-party authority (authenticated party) cannot forge. The authenticated party needs to show or prove what information they have, thereby gaining the confidence and recognition of the authenticated party and thus identity recognition. In response to the security guarantee requirements of the Internet of Things, identity authentication and identification technology integrates various security authentication technologies, such as RFID smart card authentication, the username/password method, dynamic passwords, USB KEY authentication, biometric recognition, gait recognition, etc. For different application scenarios and security guarantee levels, suitable or multiple integrated authentication technologies are chosen to ensure safe, stable, and efficient access to the IoT platform.

Efficient Thread Scheduling: In the face of massive data collection, transmission, and processing from numerous devices, computational tasks exhibit characteristics of multiple types, with serial and parallel considerations. To address this situation, efficient thread scheduling technology is established. Multithreaded task scheduling refers to the allocation of processor time and thread priority in a multithreaded program according to a specific scheduling algorithm, aiming to increase the efficiency of program execution. The fundamental principle of multithreaded task scheduling is to distribute several threads across different cores of the processor and reasonably schedule the execution order and time slices of these threads through various scheduling algorithms. This principle ensures that all threads have a fair chance of execution, thereby improving the overall performance of the program. Efficient thread scheduling technology integrates static and dynamic scheduling algorithms and selects the appropriate multithreaded task scheduling algorithm based on the specific scenario to better enhance the performance and response speed of the program.

Cloud-Edge-End Collaboration Technology: In response to the insufficient data, algorithms, and computational power coordination brought about by massive device networking, data collection, and data transmission and processing, the Internet of Things may fail as an information processing relay. Cloud-edge-end collaboration technology is a key technology for coordinating computational tasks and resources. First, by building cloud network data storage, backup, and integration, parallel computing is performed on the edge-side business model. The cloud network uniformly schedules computational resources, data resources, and algorithm resourcesusing methods such as computational resource slicing and stacking, peak-valley allocation of computational resources, on-demand dynamic GPU mounting and release, and business storage calculation power allocation. This function optimizes the utilization of computational resources, reduces the impact of computational tasks on the cloud network, enhances the utilization rate of cloud-edge computational resources, and greatly improves the performance of business operations by offloading parallel computing tasks to the edge side and building lightweight computational models there.

Video Surveillance Image Recognition Technology

With the significant development of smart cities and industrial digitization, video surveillance, as an essential tool for comprehensive monitoring, factual control, and anomaly management, has become a must-have for smart hardware. The demand for intelligent analysis of collected video data is growing exponentially, exhibiting characteristics of fragmentation, high customization, and refinement. As a foundational technology for intelligent analysis, machine vision technology can effectively extract key points for anomaly warnings, factual representations, and process optimization from video data. By building video surveillance image recognition technology, technical applications for people, vehicles, and objects, including object detection, event detection, and behavior analysis, such as facial recognition, object tracking, and smoke and fire detection, can be realized.

Data Resource Management: To achieve artificial intelligence, it is necessary to teach computers the ability to understand and judge things, and data annotation is the first step in this process. Data annotation labels images that computers need to recognize and distinguishes features, allowing computers to continuously recognize these image features, ultimately enabling autonomous recognition by computers. Data annotation provides a large amount of labeled data for AI companies for machine training and learning, ensuring the effectiveness of algorithm models. A data resource management platform is established with multiperson annotation, intelligent annotation, and open-call data annotation tools, thereby creating high-quality data resources. Within these data resources, data analysis can be performed directly using preannotated data, thereby reducing the difficulty of data acquisition, annotation, and invocation.

Video Decoding: Traditional video encoding and decoding technologies often face significant computational complexity when processing high-definition video data and thus require stronger computational performance and more efficient algorithms to achieve high-definition video processing. Moreover, the transmission and storage of video data also face issues such as bandwidth and storage capacity. To address these issues, distributed video decoding technology is established, dividing the video encoding and decoding process into multiple subtasks distributed across multiple computing nodes for parallel processing, thereby improving the computational efficiency and transmission and storage efficiency of video encoding and decoding. Compared with traditional video encoding and decoding technologies, distributed video encoding and decoding technology has advantages such as scalability and high fault tolerance and can be applied to large-scale video processing platforms, video surveillance systems, video conferencing systems, and other scenarios.

Machine Vision Model Transfer Learning Technology: The increasing diversity of application scenarios highlights the limitations of single-class models adapted to specific scenarios. In addition, the data quality sensitivity of deep learning models reveals the unsatisfactory state of general intelligence. For more complex and diverse task scenarios, building machine vision model transfer learning technology is a key technology for quickly achieving suitable AI model construction, training, and deployment. Pretrained models are an application of transfer learning and obtain task-independent pretrained models from large-scale data through self-supervised learning. In the field of computer vision, when large-scale image data from various scenarios are used, an initial model encompassing nearly all the entity and semantic features can be trained. When transferred to a specific scenario, relatively accurate results can be obtained with only a small amount of data training. On the other hand, the complexity of deep learning models increases with increasing model accuracy; the latter is achieved mainly by significantly increasing the computational load, which in turn leads to difficulty in deploying highly complex models on hardware-limited devices. By building a model lightweight deployment technology system that includes knowledge distillation, model pruning, model structure optimization, sparse storage and computation, and general model transformation methods, the lightweight deployment of AI models is achieved to adapt to edge computing devices while improving model computational efficiency.

Machine Vision Model Training and Deployment Technology: As the requirements for precise, rapid deployment, and multimodel collaboration of artificial intelligence models become increasingly high with the digitalization of scene applications and industries, the keys to solving this challenge are the collaboration of the entire process of AI model development, training, and deployment, as well as multimodel collaborative computing. To this end, we have established a unified technology for machine vision model training and deployment, providing a full lifecycle toolchain for AI development and application, from data management and processing, algorithm development testing and iteration, model deployment and application, to system operation and maintenance. The powerful performance of this technology in terms of unified training and deployment, open compatibility, and flexible efficiency can significantly lower the threshold for algorithm development and application, helping users quickly and cost-effectively establish algorithm development and application capabilities. By deploying a unified environment for training and inference, we can continuously optimize models to maximize the satisfaction of industry AI application and development scenarios and achieve the deep integration of AI infrastructure and industry needs.

Deep Learning Platform: The development of smart cities requires a smart engine to handle massive amounts of data, business collaboration, and innovative public services. The digital transformation of industries also requires a smart brain to achieve the synergy of business systems and industry operations. Building a deep learning platform that integrates data resources, model development, computing platforms, and AI capability output will be one of the foundations of digital development. The deep learning platform consists of modules such as a data resource management platform, a data annotation platform, a model building and training platform, a model deployment platform, a model invocation and management platform, and an access device management platform, thereby quickly matching the needs of model building, training, deployment, and computing to business scenarios.

High-precision Indoor Location Service Technology

In the era of the Internet of Things, most applications are related to location services, especially for mobile objects, and the demand for positioning is more obvious. In practical applications, how to seamlessly connect indoor and outdoor positioning is more meaningful because only when indoor and outdoor positioning can be seamlessly connected can we truly achieve all-round real-time positioning. To construct precise indoor location services, we have developed high-precision indoor location service technology. Based on powerful indoor capabilities such as 3D indoor maps and indoor–outdoor integration and the integration of multiple indoor positioning technologies such as 5G + Bluetooth + UWB,[1] the indoor positioning accuracy can reach 1–5 m. To solve the problems of the need for destination navigation for the last 50 m and the inability of satellite positioning to accurately cover indoor scenarios, as well as the demand for precise positioning management of indoor personnel and assets, we achieve high-precision positioning and navigation in complex indoor environments, personnel and asset positioning management through "5G + Bluetooth + UWB" collaborative positioning, which enhances the efficiency of user affairs, the level of intelligent management, service efficiency, and digital and information processing capabilities.

Multisource Collaborative Positioning: For positioning services in multiple scenarios, a combination of multiple devices, multiple positioning signals, and multiple signal receptions is often used to meet the requirements of high reliability, high precision, and low latency. We have built a multisource collaborative positioning technology that combines 5G, Bluetooth, UWB, inertial navigation, and machine vision to create a multimodal positioning solution, which further enhances the accuracy and reliability of the positioning system. Given the need to process multimodal information from various positioning technologies, we have built an attention-based feature fusion and time delay difference-based positioning algorithm. Through time difference measurements between multiple nodes and feature-level fusion of various positioning information, we can achieve more accurate positioning results.

[1] UWB (Ultra-wideband) is a wireless carrier communication technology known for its high positioning accuracy, making it particularly well-suited for high-speed wireless access in indoor environments with dense multipath propagation.

Scene Reconstruction: Synchronous positioning and scene reconstruction have always been the research frontiers of multidisciplinary intersection. Building vision-based synchronous positioning and scene reconstruction is one of the basic technologies used to achieve positioning and autonomous navigation. Visual sensors are used to perceive the scene environment, which provides many advantages, such as high accuracy and low consumption. Compared with methods that use traditional sensors, the algorithm for visual positioning reconstruction can not only remain unaffected by some unfavorable factors of the scene and the system itself but also accurately locate the current position of the sensor by optimizing cumulative errors. In terms of map construction, the map composed of sparse point clouds does not have obstacle information and cannot be used for robot navigation and higher-level applications, the dense reconstruction of the entire large scene will place high demands on the robot's operating memory and computing power, and the subsequent processing of the dense reconstruction results will also be very tricky. Hence, we developed a method to densely reconstruct the specific color road lines in the environment into a two-dimensional grid map. Specifically, the map of the landmark line is reconstructed by color domain division, many scenes do not have color-separated road sidelines, and it is very time-consuming and laborious to manually paste landmark lines for larger scenes. Therefore, we use a plane extraction algorithm through the hierarchical clustering of depth images, combined with the gravity direction information of the IMU inertial unit and normal vector screening, to select only the ground for reconstruction.

Information Noise Filtering: One problem to be solved in multisource cooperative positioning is the delay and offset caused by redundant information from multiple sources. By establishing information noise filtering techniques, we construct methods of information noise filtering that include filtering, smoothing, peak detection, signal reconstruction, signal shifting and scaling, data fitting, etc.. The appropriate noise filtering method is selected based on the type of information noise, noise intensity, signal characteristics, etc., of different devices, combined devices, and combined technologies. Moreover, a combination of various methods can be used when processing signal noise to achieve the optimal noise filtering effect, ensuring low latency and drift-free positioning services.

Positioning and Autonomous Navigation: To meet the needs of abnormal patrol inspection in data center machine rooms, facility horticultural crop inspection, large-scale livestock farm biological group inspection, and smart park security patrol, we construct positioning and autonomous navigation technology based on SLAM[2] technology. The robot's positioning on the map is determined by the relocalization algorithm provided by AprilTag.[3] Global path planning uses the A* algorithm[4]

---

[2] SLAM (Simultaneous localization and mapping) is a technology that enables the construction of 2D or 3D maps of an environment while simultaneously determining the position of the device within it.

[3] AprilTag is a visual fiducial system used for a variety of tasks, including augmented reality, robotics, and camera calibration.

[4] *The A** algorithm is a path planning algorithm used within the open-source robotics framework ROS (robot operating system).

provided by the robot operating system (ROS) planner, and local path planning uses the dynamic window method to jointly perform navigation to the designated point on the cost map. The pose correction detected by AprilTag can correct the position error to less than 10 cm. Combined with the ROS-Navigation[5] navigation framework, basic path planning and obstacle avoidance can be completed within the park, thereby obtaining relevant information at specific locations in the park when other sensors are loaded.

Hierarchical Distributed Integration Technology

To build a new type of smart city that includes device interconnections, data interconnections, business interconnections, industry interconnections, and industry–city integration, achieving synergy between devices, data, technologies, business systems, and business models is necessary. A secure and stable digital system requires the construction of a complete set of integrated technology systems to ensure effective coupling between devices, data, core technologies, business systems, and between the technology system and business needs. In response to the integration requirements established for the digital system, we have established a hierarchical distributed integration technology. This technology adopts system integration principles, which aim at subsystem integration, function integration, network integration, and software interface integration. It performs on-demand integration at the device layer, monitoring layer, and information layer, thereby achieving information sharing, centralized management, and distributed control. This technology system has the advantages of high reliability, strong implementability, and a clear design hierarchy. It includes device integration, data integration and sharing, technology system construction and collaboration, and information system integration capabilities. Thus, it is an effective information integration technology for the construction of digital systems in smart cities and smart industries.

Device System Integration: With the aim of building an information management support platform within the organization, related devices and software are integrated in design, installation and debugging, interface customization development, and application support. Based on the AIoT as the foundational technology, a new IoT gateway with new functions is developed, which requires only secondary development for the integration point, thus greatly improving the efficiency and reliability of gateway development. This technology includes the following features: southbound device integration: interfacing with devices of various models and different protocols, collecting their data, and sending control commands to them; northbound cloud integration: the gateway converts the collected data into different protocol formats and sends them to different cloud services; and custom business logic: according to requirements, custom development of the gateway's GPIO (general-purpose input/output) interface, reset button, time synchronization, dip switch, network connection, and other functions.

Data Integration and Sharing: In response to the full business, multiterminal, and multiform characteristics of data presentation in the digital system, we construct data

---

[5] Navigation is a 2D navigation package within the ROS (robot operating system) framework.

integration and sharing technology. Specifically, we structure and standardize data indicators to unify the caliber of indicators, storing them in various databases, data warehouses, or data lakes to achieve data asset management. Various data services are provided upward, building a unified data service interface and data query logic for businesses, facilitating data analysis and display, forming a connection and label system centered on business core objects, and deeply extracting data value. This technology has four major applications: data storage, data fusion, data processing, and data sharing. It integrates the two technical evolution directions of data lakes and data warehouses, providing enterprise users with a cloud-native warehouse-lake integration solution.

Technical System Construction and Collaboration: A digital system often requires the use of various foundational technologies to enable effective interconnection, integration, and interaction among objects such as devices, data, businesses, industries, and organizations. To quickly build a suitable digital system that empowers traditional business processes, it is necessary to select appropriate technologies and ensure effective collaboration among them. In the construction of a digital system, the concept of a layered distribution is adopted to build an advanced and suitable technical system and collaboration method, which fully leverages integration benefits and provides intelligent digital services. The construction of this technology's open technical system includes a foundational layer, a capability layer, and an application layer. The foundational layer is the basis for the operation of the digital technology system, with the core being the virtualization and partitioning of physical hardware infrastructure resources such as servers, storage devices, network devices, and security devices through virtualization technology. Thus, this layer provides scalable, on-demand secure cloud resources and services, covering various types of IT infrastructure such as computational foundations, storage foundations, network foundations, and cloud operating systems. The capability layer focuses on the deployment of the digital technology system business; connects all industrial data through the foundation layer; supports the development, deployment, and operation optimization of the application layer; facilitates data interaction, application development, testing, and operations; and employs cloud computing to transform basic resources into a platform environment for users and applications, including data sharing, business services, data development, operation support, general management, and other functions. The application layer is key to the service provision of the digital technology system. It provides personalized application services for different business scenarios in the smart city field, such as cost management, project management, equipment management, energy consumption management, and other specific needs, through software cloudification and mobile end development, achieving the softwareization, modularization, platformization, and generalization of businesses, technologies, data, resources, and other attributes.

Information System Integration: System integration refers to the process of combining software, hardware, and communication technologies to solve information processing problems for users. The separate parts of the integrated system are originally independent systems. After integration, the parts of the whole can work organically and harmoniously with each other to achieve overall benefits and overall optimization.[6] To ensure effective collaboration among various information systems within the digital system, information system integration technology, which includes hardware integration, software integration, and data information integration, is built based on the concept of layered distributed business system integration, addressing the difficulties of interconnection, interaction, and coupling between systems in traditional business processes. This technology can provide an optimized comprehensive design and large-scale integrated digital system for various applications within smart cities; maximize the organic composition, efficiency, integrity, flexibility, etc., of the system through the integration of products from different manufacturers; and simplify the complexity of the system.

Data Visualization and Analysis Technology Based on GIS

To date, GIS has been widely and successfully applied in domestic and foreign fields. It has not only been successfully used in surveying, mapping, resources, and the environment but also become an important tool for urban planning, public facility governance, engineering construction, and other applications. With the development of the Internet of Things, artificial intelligence, digital twins, and other technologies, building a visualization space that integrates advanced technologies and GIS is an important part of the digital system. By deeply integrating data storage management, computation, and spatial visualization technologies, static geographic spatial data or real-time dynamic location data can be presented in a more intuitive and efficient way, extracting and analyzing information about the spatial location, distribution, morphology, formation, and evolution of geographic objects from spatial data. By accurately combining massive amounts of data such as population, housing, and component data with maps, a more vivid spatial big data application is formed, which integrates cutting-edge technologies and excellent components and enhances the technical performance of the GIS platform. It combines cross-platform, distributed, data storage, data computation, and other technologies into existing technologies and enhances spatial data operation analysis. Given the problems of data integration and interoperability between different data sources in the digital space based on GIS, such as data format conversion, coordinate system conversion, and data consistency, data visualization and analysis technology, which performs group knowledge perception, spatial analysis, spatial information visualization, collaborative decision control, and group safety maintenance, is established to support the construction of the digital system.

Group Knowledge Perception: The quality of GIS data is crucial to ensure the accuracy and reliability of analysis and decision-making. However, the quality of geographic spatial data may be affected by multiple factors, such as data sources, collection methods, and update frequency, leading to uneven data quality, including aspects such as accuracy, completeness, consistency, etc. When processing and managing a large amount of geographic spatial data, the uncertainty of data quality

[6] Hohai University. Compiled by the Editorial and Revision Committee of the Hydropower Dictionary. *Hydropower Dictionary* [K]. Shanghai Lexicographical Publishing House, October 2015: p. 528.

may lead to errors or misguidance in terms of analysis and decision-making. On the other hand, GIS usually needs to integrate geographic spatial data from multiple sources, including data of different formats, different coordinate systems, different accuracies, etc. Data integration and interoperability are the pain points of GIS because they may face problems when working with different data sources, such as data format conversion, coordinate system conversion, data consistency, etc., thus requiring considerable time and resources to solve. By building group knowledge perception technology oriented to GIS space establishment, multisource information is collected and accessed with the help of intelligent networking technology, and multisource data are managed uniformly with the help of big data technology, ensuring the consistency of the data accessed via the GIS platform.

Spatial Analysis: To provide convenient information services for the digital system, it is necessary to use spatial analysis techniques to obtain the spatial location and distribution of key targets in the spatial data of the Internet of Things, revealing the relationships and patterns among targets, geographic data, and spatial locations. Spatial analysis integrates spatial queries, overlay analysis, buffer zone analysis, etc., to reveal the relationships and patterns among geographic data. By performing geographic data analysis based on attribute analysis, the attributes of the data are statistically analyzed, such as via data classification, cluster analysis, and data filtering. Additionally, technologies such as knowledge engineering, planning, decision-making, and automatic reasoning in artificial intelligence are used to perform complex relationship analysis and achieve the dynamic display of spatial data.

Spatial Information Visualization: In response to the needs of spatial geographic information visualization, a variety of spatial information visualization technologies have been developed, including dynamic maps, interactive fusion maps, and city information modeling (CIM). Using GIS mapping, the CIM model framework and other tools, geographic information, demand target information, spatial analysis information, etc., are displayed in multiple dimensions. Moreover, data interpolation, heatmaps, computational fluid dynamics and other methods are used to estimate the entire 2D/3D spatial information with locally collected data. Based on multiresolution data expression, multidimensional table data display, real-time dynamic processing, parallel technology, interactive feedback technology, and GIS information visualization services for multidimensional space are provided.

Collaborative Decision Control: In a standard digital system, collaborative decision control is an extremely important part of the information acquisition–calculation–decision–feedback process. Only when the collection, analysis, and processing of information are mapped to industrial operation monitoring, control, and decision-making can the digital system truly provide effective services for the modeling object. In response to the visualization needs of industrial operation feedback decision-making, a collaborative decision control technology based on GIS spatial information is established. In the constructed GIS space, on the one hand, important information monitoring and early warning are realized by reflecting the changing trend of spatial information, facilitating the discovery and prediction of abnormal phenomena in multidimensional space, local space, and fine targets. On the other hand, combined with the theoretical knowledge and control system of the target business process, a feedback decision mechanism is established to facilitate the real feedback of decision information and the precise issuance of decision instructions, thus enabling fine control of the space process and the object process.

Group Safety Maintenance: Geographic information is an important basic and strategic resource of a country, is directly related to national sovereignty, security, and interests, and reflects insights about the resources, the environment, the population and other economic construction and social development. Digital spatial information security has two parts: the data security of geographic information and the security of target information processing. To ensure geographic information security, spatial geographic information development technology oriented to domestic GIS basic software is constructed, quickly adapting to the domestic GIS basic software framework. At the same time, a full-process security isolation mechanism for geographic information collection, processing, and display is constructed, ensuring the security of geographic information. In response to the security needs of spatial target information processing, a customer-oriented data security, data backup, and data display system is established using network security, big data security, the Internet of Things privacy computing, and trustworthy deployment of artificial intelligence models, thus maximizing the security of the spatial information library and disaster recovery capabilities.

BIM Visualization Technology

In the context of information technology development, with the application of cloud computing and big data technology and the significant improvements in computing resource performance, BIM technology is becoming one of the core technologies driving the development of the construction industry. The mature development of BIM technology has improved the degree of information integration in construction engineering, deepened its application throughout the entire life cycle of the construction industry, and ultimately formed a development pattern of high efficiency, energy savings, and cost reduction. Currently, intelligent building digital systems based on BIM technology present trends such as multitechnology collaboration, multisource information fusion, large computational volumes under complex multicomponents, and increasing requirements for the timeliness of spatial information calculations. In response to the needs of constructing an intelligent building digital system and the basis of the multitechnology development of new technologies, a digital system visualization technology for intelligent buildings based on BIM technology is established. Based on BIM and Internet of Things technology, a new three-dimensional simulation intelligent park equipment management system is built, which achieves structural interaction simulation and related management equipment and facility data integrated management and the linkage of the entire park and buildings and rooms and bridges the underlying data of intelligent subsystems, achieving design visualization, electromechanical pipeline comprehensive visualization, collision detection visualization, technical disclosure visualization, space visualization, construction organization visualization, construction progress visualization, three-dimensional rendering effects, etc.

Spatial Information Perception: During the construction design phase, the integration of BIM and Internet of Things information allows engineers to fully understand the comprehensive situation of the building, thereby enabling them to propose more reasonable design and construction plans. By setting controllers and sensors in a distributed manner in the building; monitoring the temperature, air quality, and humidity in the building; combining heating, ventilation, water supply and other control information; and using intelligent Internet of Things technology to collect the above information in real time, a digital information collection system for intelligent buildings is constructed.

Three-Dimensional Space Reconstruction: Intelligent buildings are complex systems that encompass the building structure, energy supply, production and living spaces, and supply of production and living materials. Constructing a three-dimensional space that aligns with and integrates reality is a key aspect of building information modeling (BIM) technology. Moreover, the reconstruction of a building's three-dimensional space is a collaborative project involving multiple technologies. On the one hand, rapid architectural modeling can be achieved using existing building component models in BIM development software. On the other hand, information about bridges, roads, pipelines, and living spaces can be generated using laser scanners and generative artificial intelligence technologies. Through point cloud modeling and the efficient production of characteristic component models using artificial intelligence 3D vision technology, precise, efficient, and detailed reconstruction of three-dimensional space can be achieved.

Spatial Simulation Calculation: To achieve detailed management throughout the lifecycle of a building, including surveying, design, construction, supervision, acceptance, and operation, a significant number of computational resources and numerous simulation calculation models are required for spatial information collection, spatial reconstruction, model rendering, and digital reality capture. The powerful computational capabilities of cloud computing technology can solve the challenges associated with architectural structure and energy consumption calculations. Its rendering and analysis processes can achieve real-time computation, helping designers make accurate choices among various designs and solutions. Moreover, a collection of multisource information fusion analysis models for the digital system of intelligent buildings, such as data analysis models for energy supply and consumption, is established to optimize the energy supply.

Intelligent Human–Machine Interaction Technology

Human–machine interaction refers to the process of information exchange between humans and computers using a certain dialog language and interaction method to complete specific tasks. This interaction is facilitated by input–output devices and corresponding software. Devices used for human–machine interactions include keyboards, mice, and various pattern recognition devices. The software corresponding to these devices is part of the operating system that provides human–machine interaction functionality. By constructing a human–machine interaction system oriented toward information processing and feedback and integrating information technology, multimedia technology, network technology, and other means, devices such as high-definition cameras, ID card readers, printers, surveillance cameras, QR code scanners, AR glasses, etc., can be integrated. By using artificial intelligence technologies such as image recognition, voice recognition, and natural language processing (NLP), graphic and voice interactions and Q&A sessions with users can be conducted, providing users with intelligent guidance and matching services. This technology is applied mainly in government and corporate consultation service halls, where it provides 24-h one-stop consultation and services and replaces manual work to complete user consultation, replies, operation guidance, appointments, and other daily user services in a quick and timely manner.

Multisource Data Fusion: Current human–machine interaction systems address the needs of massive data collection, multidimensional data analysis, and more complex scene fusion applications. Humans often require the input of massive amounts of multisource information to make quick and accurate decisions. For machines, the information to be processed is even more massive and multidimensional. Therefore, a multisource data fusion technology oriented toward intelligent human–machine interactions is established. Through multisource data collected by interactive software front and back ends and front-end devices, such as images, voices, structured data, and text, a multisource data collection, storage, backup, and processing fusion system is constructed. By using big data technology to classify, store, backup, and mine the relationships of multisource data, the layered consistency of multisource data can be ensured. By using cloud computing technology to uniformly schedule multisource data, human–machine information interactions, and machine-to-machine information interactions, the stability of information transmission and computation can be ensured.

Multitechnology Collaboration: A human–machine interaction system is a process in which machines perceive and interact with humans through perception technologies such as vision, hearing, and supersensing, as well as cognition and decision-making technologies. The multisource data origin determines that human–machine interaction is a system based on multitechnology collaboration. To meet the collaboration needs of multiple technologies in one or more human–machine interaction systems, a multitechnology collaboration system that integrates the intelligent Internet of Things, big data, artificial intelligence, GIS, digital twins, etc., is constructed. This system includes AIoT device management, computational and resource management, artificial intelligence algorithms and engines, human–machine collaborative perception APIs, model training and learning, development tools and API services, data lake analysis, business process engines, and knowledge decision systems as core modules. By establishing data access, cloud-edge integrated collaborative computing, high-performance file storage, real-time task scheduling, and other basic capabilities, the automatic deployment, monitoring, and control of edge perception devices for information interaction capabilities can be realized.

Intelligent Algorithm Coupling: To continuously optimize the human–machine interaction experience, multimodal perception data and business data inference and decision-making are constructed based on natural language processing technology, intelligent voice technology, and big data analysis technology. Through the underlying AI business process engine and knowledge decision system, intelligent transformation is carried out in depth for each node of the business process in industry application scenarios, optimizing the human–machine interaction experience and improving the efficiency of business process execution. Additionally, to meet the needs of intelligent and interconnected information processing between business systems, a business data analysis algorithm library is constructed based on the two core information processing engines of the data middle platform and the AI middle platform. By coordinating and scheduling cloud-edge computing resources with cloud computing technology, the intelligence level of business data processing is improved, and the pressure on computational resources by information processing tasks is reduced, ensuring the timeliness of human–machine interactions.

Intelligent Hardware–Software Integration: The intelligent entity of hardware–software integration is the core of human–computer interaction and information exchange. The development of intelligent hardware-software integration technology combines three major elements: humans, interactive devices, and interactive software. The human element is indispensable in the process of human–computer interaction; that is, the user cannot be absent. This aspect of the human element involves the user operation model, which is related to various characteristics and preferences of the user. The task is to organically combine the various behaviors of the user and the computer. Interactive devices, such as graphics, image input and output devices, sound, posture, tactile devices, three-dimensional interactive devices, etc., are also indispensable in the process of human–computer interaction. These interactive devices are also constantly being integrated into various business scenarios, achieving the best state and effect in the interaction process. Interactive software is the core of information collection, processing, and feedback. With the help of multisource information collection, information transmission, information analysis, information feedback, and effective exchange of information between humans and machines, interactive software deeply couples the above processes and technologies to support industrial monitoring and decision-making.

General Basic Office Platform Technology

To meet the needs of customers for a large number of business systems, high interconnectivity requirements, unified identity verification, and collaborative management of application permissions in the construction of digital systems, we build a general basic office collaboration platform that can conveniently connect people, data, information, and business processes. Through the portal, the functions that users need to operate are unified and aggregated to form a one-stop workstation. Integration technology extracts and integrates information and data and displays them uniformly through the portal. Business collaboration reshapes employee collaboration through business processes, facilitates the collaboration of internal and external platforms and users of the enterprise, and improves business execution efficiency. Based on microservices and distributed architecture, it adopts a service-oriented strategy, with unified gateways, distributed logs, system monitoring and other basic components, and has functions such as the unified organization of users, permission management, login verification, application registration, data exchange, document management, and application integration. This technology is used mainly in basic office scenarios. It unifies the technical framework and process management, improves data consistency and resource utilization, enhances the standardization capabilities of services and technologies, simplifies the development of upper-layer applications, and reduces operational and maintenance costs.

Unified Identity Authentication Management: By integrating the single sign-on module and calling the unified identity authentication platform service, different content can be displayed for different user logins. This system can provide users with the function of customizing the desktop according to different user focus points. All authorized application systems share an identity authentication system. Moreover, all application systems can recognize and extract ticket information. To achieve the single sign-on (SSO) function, users need to log in only once, so the application system must be able to recognize users who have already logged in. The application system is able to recognize and extract tickets and automatically judge whether the current user has logged in through communication with the authentication system, thereby completing the single sign-on function. On the other hand, centralized authentication management provides unified identity authentication for IT systems, which is a group security portal. Only a secure authentication mechanism can ensure that the institution's gate is not entered by illegal third-party platforms; the platform provides unified authentication management, integrates single sign-on modules, calls unified identity authentication platform services to achieve authentication for different users, and authenticates third-party platforms, establishing a secure identity authentication service platform with a unique identity identification secret key digital certificate. The service objects of the entire authentication system include all business systems, management systems, and application systems of the group that access the unified authentication third-party platform. The unified authentication system can provide fast, efficient, and secure services. The application system offers small access transformation, and flexible scalability and high availability.

Security Audit and Compliance: This feature implements the log recording and audit functions of the identity management platform. The information recorded in the log includes administrator operation logs, data synchronization logs, function operation logs, system access records, etc.; the final user login authentication record audit function, which records the final user's login IP, time, login user, login status and other information; and password policy configuration management, which can be configured by the administrator system password policy, username policy, internal and external network access policy, mobile security policy, etc.

Unified Gateway Service: In the process of developing a digital system business system, as the complexity of device access, data interconnections, and data exchanges between business systems increases, the requirements for interface scheduling and maintenance also increase. Numerous API calls urgently need a unified entrance to support the call of various business systems. In this case, the gateway is responsible for access, routing, and flow control, and each service provider focuses on the implementation of business logic, thereby providing a robust service call environment for business system calls. Later, under the condition of a large number of gateway calls, it can also ensure that the gateway is degradable, flow-controllable, and isolatable, along with other fault-tolerant capabilities.

Multiapplication Collaboration: In modern society, with the explosive growth of information, users need to quickly and accurately obtain information. Consequently, information service platforms have become indispensable tools. However, there are too many information service platforms; thus, users need to remember many websites, accounts and passwords, which not only wastes time but also poses many management and maintenance difficulties for IT departments. Moreover, with the continuous development of artificial intelligence technology, how to make information service platforms more intelligent so that users can obtain more personalized and intelligent services has become a new direction and challenge. Against this background, a unified portal for multiapplication collaboration can be built. The creation of a unified application portal entrance supports users with different roles and permissions. Through a portal platform, users can view multiple applications and their messages concerning to-do items. Each function module is grouped according to its application, thus achieving unified interface integration, business integration, message integration, authorization management, interface management, etc., and providing a unified information resource access entrance for business managers, company employees, partners, and the public.

3.2 Applications of Smart Cities Based on the AIoT

3.2.1 Development and Application of an Energy-Saving Operation Management System for Large Public Buildings Based on IoT Technology

Research Background

The construction industry is the second largest energy-consuming industry after the industrial sector. Energy consumption in the construction industry is approximately divided into five categories: electricity, water, gas, centralized heating, and centralized cooling. In the context of China's current "dual carbon" goals, the energy consumption and energy-saving management of buildings are particularly important.

However, traditional energy management systems collect energy consumption data from only some electromechanical equipment. They do not consider the comprehensive collection of energy consumption data from all energy-consuming equipment (including electromechanical and weak current) in the building, as well as the environmental factors and building components that affect energy consumption. At present, in building equipment management systems, while basic dispersed monitoring, control, and management of various systems can be conducted, simple energy consumption equipment monitoring cannot comprehensively evaluate the operating efficiency of equipment and help uncover energy-saving potential. Moreover, many large public buildings do not have corresponding energy-saving management methods, cannot capture overall energy consumption in a timely manner, and are unable to provide a timely understanding of the operation and energy-saving status of major energy-consuming equipment. Therefore, it has become very urgent to establish an energy-saving operation management system within buildings.

In today's advocacy for energy conservation and emission reduction, this research not only responds to the needs of national development strategies but also helps to improve the low-carbon energy-saving emission reduction technology in the construction industry and to promote the application of this technology in construction engineering.

Research Content

For large public buildings, we aim to collect data from various equipment and systems that significantly impact building energy consumption, such as intelligent lighting control systems, air conditioning systems, elevator systems, hot water systems, and electric sunshade equipment, with Artificial Intelligence of Things (AIoT) technology. Through the storage and optimization of massive dynamic real-time energy consumption data, we aim to achieve intelligent online monitoring and processing of the overall operation of the building from an energy efficiency optimization perspective, establish an intelligent decision support model, form scientific and feasible energy-saving strategies, and significantly improve the operation management and energy efficiency utilization of the entire building.

The technical route of the energy-saving operation management system for large public buildings based on IoT technology is shown in Fig. 3.2:

Building Energy Consumption Measurement and Control Perception

Low-power measurement and control terminal: Research on low-power measurement and control terminals based on AIoT technology, using low-cost, low-power, miniaturized design and other hardware development technologies, as well as dynamic loading, online debugging and other software development technologies, is needed to develop measurement and control terminals for building applications, such as temperature and humidity sensors, human–machine interactions, air quality sensors and other measuring devices, as well as air conditioning fan coil controllers, lighting controllers, switch controllers and other control devices.

Multinetwork collaboration and adaptive communication: In response to the needs of real-time communication of self-configuring networks in public building environments, the low-power operation of measurement and control terminal equipment in large buildings, the coexistence of building measurement and control wireless networks with other wireless networks, and the interconnection and interoperability of building measurement and control networks with existing building intelligence systems, we aim to develop multinetwork coexistence and adaptive communication technology via the networking of large-scale sensing devices, design protocol

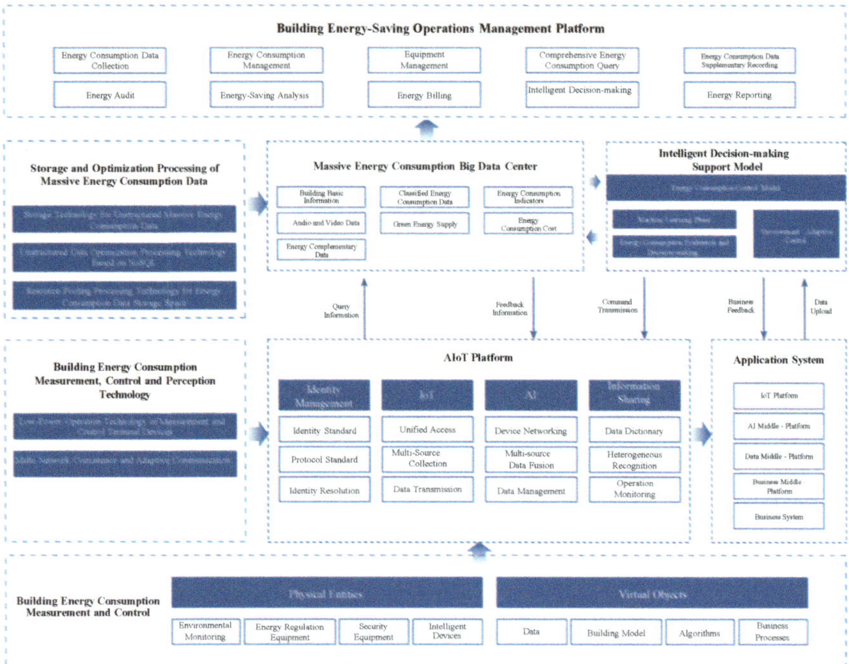

Fig. 3.2 Technical Route of the energy-saving operations management system for large public buildings based on IoT technology

stack software to support the operation of building wireless networks, and develop wireless communication modules, wireless communication gateways, and wireless communication adapters.

Massive Energy Consumption Data Storage and Optimization Processing Based on Cloud Computing

Large public buildings have a wide variety and large number of energy-consuming devices. Moreover, the data generated by different energy-consuming devices have problems such as inconsistent standards and pose difficulties in storage and processing in terms of type, quantity, coding, and structuring. Traditional data storage and processing technologies have difficulty adapting to the characteristics of energy consumption data in large public buildings. We aim to develop massive energy consumption data storage and optimization processing technology based on cloud computing and innovative applications, employ cloud computing technology for the storage and optimization of energy consumption data in large public buildings, and adopt unstructured massive energy consumption data storage technology, NoSQL-based unstructured data optimization processing technology, and energy consumption data storage space resource pooling processing technology to overcome the performance bottleneck of traditional technologies.

Intelligent Decision Support Model

Machine Learning Phase: In the initial stage of product deployment, the learning phase of intelligent decision support needs to be completed. First, energy consumption data samples are collected through the normal operation of large public buildings, forming a time series of energy consumption data. After expert review, these data serve as the basis of the energy consumption model, constituting the energy consumption model library, which subsequently serves as the basis for evaluating subsequent energy consumption.

Energy Consumption Assessment and Energy-saving Decision Phase: During the operation of large public buildings, energy consumption data are collected to form an energy consumption time series. These data are compared with the energy consumption model in the energy consumption model library for inference and model fitting. An energy consumption assessment report is then produced, and based on the degree of fit, an energy consumption anomaly report and energy-saving strategy are created.

Model Fitting and Adaptive Phase: Periodically, expert reviews are conducted on the collected energy consumption data to form a new energy consumption model and update the energy consumption model library. The more energy consumption models there are in the library, the better the energy consumption patterns of large public buildings can be described, and the more accurate, scientific, and feasible the energy-saving strategies will be.

Building an Energy-saving Operation Management Platform

The Building Energy-saving Operation Management Platform is designed and developed using a combination of C/S and B/S architectures. It follows object-oriented, modular design principles, integrating functions such as data collection, extraction, filtering and cleaning, business transformation, analysis and mining, and intuitive display. It can meet the various needs of user business analysts, managers, and decision-makers for building energy management, providing energy decisions for enterprise managers and decision-makers. Its main functions include energy consumption data collection, energy management, equipment management, comprehensive energy consumption queries, energy consumption data supplementation, energy audits, energy-saving analysis, energy billing, intelligent decision-making, and energy reporting, thus facilitating decision support and energy consumption evaluation.

Research Results

In response to the problems of high energy consumption in large public buildings, traditional energy management only involves simple energy consumption data. It does not involve comprehensive data collection and analysis of energy consumption equipment and lacks energy-saving strategies. The innovative application research results based on AIoT technology has led to the development of an energy-saving operation control technology system and management platform for large public buildings, achieving intelligent online monitoring and processing of the overall operation of the building, thereby greatly improving the operation management and energy efficiency utilization of the entire building. An energy consumption measurement and control technology for large public buildings has also been established, and energy consumption measurement and control terminal equipment integrating multiple communication modes has been developed to achieve efficient, accurate, and low-power consumption data collection. In response to the energy-saving management needs of public buildings and the storage and processing needs of massive multimodal energy consumption data, cloud computing-based massive energy consumption data storage and optimization processing technology has been developed. Based on public building energy management systems, expert theoretical knowledge, and energy supply and consumption trends, a machine learning-based energy consumption monitoring and energy supply regulation model has been established to achieve efficient energy supply and regulation. Finally, a building energy-saving operation management platform has been developed for energy management personnel, providing energy decisions for enterprise managers and decision-makers.

The research results have been successfully applied at the Guangzhou Industrial Intelligence Research Institute, Nanning Energy-saving Supervision Center, and Guangxi Zhuang Autonomous Region Public Security Department. The system operations are stable and reliable, and the energy-saving data in some areas have achieved the expected results. For example, after implementing energy-saving strategies at the C509 area demonstration site of the Guangzhou Industrial Intelligence Research Institute, during the statistical period, the total energy consumption during working hours decreased from 205.5 to 62.9 kW·h, a decrease of 69.39%; the total energy consumption during nonworking hours decreased from 217.86 to 65.91 kW·h, a decrease of 69.75%; the total energy consumption decreased from 423.36 to 128.81 kW·h; the average daily energy savings decreased from 42.34 to 23.57 kW·h; the energy-saving ratio was 69.57%; the amount of energy waste was essentially eliminated; the efficiency of equipment use improved; and the energy consumption level of this area significantly decreased.

3.2.2 Research and Application of Key Technologies of Building Robots Based on Building Information Modeling and Artificial Intelligence Technology

Research Background

The traditional construction industry construction process has problems such as labor shortages, high safety risks, high work intensity, low work efficiency, and difficulty in systematizing, standardizing, and normalizing construction work. Various types of single-process construction robots have appeared abroad and are widely used in the construction industry, but problems still remain in terms of multiple scenario applications, scene adaptive applications, and real-time perception interactions of robots. Domestic construction robots also have applications in the construction industry, but there are also problems such as single-application scenarios, a lack of high-level robot intelligence, insufficient human–machine interaction capabilities, and insufficient integration of AR/VR, artificial intelligence, automatic positioning and navigation technologies.

Therefore, this research is based on building information modeling (BIM) and artificial intelligence of things (AIoT) technology, which involves the creation of various types of construction robots; focuses on solving technologies such as multirobot and multiscenario coordinated application, robot automatic navigation, real-time perception, and precise operation; is applied to the entire process of traditional intelligent building design, construction, and operation; and promotes the in-depth application of multitrade construction robots in the construction field. The technology ultimately achieves the goals of replacing a portion of the labor in the intelligent building field, improving construction safety, improving construction efficiency, and reducing costs.

Research Content

Based on BIM and AIoT technologies, we aim to develop intelligent construction robots for multiple application scenarios and professions. This endeavor includes intelligent perception technology based on BIM, which allows the robot to collect a large amount of sensory data during its movement and grasping processes, enabling automatic scene recognition, target recognition, and target positioning. We also focus on multisource information fusion analysis and control technology based on BIM, which allows the robot to plan its path, navigate autonomously, and correct its course in real time. This technology also enables high-precision, unordered grasping techniques adaptable to different tasks, application scenarios, and technical requirements. Furthermore, we aim to develop an integrated human–machine interaction platform for intelligent construction robots based on BIM and AIoT technologies, which facilitates coordination between humans and machines and between machines themselves, enabling intelligent scheduling, management, and closed-loop management decision-making for intelligent operations.

The key technical architecture of construction robots, which is based on building information modeling and artificial intelligence technologies, is shown in Fig. 3.3:

Intelligent Perception of Robots Based on BIM

By using AI-based deep learning for target detection, multiscale scene modeling, and SLAM technology, we can identify, understand, and model scene targets. Simultaneously, by effectively integrating BIM cloud information with intelligent sensor algorithms, we can correct sensor data in a timely and effective manner, thus achieving high-precision recognition and positioning of targets.

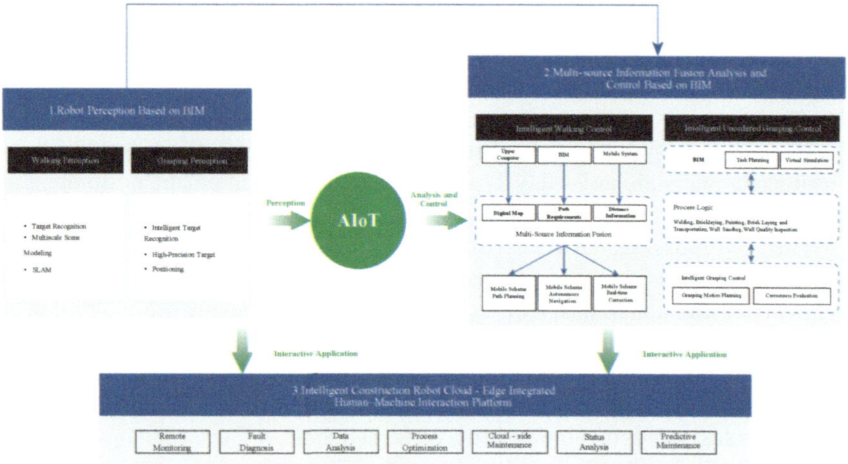

Fig. 3.3 Key technical architecture of construction robots based on building information modeling and artificial intelligence technologies

Multisource Information Fusion Analysis and Control Based on BIM

By adopting multisource information fusion methods, we can achieve autonomous planning and real-time correction capabilities in walking control by analyzing and calculating digital maps, path requirements, and distance information from the upper computer, BIM, and mobile systems. In terms of the grasping control, based on intelligent construction task planning in the cloud-based BIM, we can calculate and optimize the precise placement position of each material through intelligent target recognition and high-precision target positioning, combined with process requirements. We can also achieve 3D simulation modeling of each material in BIM, plan the grasping path with constraints, and ensure automatic obstacle avoidance in dynamic environments. After the grasping motion is completed according to process requirements, we use visual assessment to verify the correctness of the grasp, promptly detect errors that occur during the construction process, and ensure high system availability.

Intelligent Construction Robot End-Cloud Integrated Human–Machine Interaction Platform Based on BIM

By collecting the robot's visual data, laser ranging data, and multimedia voice data through artificial intelligence of things (AIoT) technology and recognizing, analyzing, and processing these data, we can optimize the robot's image data processing and voice semantic processing capabilities. By collecting the robot's coordination control data and modeling it with the BIM model, we can optimize the robot's adaptive ability to the environment and its choice of optimal solutions. Ultimately, we aim to provide unified monitoring, management, and operation and maintenance services for intelligent construction; reduce operational costs; and perform remote monitoring, fault diagnosis, data analysis, process optimization, cloud maintenance, status analysis, predictive maintenance, and other functions for production lines and automated production robots.

Research Innovation Points

We aim to establish a new intelligent environmental perception model based on BIM and AI, using BIM information to solve the "uncertainty problem" in traditional SLAM technology. We also aim to establish a new BIM-SLAM intelligent robot positioning and navigation technology. Moreover, we use AI technology to solve the "adaptive difference" problem in traditional methods based only on BIM through intelligent perception, achieving effective path planning, timely discovery, and handling, thereby avoiding and eliminating hidden dangers.

We aim to develop a fully automatic mobile system for robots based on BIM, surpassing high-passability, high-speed, high-stability, and other robot mobile technologies. This system can ensure the smooth completion of related construction engineering tasks and perform tasks such as automatic robot charging.

We aim to develop a high-precision grasping system adaptable to complex construction environments based on machine learning and intelligent big data vision algorithms. This system can solve the problem of visual detection and high-precision positioning under conditions of few samples and the problem of traditional convolutional neural networks being easily disturbed by local high-frequency signals in images. We aim to organically combine BIM with the ROS system to achieve automatic planning and real-time interaction during project construction.

By utilizing BIM and AIoT technologies, we aim to combine advanced software tools that integrate intelligent building design, construction, and operation with various construction robots and artificial intelligence technologies. This approach will enable the interconnection and two-way interaction between the multiscenario construction robot end and the cloud-based platform end with distributed storage based on big data, thereby achieving functions such as intelligent building transportation, intelligent navigation, intelligent positioning, and intelligent building. It will also facilitate cloud-based data analysis and unified management of integrated robots, improving their management capabilities and efficiency.

3.2.3 *Research and Demonstration of Key Technologies for Intelligent Group Control of Multiparameter Greenhouses for Fruits and Vegetables*

Research Background

In response to the difficulty of implementing intelligent measurement and control in existing fruit and vegetable greenhouse groups, this research focuses on the sensing and coordinated control technology of multiple parameters and variables in individual greenhouses and the analysis technology of group control strategies for multiple greenhouses to form a cloud platform for the management of fruit and vegetable greenhouse groups. This research can effectively solve the problems of the intelligent management of greenhouses in southern China, labor shortages, and high labor costs. Through adaptive and self-learning group control strategies, it can be applied to different greenhouse systems for seedlings, vegetables, flowers, and herbs.

Research Content

The architecture of the key technology for intelligent group control of multiparameter greenhouses for fruits and vegetables is shown in Fig. 3.4:

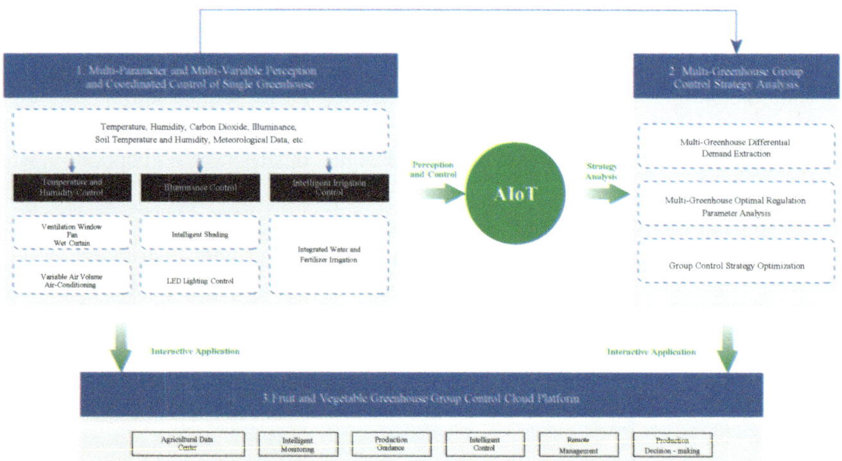

Fig. 3.4 Architecture of the key technology for intelligent group control of multiparameter greenhouses for fruits and vegetables

Sensing and Coordinated Control of Multiple Parameters and Variables in a Single Greenhouse

Based on the improved particle swarm optimization (PSO) algorithm, the method of variable coordinated control for the Internet of Things (IoT) group control terminal in multiple greenhouses intelligently processes collected data such as temperature, humidity, carbon dioxide, light intensity, soil temperature and humidity, and meteorological data. It provides control signals for the automatic control system to control ventilation windows, fans, wet curtains, variable air volume (VAV) air conditioning, shading, LED lights, and integrated irrigation of water and fertilizer, achieving automatic adjustment of the microclimate within the greenhouse. This method meets the suitable conditions for the growth of fruits and vegetables, thus improving the yield and quality of different types of fruits and vegetables.

Analysis of Group Control Strategies for Multiple Greenhouses

After the coordinated control of a single greenhouse is completed, it is necessary to implement group control for multiple greenhouses. This process is achieved through centralized control, namely, by extracting the differential requirements of multiple greenhouses, analyzing optimal control parameters, and optimizing strategies to form group control strategies for multiple greenhouses. If each greenhouse grows the same variety of fruits and vegetables, different parameters can be controlled in multiple greenhouses, and the growth conditions of fruits and vegetables at the corresponding times can be observed. The optimal environmental parameter range values can be obtained for fruits and vegetables at the same time. Through centralized control of the greenhouse group, the optimization control of greenhouse environmental parameters can be achieved, realizing the efficient, energy-saving, and high-yield production of fruits and vegetables in the greenhouse.

Cloud Platform for the Management of Fruit and Vegetable Greenhouse Groups

The cloud platform for the management of fruit and vegetable greenhouse groups, which is based on agricultural big data, integrates agricultural characteristics such as regionalism, seasonality, diversity, and cyclicity. It generates a wide range of sources, diverse types, complex structures, and potential value datasets that are difficult to process and analyze with conventional methods. Through key technologies of multi-source heterogeneous data and cross-domain knowledge aggregation, it achieves precise production guidance and intelligent production decisions for different crops in the greenhouse group. Consequently, it realizes efficient unified management of facility horticulture greenhouse groups, improving the level of fine production control.

Research Results

To address the existing problems of greenhouse control methods, such as information transmission delay, single controlled environmental variables, and difficulties in the coordinated control of multiple parameters and variables, we developed IoT group control technology for multiple greenhouses. Specifically, we propose a method of variable coordinated control for the IoT group control terminal in multiple greenhouses based on the improved PSO algorithm. This method achieves monitoring and coordinated control of air temperature, relative humidity, light intensity, and carbon dioxide concentration in multiple greenhouses, greatly improving the level of intelligent coordinated control of the greenhouse group. Based on this method, we conducted research on the coordinated control of multiple parameters and variables in a single greenhouse for fruits and vegetables, intelligent group control of fruit and vegetable greenhouses, big data analysis of fruit and vegetable growth, and comprehensive analysis of data from multiple sensors in the greenhouse. We established a layered distributed intelligent group control integration system based on the fruit and vegetable greenhouse group. This system effectively solves the problems of intelligent management of greenhouses in southern China, labor shortages, and high labor costs. After researching and optimizing the coordinated control strategy of multiple controllers in the greenhouse environment for fruit and vegetable growth, we achieved precise guidance for the growth of seedlings, vegetables, flowers, and herbs.

We developed a layered distributed intelligent group control integration system based on the fruit and vegetable greenhouse group. This system can achieve intelligent measurement and control of the greenhouse group and can monitor and analyze the growth of fruits and vegetables. It can manage more than 10 individual greenhouses in the greenhouse group.

The developed application platform can monitor environmental parameters for fruit and vegetable growth in greenhouses, including temperature, light intensity, humidity, and carbon dioxide concentration. The measurement error for temperature and light intensity is $\pm$ 1%, and that for humidity and CO_2 concentration is $\pm$ 1%.

The control accuracy of environmental parameters (temperature, light intensity, etc.) for fruit and vegetable growth in greenhouses is within $\pm$ 5% of the normal growth range.

3.2.4 Research and Demonstration of Key Technologies for the Precision Control of Horticultural Crops in Facilities

Research Background

Facility horticulture refers to the creation of an environment suitable for crop growth in seasons or regions where open-field cultivation is not suitable. It uses specific facilities to produce high-quality, high-yield, and stable vegetables, flowers, fruits, and other horticultural products. There is extensive research on facility horticulture both domestically and internationally. For example, China has the largest area and variety of facilities; and Japan, the United States, and the Netherlands have conducted advanced research on fully automated facility horticulture. However, there is room for improvement in facility horticulture in both China and developed countries such as the United States.

The perception of growth information in facility horticulture is not comprehensive enough. In facility horticulture, the comprehensive perception of growth information provides data support for crop growth control models. The environmental parameters required for crop growth need to be collected, and crop types and their growth stages and conditions need to be identified. Currently, the United States has applied wireless sensing technology to greenhouse monitoring, and the School of Mechanical Engineering at Beijing University of Technology has monitored tomato cracking via machine vision; however, neither has used image recognition technology for research in facility agriculture.

Lack of a crop growth control model: A crop growth control model is the basis for integrated management. It includes the identification of crop growth characteristics and the supply of coupled nutrients. The Netherlands and Israel have jointly developed the HORTISTM horticultural simulator to simulate crop growth to study crop growth characteristics. Domestic research has accumulated a certain foundation with the development of facility agriculture, but there is a lack of sufficient research on nutrient coupling supply, and establishing a precise control model for the production of various specific crops is impossible.

Lack of integrated management of data collection, diagnosis, and control of facility horticulture crops: The precision control of facility horticulture crops needs to be achieved through data collection, diagnosis, and control of facility horticulture crops. Japan has essentially achieved full automation in the facility agriculture of planting tomatoes and leafy vegetables, but there is still room for research on fruit seedlings and edible fungi. China has the largest volume of facility horticulture, but it lags behind in the integrated management of data collection, diagnosis, and control.

Therefore, this study highlights the urgent need for precision control of facility horticulture crop production, with cherry tomato, orchid, citrus, and straw mushroom cultivation as the main objects, to develop a precision control system for facility horticulture crops, improve the level of fine production control, and achieve efficient operations.

Research Content

This research develops multimodal fusion perception technology for facility horticulture crops, achieve environmental ecological monitoring, crop status analysis, and nutrient element content monitoring of facility horticulture crops. Through in-depth research on key technologies such as growth characteristic identification and nutrient coupling supply, a precision control model based on crop growth was proposed, precision control of the crop growth environment and nutrient supply according to crop

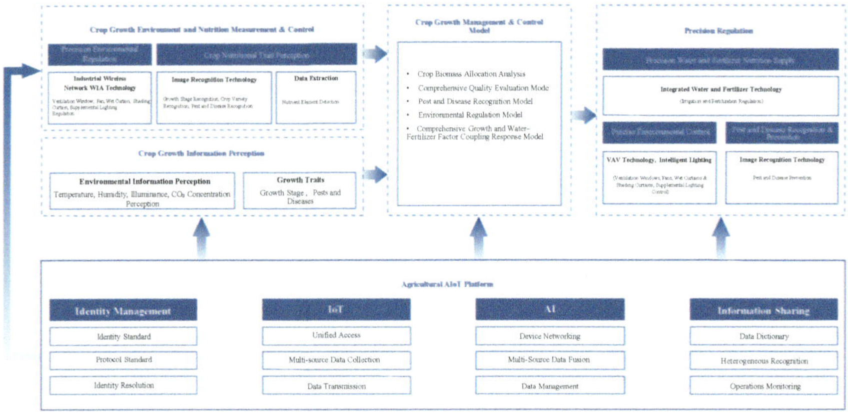

Fig. 3.5 key technology route for the precision control of facility horticulture crops

growth needs was achieved, the quality and yield of agricultural products was effectively improved, and the comprehensive utilization rates of water and fertilizer were improved.

The key technology route for precision control of facility horticulture crops is shown in Fig. 3.5:

Multimodal Fusion Perception Technology

The perception of growth information of facility horticulture crops includes two parts: crop environment perception and crop nutrient characteristic perception. Multimodal data collection of environmental data (temperature, humidity, light intensity, CO_2 concentration, etc.), crop growth status data, and crop nutrient element content data was performed.

Ecological Environment Perception

Based on Internet of Things technology, wireless environmental sensing devices collect environmental data such as temperature, humidity, light intensity, CO_2 concentration, pH value, conductivity, soil moisture, etc., and then transmit the data to the cloud platform through a wireless collection gateway.

Growth Characteristic Perception.

Crop growth status perception based on image recognition technology

To address the drawbacks of the manual detection of crop growth status, a deep learning-based intelligent detection algorithm for crop images was designed. First, a crop image dataset was constructed. Then, image preprocessing and data augmentation were performed, followed by training the dataset using a target detection algorithm model and optimizing the model, ultimately achieving intelligent monitoring of crop growth status information.

Perception of crop nutrient content via hyperspectral monitoring technology

By analyzing the relationship between hyperspectral reflectance and crop nutrient content, a nutrient estimation model was established using spectral feature parameters, achieving rapid and nondestructive detection of nutrient content.

Key Technology for Nutrient Coupling Supply

Based on the growth characteristic data collected from crops under different environmental and nutrient regulation conditions, the relationship of nutrient coupling supply was further studied through the fusion analysis of multimodal data. A growth control model for crops was established, which includes an analysis of crop biomass allocation, a comprehensive quality evaluation model, a pest and disease identification model, an environmental control model, and a comprehensive growth and water-fertilizer factor coupling response model.

Precision Control System for the Environment and Nutrition

The precision control system for environment and nutrition includes a precise environmental control system, a precise water and fertilizer nutrient supply, and pest and disease identification and prevention. By comparing multimodal real-time data with the optimal parameters of the growth control model, a control strategy was formed. Using VAV technology, intelligent lighting technology, and integrated water and fertilizer technology, precise control of ventilation windows, fans, wet curtains, shading curtains, supplementary lights, and integrated irrigation systems for water and fertilizer was achieved, thus realizing real-time matching of environmental and nutritional conditions with crop growth needs.

Research Achievements

Some of the results of this research have been applied to several bases in Guangdong Province, including the Foshan Zhihua Mushroom Production Base, Meizhou Fengxia Citrus Production Base, Guangzhou Dafeng Tomato Production Base, and Shaoguan Wengyuan Orchid Production Base.

The research formed a precision control technology system for facility horticultural crops. Based on Internet of Things, big data, artificial intelligence technologies, and planting theory, a group of precision control models for facility crops was established, including precise environmental control, nutrient coupling control, pest and disease identification, and multisource data fusion. With the help of connected environmental sensors, integrated water and fertilizer, video surveillance and other equipment, unified integrated management of facility horticulture data collection, diagnosis, and control was achieved, including the following:

The construction of machine vision technology for facility agricultural crops, which includes the identification of types of pests and diseases, estimation of disease severity, and recognition of crop growth stages.

According to the growth characteristics of crops, environmental data models, nutrient coupling data models, and appearance data models were established, and a crop growth control model was established.

The construction of a comprehensive control platform for facility horticulture based on digital twins, achieving remote environmental perception and equipment control.

The research results show effective improvements in crop quality and yield. This project relied on crop growth control models and precision control technologies and strategies, combined with crop growth analysis, to accurately supply water, nutrients, and plant protection agents according to the needs of each growth stage of the crop. Experiments on crops such as straw mushrooms, orchids, and cherry tomatoes showed that the crop yield increased by 10–30%, and the consistency of fruit shape and the proportion of superior products increased by 20%.

The research results also show great improvements in the quality and safety of agricultural products. Precision control technology and platforms were applied mainly in the facility horticulture, which is relatively closed, effectively preventing pests, diseases, and weeds and reducing environmental pollution. Through the application of precise plant protection measures and fertilization strategies, the use of pesticides and fertilizers can be reduced. In particular, according to data provided by the Agricultural Academy, the use of pesticides in the cultivation of straw mushrooms has been reduced by 100% using precision control technology. In addition, the platform has data collection and analysis capabilities. Based on real-time environmental and crop data, predictions and optimization decisions can be made, such as irrigation scheduling, fertilization plans, and pest and disease prevention.

3.2.5 Construction of a Smart Park Platform Based on Big Data and Distributed Cloud Storage

Research Background

Currently, various parks are still in the process of system and smart scenario construction and upgrades. The following pain points are commonly observed in the construction process of smart parks:

Information islands, low operational efficiency, and a lack of unified system control in traditional industrial parks lead to "chimney-style" system deployment, resulting in data not being shared and business operations not being linked. These issues make it difficult to achieve information management, support interconnections within the park, and break from traditional leasing and selling models. Most parks consider only the need for information technology to cover some park functions. There are information barriers between systems, functions cannot be interconnected, and data cannot be shared. Single-function services cannot fully meet the needs of park users, and the transmission of information between different departments (systems) consumes considerable labor and other resources.

Park construction lacks intelligent design. Traditional park construction lacks intelligent design based on the type and nature of the park. Limited funds and restrictions on preinstalled hardware make it difficult to implement intelligent transformations.

The park ecosystem has not yet been built. The park itself has various industrial ecosystems, such as enterprise resources, services, and data ecosystems. These ecosystems need to be interconnected. Currently, there is no closed-loop ecosystem at the application, management, and analysis levels.

Smart parks urgently need to deeply integrate new generation information technologies such as cloud computing, big data, the Internet of Things, artificial intelligence, 5G, and digital twins with industries; integrate park resources and third-party service capabilities; achieve layered resource sharing, industry-linked development, real-time environmental perception, full-process event visibility, automatic production adaptation, full-time equipment utilization, and community value association; promote the extension of the industrial value chain; enhance the intelligent management of the park and social integration capabilities; and build a closed-loop ecosystem.

Research Content

To meet the needs of efficient management, industrial ecosystem integration, and energy savings and carbon reduction in industrial parks, a smart park platform based on big data and distributed cloud storage was built to achieve interconnection and interoperability of management and operation data in industrial parks. The construction and development of industrial internet identification resolution and park energy, assets, operations, etc., was carried out to achieve standardized and intelligent management of park comprehensive security, energy monitoring, equipment and facilities. By docking with national, industry, and technical standards, a standard system for identification resolution, application development, and algorithm development can be established. In the research and development of key technologies such as identification coding and resolution, synchronous identification resolution and information collaboration, multiprotocol adaptation, equipment safety certification, deep learning algorithms for industrial parks, multisource data access and integration, batch data rapid processing, business engines, etc., ensure the advancement and feasibility of technology through standard resolution, theoretical verification, simulation, business process design verification, etc.

The technical architecture of the smart park cloud platform, which is based on big data and distributed cloud storage, is shown in Fig. 3.6:

Multisource Heterogeneous Information Perception in Parks

With the rapid expansion and intelligent development of various park formats, the data generated through park construction, operation, and management present multisource, multidimensional, multimodal, and massive characteristics. The data types include building automation system (BAS) data, security data, environmental data,

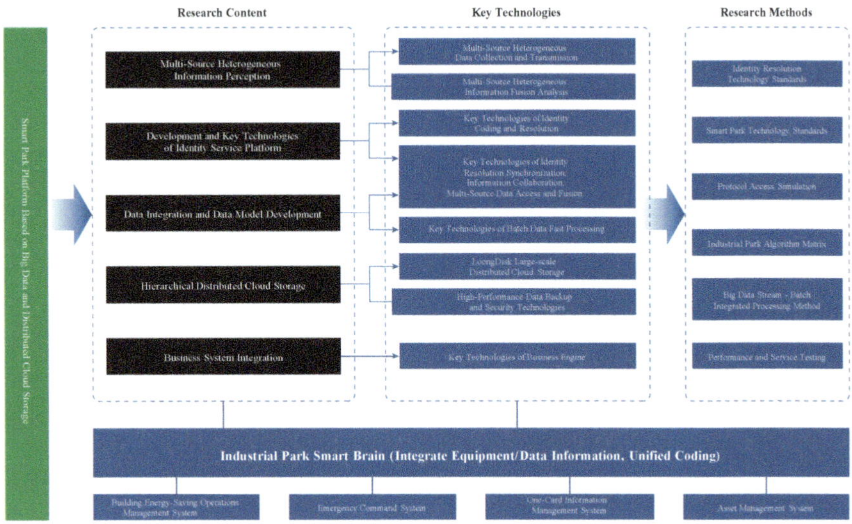

Fig. 3.6 Technical architecture of the smart park cloud platform based on big data and distributed cloud storage

energy consumption data, asset data, operational data, video data, and other types of data. These data include both structured data such as BAS data and environmental data and unstructured data such as video data and image data.

To meet the needs of multisource heterogeneous information perception in parks, an information perception technology and platform for full-process and full-scene control of parks based on AIoT technology was developed. Among them, BAS data are obtained through the system field bus or TCP/IP network; environmental data are collected through IoT technology (a combination of wired and wireless); equipment asset data are obtained through a combination of RFID technology and QR codes; energy consumption data are obtained through the use of smart meters (a combination of wired and wireless); and video conference data are obtained through video conferencing systems or visual network perception.

Identification Resolution

The main services included in identification resolution include real-name verification, identification registration, identification resolution, and data management. They provide APIs to various business systems for easy identification management and query software development. The identification registration system establishes a mapping relationship between the industrial internet object identification and its corresponding information address. After the identification is assigned to a specific object, before the object information is recorded, the location of the information record needs to be determined, that is, the addressable identification, to form a mapping relationship between the object identification and the addressable identification. This mapping relationship is submitted to the superior identification service provider for registration, and the superior identification service provider records and maintains this mapping relationship.

When different applications require information collaboration and sharing, the platform can provide identification resolution functions for related applications or users. For the unified identification allocated by the platform, this platform should include root servers and various levels of servers, and after resolution, it can return the communication address; for the identification allocated by other institutions, the platform establishes a root server, and after resolution, it can return the address of the resolution system within the industry to which the identification belongs. If the resolution system within the industry can communicate with this platform, the platform can directly return the resolution result of the resolution system within the industry. The resolution subsystem of the platform consists of a public resolution access point and a public resolution system, facing various triggering mechanisms such as barcodes, RFID, IC cards, intelligent systems, etc. It is a compatible public resolution system that has cross-industry, cross-platform, and multiapplication features.

Data Integration

Data integration capabilities solve data integration and synchronization issues in various scenarios, such as data mid-end construction, database migration backup, business upgrades, integration, data access acceleration, and full-text search, by quickly connecting and merging various data from self-built cloud or noncloud sources. Data integration supports three core capabilities: offline synchronization, real-time synchronization, and solutions. Offline synchronization involves mainly the periodic batch active extraction and synchronization of data sources; it currently supports 23 types of data source types, including MySQL, EMR Hive, Kafka, HDFS, Redis, Elasticsearch, etc., and 23 types of writing ends. Real-time synchronization monitors data changes passively based on database logs and refreshes data dynamically synchronized to the target end in real time; it currently supports MySQL, Kafka, and Hive data source types. It also supports multidata source collection, external file processing, concurrent data processing at multiple execution ends, and the fulfillment of TB-level data interface collection requirements.

Layered Distributed Cloud Storage

Distributed cloud storage is a distributed system that aggregates the storage space of multiple storage devices to provide data storage and business access functions through cluster applications, grid technology, distributed file systems, etc., and coordinates a large number of different types of storage devices in the network to work together through application software. In the smart park platform, a large-scale distributed cloud storage system is used, which can effectively provide high-performance data access interfaces and ensure high-speed access when multiple business systems and multiple client terminals share access to the data pool. On the other hand, it provides a high-security protection mechanism for the core data of the smart park platform. Based on the replica mode and LeoRaid[7] protection mode security mechanism, it can ensure that data automatically migrate and rebuild when any hard disk or device fails, ensuring the safe and continuous operation of the core business system.

Smart Park Innovation Application

As a microcosm of a smart city, the smart park covers various construction functions. Through overall intelligent transformation, it comprehensively enhances overall management and service capabilities, thereby achieving visual real-time equipment monitoring, improving park work efficiency, enhancing park service levels, and reducing manual costs. To address the needs of fine management and efficient operation of the smart park, based on the fusion analysis technology of multi-source heterogeneous information perception in the park and using the layered distributed integration concept as a method, a smart park innovation application system was developed. The system includes a building energy-saving operation management system, an emergency command system, a one-card information management system, and an asset management system. This innovative application system provides a digital infrastructure for the park, connects various smart devices and facilities, achieves horizontal and vertical penetration, offers a comprehensive service platform for the smart park, and relies on advanced and applicable technologies such as big data, cloud computing, artificial intelligence, and the intelligent Internet of Things to guide the fine and intelligent operation and maintenance work of the park.

Research Results

To address the needs of efficient management, industrial ecological integration, and energy savings and carbon reduction in industrial parks, a smart park platform based on big data and distributed cloud storage is built to achieve the interconnection and interoperability of the management and operational data of industrial parks.

The unique data recovery algorithm increases the utilization rate of the entire storage space as high as 80%, and the data recovery speed is more than 5 times higher than that achieved with the traditional RAID technology data recovery mechanism.

Users are provided with more than 120 GB/s aggregate bandwidth and 1 PB of actual usable space.

The storage system has no capacity limit. It effectively supports a capacity of more than 300 PB and balances the data server load and data consistency.

The platform demonstrates functions such as fault recovery, automatic fault switching, and load balancing.

[7] LeoRaid is the collective term for the data protection technologies used in LeoStor, a distributed storage system.

3.2.6 Smart City's IoT/Visual Networking Information Integration Platform

Research Background

In 2010, we innovatively divided the concept of a "smart city" into three levels—intelligent buildings, smart parks, and smart cities—and noted that smart parks are the breakthrough and entry point for implementing smart cities and the key link for making a qualitative leap toward smart cities. We also innovatively proposed the construction path of the IoT/visual networking multimedia information integration platform for smart buildings/smart parks/smart cities. We proposed that smart cities are based on intelligent buildings and layered distributed integration technology to create a digital ecology between devices, data, technology, businesses, and organizations for building intelligence. Moreover, based on the traditional technology, business, industry, terminal, and network integration model centered on three-network integration (telecommunications network, broadcast television network, internet), we innovatively proposed a new three-network integration model centered on the Internet of Things, visual networking, and intelligent networking. To adapt to the trend of the diversified business development of smart cities owing to intelligent building technology, we have extended the layered distributed integration technology and the new three-network integration concept to smart cities and studied the common technology, path, and mode of building the IoT/visual networking information integration platform for smart cities.

The concept of smart city information integration under the new three-network integration is shown in Fig. 3.7:

Research Content

This project starts from practical applications, which are based on the construction of smart buildings/smart parks, and targets smart cities. It applies technologies such as intelligent buildings, new triple network convergence, cloud computing, and Internet of Things/Visual Networking to research and develop an information integration platform for smart buildings/smart parks/smart cities. During the project implementation, comprehensive perception, high-definition visualization, and data mining are used to make the leap from intelligent buildings to smart buildings. Moreover, dozens of information systems in intelligent buildings are integrated into one platform, namely, the smart building information integration platform. The overall plan for smart city integration, industry interface protocols, and standards encompasses the entire process of project research and implementation.

The research technology route of the Internet of Things/visual networking information integration platform for smart cities is shown in Fig. 3.8:

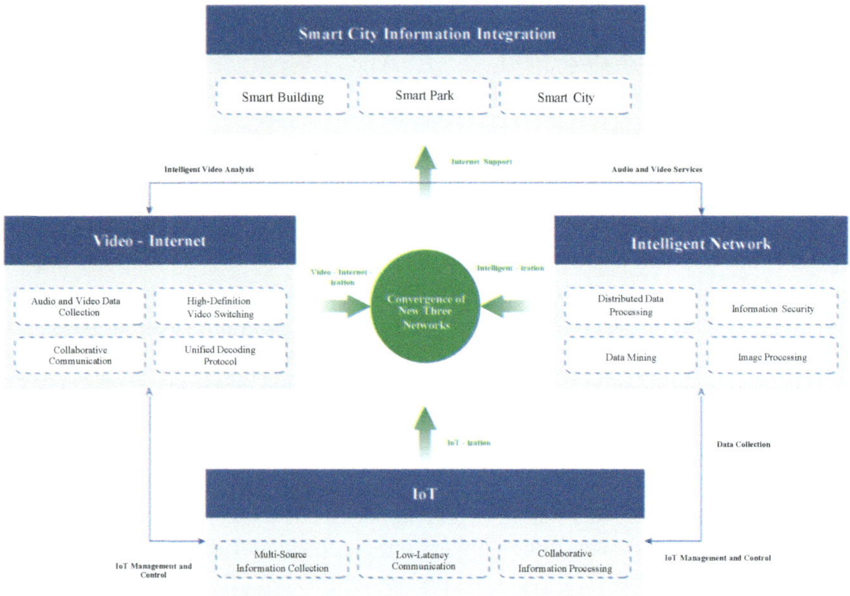

Fig. 3.7 Concept of smart city information integration under the new triple network convergence

Overall Plan for Smart City Information Integration

The smart building information integration platform is the cornerstone for achieving the expected goals of smart cities. Based on the foundation of doing well and refining the smart building information integration platform and on core technologies such as big data and cloud computing, a smart park information integration platform can be built. A breakthrough in quality can lead to the smart city information integration platform. The overall plan for the smart city information integration platform is divided from bottom to top into the perception layer, network layer, platform layer, and application layer. The network layer uses TCP/IP networks and visual networks as the basic communication infrastructure for smart buildings, establishes a cloud data center, and provides real-time, massive cloud storage environments for smart buildings. The platform layer collects, classifies, integrates, and performs other data preprocessing as well as intelligent processing and analysis of the collected information. The application layer provides various city function management services for end users and is compatible with existing application systems.

Smart City Information Collection and Comprehensive Perception Technology Based on the Internet of Things

The Internet of Things is first reflected in the asset management of buildings and parks in the application of city information collection and comprehensive perception. Starting from the subtleties of city information collection and comprehensive perception, research on the application of the Internet of Things represented by RFID

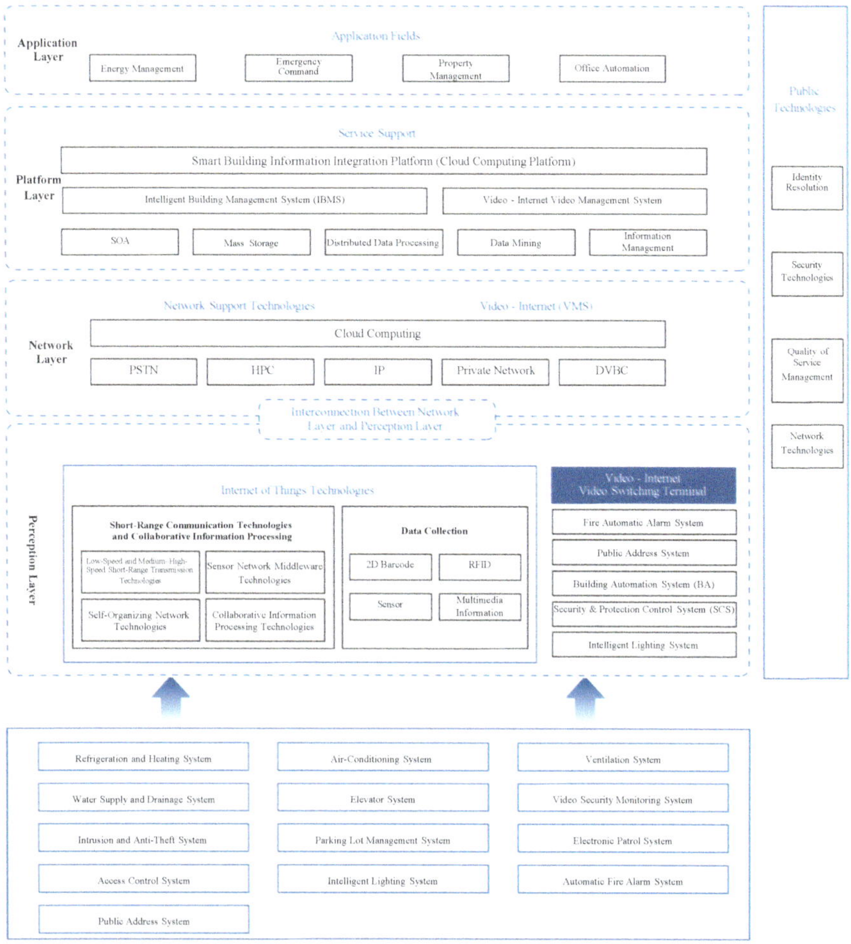

Fig. 3.8 Research technology route of the internet of things/visual networking information integration platform for smart cities

in asset management is conducted. An asset management system based on the Internet of Things, reliable wireless network transmission and anti-interference technology, user self-modeling technology based on OPC specifications, and data visualization technology for buildings, parks, and cities based on virtual reality is constructed. The user self-modeling technology uses a new configuration software architecture (OLE for process control) and a client/server model to establish a graphic configuration subsystem to support asset management and environmental monitoring customization. Through the configuration controls, graphic drawing tools, etc., provided by the subsystem, personalized customization of the variable monitoring environment is carried out. Virtual reality technology is used to construct 3D scenes of buildings, parks, and cities; establish mathematical models of important assets; and present data from various sources of buildings, parks, and cities in a vivid, intuitive, and perceptual way.

Smart City Information Exchange and Video Service Technology Based on Visual Networking

Video data communication is the main data flow of buildings, parks, and cities, accounting for more than 90% of the total communication flow and consuming a large amount of bandwidth in the TCP/IP network. Based on the wired broadcast television network as the transmission channel, visual network technology and its product services are used for the information exchange of smart buildings, providing video surveillance, video conferencing, remote training, emergency command, media release and other video data services; event analysis functions such as intrusion, departure, tripwire, directional tripwire, loitering, sudden change, etc.; and statistical analysis functions of the flow of people in designated areas.

Data Fusion and Massive Data Intelligent Processing Technology for Smart Cities

The forms of perception data in smart cities are diverse. Research is necessary to understand how to manage and utilize these data and extract useful information for building smart cities. Cloud storage technology for massive data of smart cities should be constructed. Data mining, knowledge discovery, and other means can be used to research the intelligent processing technology of massive data in smart cities, form knowledge from data, and provide scientific and accurate decision-making support.

Development of an Information Integration Platform for Smart Buildings/Smart Parks/Smart Cities

Based on the above research, a cloud computing-based information integration platform is designed and developed, data access protocols for various industry information systems are formulated, service interfaces are provided for various functional departments of the city, and ultimately, an information integration platform for smart buildings/smart parks/smart cities is developed. The platform is based on the principle of intelligent building layered distributed information integration. It adheres to the principles of data centralization, management centralization, and decentralized control, integrating the information systems of fire protection, HVAC, security, water supply and drainage, power supply, lighting, elevators, parking, etc., within the building into a single integrated platform.

Research Outcomes

Focusing on the needs of massive amounts of information processing, video information transmission processing, and information system integration for smart cities, with layered distributed technology and the concept of new triple network integration as the core foundation, a construction plan for a system and platform for the Internet of Things/visual networking information integration of smart cities was formed. The research results constitute the foundation of an information integration platform for smart parks and smart cities and can be gradually updated to promote advances in information integration platforms for smart parks and smart cities.

The project results have several impacts on the development of related R&D work and of this discipline and related disciplines.

The introduction of visual networking technology into smart buildings provides a reference for data processing and transmission in smart buildings, smart parks, and smart cities. The volume of video data inside and outside buildings is very large, and the redundancy of video data is very high. Using the current TCP/IP network transmission will severely consume limited bandwidth resources, which will obstruct the communication of city emergency data and even cause data loss. This project thus diverts video data to the visual network, reducing the load on the TCP/IP network, and provides a reference for data transmission in smart cities, technical research in related fields, and product development.

A layered distributed construction approach for smart cities is proposed. This project starts with smart buildings and explores the construction approach of the smart building information integration platform. Based on this approach, it expands the number of smart buildings to the scope of smart parks, then achieves a qualitative leap, expanding from the smart park to a city via step-by-step implementation, and finally constructs a smart city information integration platform.

Several experiences and reference cases of smart building construction have been accumulated. This project integrates the construction ideas of smart buildings, smart parks, and smart cities into actual engineering cases, achieves several results, accumulates experience, and provides reference cases and reference value for other smart constructions, smart parks, and smart city engineering construction.

3.3 Cloud-Edge-End Collaboration Technology System and the Concept of Application Innovation

3.3.1 Background and Goals of Cloud-Edge-End Collaboration Technology and the Concept of Application Innovation

The "Fourteenth Five-Year Plan" proposes to "coordinate the development of cloud services and edge computing services". In this national top-level planning document, the idea of the coordinated development of the cloud and the edge, which is far-sighted and profound, is proposed.

The past decade has witnessed rapid progress in cloud computing, with the global cloud computing market expanding several times and China's cloud computing market growing from billions to the current scale of hundreds of billions. In the future, cloud computing will usher in the next golden decade, marked by a period of inclusive development. First, with the advancement of new infrastructure, cloud computing will accelerate the application landing process and achieve rapid development in different fields. Second, in the context of the global digital economy, cloud computing has become the inevitable choice for enterprise digital transformation, and the process of enterprises moving to the cloud will further accelerate.

Edge computing is also receiving increasing attention, and cloud-edge collaboration has become an important trend. With the rapid development of technologies such as 5G, IoT, and AI and the promotion of cloud services, edge computing has received industry attention. However, only through close collaboration between cloud computing and edge computing can the needs of various scenarios be met, thereby maximizing the application value of cloud computing and edge computing. With the continuous implementation of new infrastructure, building an end-to-end cloud-edge-end collaborative architecture will be important to achieve the high-speed interconnection of data across the entire domain, application integration scheduling distribution, and full coverage of computing power. Future computing will not only be limited to large data centers but also be distributed on a continuous spectrum of cloud-edge-end collaborative architecture. Therefore, the integration of the cloud-edge-end concept will drive the diffusion of computing processing to the edge, and the ubiquitous development of computing power is the inevitable direction and trend of future technology development.

In response to the new demands in the construction of smart cities and the rapid development of new generation information technology, we apply technologies such as intelligent information sensing, the Internet of Things (IoT), artificial intelligence, big data, and cloud computing to create a new generation of smart city industry application products and solutions with cloud-edge-end collaborative control. This application promotes the digital upgrade of city construction toward smart livelihoods, comprehensive city management, and smart parks, forming modular, standardized, and systematic products and solutions. It meets the needs for massive IoT device access, spatial management and analysis, intelligent processing of large amounts of information data, and data analysis and prediction. By realizing the "Internet of Things", data collection and preliminary processing, AI deep analysis, big data analysis and mining, we can accurately predict the operating status of equipment or perform a comprehensive analysis of user behavior, making the equipment "smarter" and the user experience richer.

The research objectives of common technology development and engineering applications are shown in Fig. 3.9:

After years of accumulating technology, products, innovative applications, engineering applications, and research experience in the field of smart cities, the goal of developing a cutting-edge technology system of cloud-edge-end collaboration for smart cities is based on the already formed AIoT-based smart city basic technology system. Specifically, we propose a cloud-edge-end collaborative technology system for smart cities (as shown in Fig. 3.10), which includes intelligent information sensing and IoT technology, edge computing and artificial intelligence technology, intelligent cloud computing and big data analysis technology, and cloud-edge-end collaborative technology. Based on this system, we establish an industrialized technology research and development and service support platform for the smart city

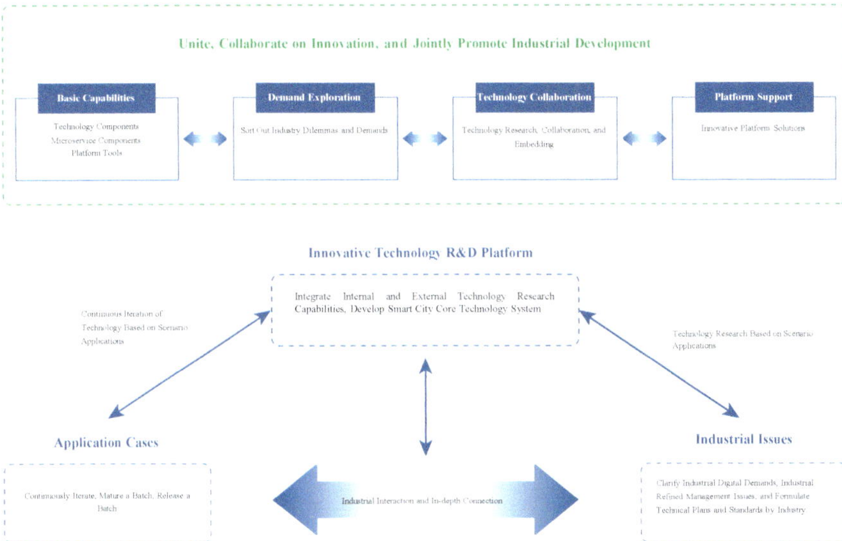

Fig. 3.9 Research objectives of common technology development and engineering applications

industry, improve the innovation ability of the smart city industry chain, effectively solve the pain points faced by enterprises in industrial informatization and digitization, and provide technical support for the high-quality development of China's urban agglomeration economy.

The cloud-edge-end collaborative technology system consists of multisource heterogeneous information sensing and intelligent IoT technology, cross-platform edge computing and artificial intelligence based on domestic AI chips, intelligent cloud computing and big data analysis technology for smart cities, cloud-edge-end collaborative technology based on domestic intelligent computing platforms, and

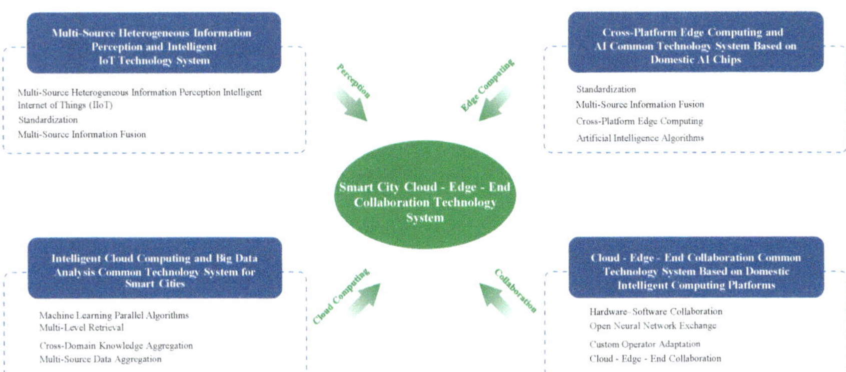

Fig. 3.10 Cloud-edge-end collaborative technology system for smart cities

system integration and application. In terms of engineering applications, it focuses on the transformation and promotion of new technology achievements in smart parks, smart energy, smart medical care, smart agriculture, and other sectors in the construction of smart cities.

3.3.2 Multisource Heterogeneous Information Sensing and Intelligent IoT Common Technology System

To address the forefront of multisource heterogeneous information sensing and IoT technology development, fully leveraging the advantages of closely combining theoretical method research with application innovation, we address the key issues urgently needed to be solved in the IoT application environment. Following the "sense, know, act" rules of understanding the real world and guided by the multi-dimensional integration of "sensing, intelligence, interconnection, collaboration", we construct a common technology system of multisource heterogeneous information sensing and intelligent IoT (as shown in Fig. 3.11) to serve pillar industries such as smart parks, smart energy, smart medical care, and smart agriculture in the construction of smart cities.

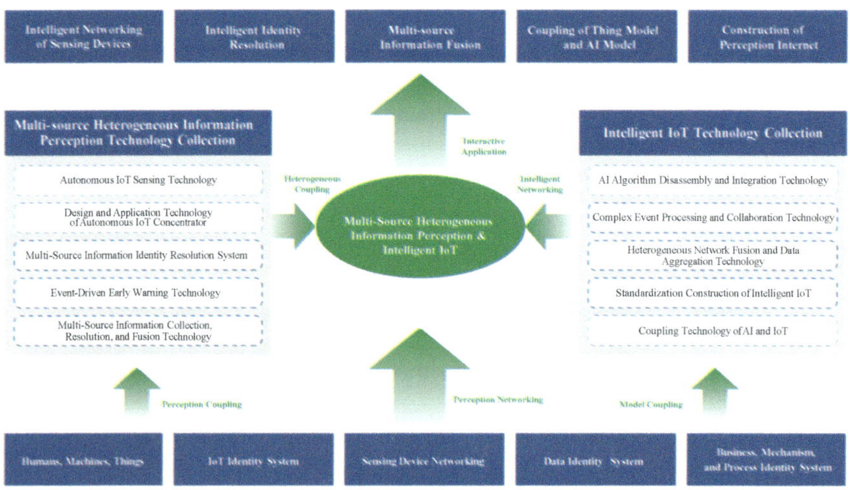

Fig. 3.11 Multisource heterogeneous information sensing and intelligent IoT common technology system

Multisource Heterogeneous Information Sensing Technology

To overcome the bottleneck of multisource heterogeneous information IoT sensing access processing technology, we propose a set of multisource heterogeneous information sensing technologies, including autonomous IoT sensing technology, autonomous IoT concentrator design and application technology, personalized configuration technology for IoT sensing applications, event-driven early warning technology, and multimodal configurable display technology. By integrating cutting-edge sensing, transmission, and parsing technologies, we address the unified access and management of massive multisource heterogeneous IoT devices. Through IoT devices, we collect, clean, parse, store, analyze, mine, and display vital sign big data, enabling the sharing and linkage of various types of data. The IoT middleware technology using the adapter pattern solves the concurrency and heterogeneity problems of shielding the underlying massive device access, thus establishing a virtual bridge between IoT devices and upper-layer applications to solve heterogeneous interface adaptation problems such as the unified access of IP cameras.

Intelligent Internet of Things (AIoT) Technology

The AIoT is the fusion application of artificial intelligence (AI) and the Internet of Things (IoT), where both technologies benefit from their integration. On the one hand, AI aids the IoT in intelligently processing massive amounts of data, thereby enhancing the level of intelligence in its decision-making process. On the other hand, the IoT, through its ubiquitous sensors and terminal devices, provides AI with a vast number of analyzable data objects. In short, AI gives the IoT a "brain," elevating "connectivity" to "intelligent connectivity," whereas AIoT provides AI with more extensive research "soil," pushing "artificial intelligence" toward "applied intelligence." AIoT-related research is still in its early stages, with many open issues in different industries and application scenarios awaiting further research and exploration. Therefore, the main problems that the intelligent Internet of Things (AIoT) technology system aims to solve are the intelligent issues of networking, transmission, and parsing. These problems include the following:

Decomposition and integration technology of AI algorithms: AI algorithms require a large amount of computing power to build AI-based intelligent terminals, perform data processing and deep analysis, and research the decomposition and simplification techniques and methods of AI models, especially deep learning models, on the edge side.

Complex event processing and collaborative technology: The network information system built by the AIoT has ubiquitous and complex characteristics. The large-scale heterogeneous network element data access, the dynamic instability of heterogeneous networks, the limited underlying resources, the massive amount of data exchange, and the lack of a centralized control structure greatly increase the difficulty of event processing. Research on complex event processing and collaborative technology is therefore being carried out.

Heterogeneous network fusion and data aggregation technology: The AIoT system integrates different types of networks (wireless sensor networks, edge servers, centralized cloud servers, etc.), and different types of networks have different characteristics and constraints (data transmission rates, energy consumption, quality of service, etc.). In response to these issues, effective heterogeneous network fusion technology solutions are being developed to ensure stable quality when tasks are offloaded and cross-network, cross-layer collaboration occurs. On the other hand, as the AIoT is implemented in different application scenarios, different levels of sensing devices will inevitably generate heterogeneous data. Therefore, heterogeneous data extraction and semantic understanding methods will be developed to achieve data aggregation, mine data value, and develop high-level AIoT applications.

AIoT standardization: Currently, the standardization of the two core technologies of the AIoT, the AI and IoT, is not very established. Therefore, as a fusion entity, AIoT standardization obviously faces many problems and challenges. Efforts will be made to unify the concepts, connotations, and application models involved in the AIoT, achieve cross-platform interoperability of data and applications, and classify the standards for evaluating AIoT solutions, among many other aspects, to help establish enterprise technology application standards and industry application standards.

3.3.3 Cross-Platform Edge Computing and Common AI Technology Systems Based on Domestic AI Chips

Owing to the development of algorithms, computing power, and datasets, AI technology has made rapid progress and has been widely applied in various industries, such as security, transportation, industry, and agriculture. With the advent of the 5G and IoT era, AI technology is gradually sinking from the central cloud to the edge, taking advantage of the convenience of edge-side data collection and real-time processing and computing. By performing collaborative inference and training of models at the edge and cloud, the "last mile" problem of AI implementation can be solved. Consequently, edge intelligence has emerged, attracting considerable attention from academia and industry. Given the current complex and changing international environment, edge computing will be based more on domestic AI chips (such as Huawei HiSilicon, Rockchip, Allwinner, Yizhi Electronics, etc.) to form an edge computing ecosystem. Therefore, we propose establishing a technology system on the basis of an intelligent edge computing platform created from domestic AI chips (as shown in Fig. 3.12), ultimately building a lightweight intelligent computing system based on edge computing and multitask collaboration with domestic intelligent computing chips as the basic platform, which provides technical support for intelligent analysis and processes tasks in industries such as smart parks, smart energy, smart health care, and smart agriculture.

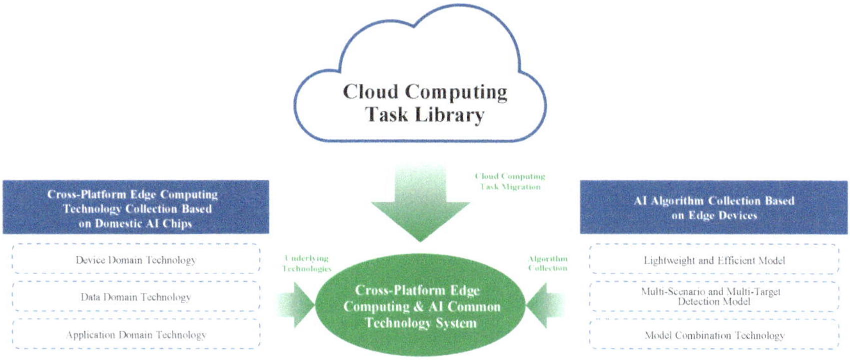

Fig. 3.12 Cross-platform edge computing and common AI technology system based on domestic AI chips

Cross-Platform Edge Computing Based on Domestic AI Chips

Device domain technology: In collaborative lightweight, low-power scenarios, zero-configuration, self-organizing network, and cross-platform technologies for edge devices are constructed. In real-time computing scenarios, multitasking and priority scheduling technology for edge devices is developed to support event response and task processing within a certain time. A security technology stack for edge devices is established to ensure that the specific design and implementation of the operating system, middleware, and upper-layer applications in the device domain meet security protection requirements, thus preventing tampering with the root keys, software, firmware, and configuration information.

Data Domain Technology: Different devices from different vendors have their own information protocols and information description formats. To meet the need for the integration of different protocols between devices, it is necessary to establish cross-protocol device integration technology. This technology integrates and consolidates information from various types of devices, thereby supporting managers in efficiently managing onsite devices globally. To achieve cross-vendor data interoperability and data parsing, the first step is to implement a unified semantic method, build a unified information model architecture, and thus achieve compatibility with various existing models.

Application Domain Technology: In the edge computing model, edge computing node devices are diverse and complex, and the operating environment for each node may not be exactly the same. Traditional programming methods cannot meet the needs of deploying applications under the edge model. Therefore, it is necessary to develop a new programming method suitable for edge computing, such as the Firework model and Docker container technology. In addition, as the scale of the data increases, relying on a single edge device for computation will lead to significant computational pressure. How to design task scheduling strategies and device collaboration schemes and work jointly with multiple devices is one of the key scientific and technical issues to be addressed.

Artificial Intelligence Algorithms Based on Edge Devices

Lightweight and Efficient Model Technology: Compared with edge intelligent computing platforms based on foreign AI chips, domestic platforms generally have problems of low memory and low computing power. How to enhance the feature extraction capability of lightweight models, improve the running efficiency of models, and build a model construction method that accurately describes the features of lightweight models are among the considerations built into the system.

Multi-Scenario Multi-Target Detection Model Technology: For the lightweight model requirements of multiscenario and multitarget detection, tracking, and structuring algorithms, a multitask learning method based on parameter sharing is adopted. This method shares shallow visual features of multiple tasks, thus making the network suitable for multiple tasks, such as visual target detection, tracking, and structuring. Considering the limited computing power of edge computing devices, a network pruning method based on weight norms is proposed. While maintaining the model's feature perception ability, it further improves the running efficiency of the model and realizes the lightweight application of the network model on the edge.

Cloud Computing Task Migration Technology: Traditional cloud computing cannot meet real-time requirements for data computation speed and data transmission because the cloud computing center undertakes the majority of the computing tasks. By adopting the software and hardware framework of edge computing architecture, a platform with preprocessing capabilities is built for the system. An efficient set of computing task migration technologies is established to migrate some of the computing tasks to the edge, thereby improving the real-time performance of the business through the additional computing power and resources provided by the edge.

3.3.4 Common Technology System of Intelligent Cloud Computing and Big Data Analysis for Smart Cities

With the continuous development of core technologies for smart cities, the Internet of Things (IoT) has formed a technology ecosystem that includes multisource heterogeneous data collection and analysis, the standardization of identification parsing systems, and the intelligent generation and combination of physical models. With breakthroughs in deep learning and reinforcement learning, the field of artificial intelligence is gradually forming a technology ecosystem centered on data, computing power, and algorithms. This ecosystem includes multitype and multimodel development, multimodal model training, and the combination and collaboration of multiple models. With the development and application of large model technology gradually evolving toward general artificial intelligence, the field of cloud computing has made significant breakthroughs in computational resource coordination, the computational power output of computational nodes, and the integration of cloud-native container technology and applications with the establishment of national supercomputing and intelligent computing networks. The IoT, artificial intelligence, cloud computing, and other core technologies are being combined to form a core technology ecosystem serving smart parks, smart energy, smart health care, smart agriculture, e-government, and digital and intelligent upgrades in other industries. Therefore, we propose a common technology system of intelligent cloud computing and big data analysis for smart cities (as shown in Fig. 3.13). This technology system includes a set of parallel machine learning algorithms against the background of big data, a set of key technologies for multilevel retrieval based on big data intelligence, and a set of key technologies for the aggregation of multisource heterogeneous data and cross-domain knowledge. It is used mainly to support functions such as IoT perception coupling intelligence; AI training and cloud computing computational power coordination; multidomain knowledge aggregation; and the combination of big data and AI multitype models, etc. Ultimately, with the common technology system at its core, it forms the ecosystem of core technologies for cloud-edge architecture in smart cities; fosters the collaboration and ecologization of IoT technology, AI technology, cloud computing technology, and other core technologies; and achieves the continuous iteration and optimization of core technologies.

Parallel Machine Learning Algorithms in the Context of Big Data

Faced with massive and diverse big data, the computational burden of traditional machine learning algorithms based on small amounts of data is increasingly apparent. Parallel computing is the mainstream method for solving this problem. The focus is on researching distributed parallel computing methods for processing big data machine learning, parallel computing methods for multicore heterogeneous computing platforms, and hybrid parallel computing methods that integrate the above two methods.

Key Technologies for Multilevel Retrieval Based on Big Data Intelligence

Research on fundamental theories and key technologies oriented toward big data intelligence should be conducted; high-increment, high-timeliness, multimodal network modeling technologies should be developed; classification methods based on graph convolutional neural networks should be studied; intelligent construction methods for knowledge graphs based on big data should be researched; and multimodal content retrieval technologies should be studied.

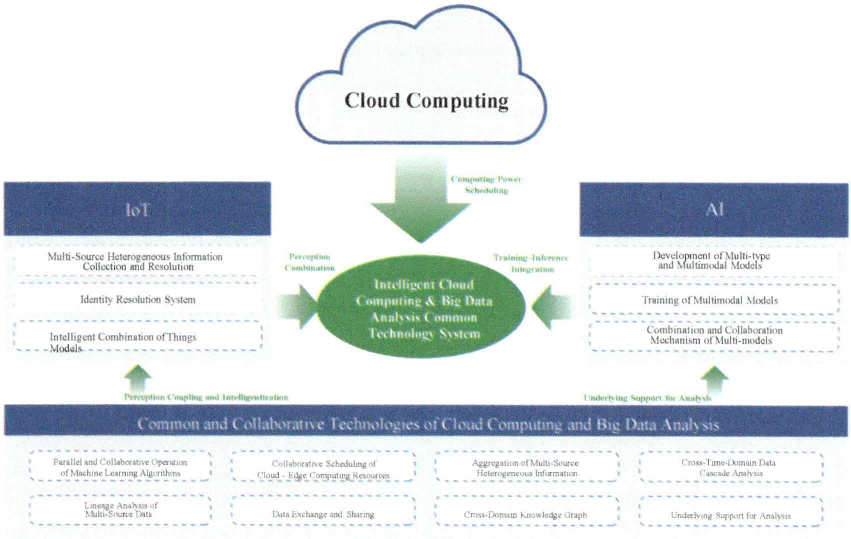

Fig. 3.13 Common technology system of intelligent cloud computing and big data analysis for smart cities

Key Technologies for the Aggregation of Multisource Heterogeneous Data and Cross-Domain Knowledge

Key technologies for the aggregation of multisource heterogeneous data and cross-domain knowledge should be developed, focusing on the efficient collection, integration, secure utilization of big data and the construction and correlation analysis of industry knowledge and datasets. A data collection system should be constructed; real-time data convergence technologies should be researched; multisource data lineage mining technologies should be explored; and cross-temporal, spatial, multiscale, and multidimensional data cascade analysis technologies should be studied. A data sharing and exchange system should be built, and multimodal data sharing and exchange mechanisms and technologies should be researched. Finally, industry data repositories and industry knowledge graphs should be constructed.

3.3.5 Common Technical System for Cloud-Edge Collaboration Based on a Domestic Intelligent Computing Platform

Owing to the development of algorithms, computing power, and datasets, artificial intelligence technology has made rapid progress and has been widely applied in various fields, such as parks, electricity, agriculture, and health care. With the

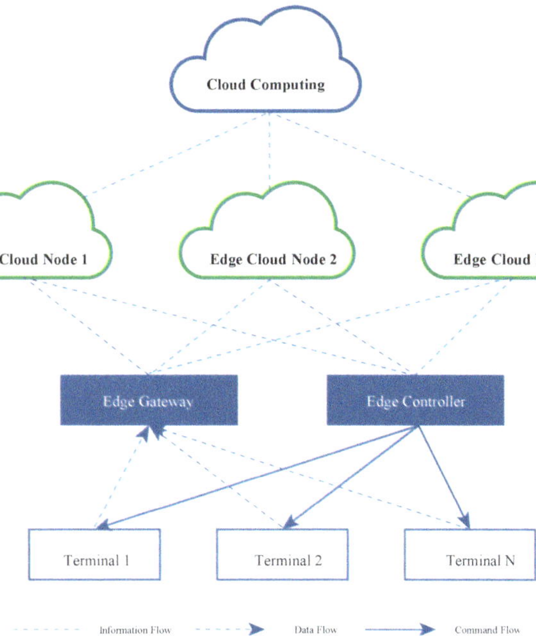

Fig. 3.14 Cloud-edge collaboration mechanism

advent of the 5G and Internet of Things era, artificial intelligence technology is gradually sinking from the central cloud to the edge, leveraging the convenience of edge-side data collection and real-time processing. By conducting collaborative inference and training of models at the edge and cloud, the "last mile" problem of AI implementation has been solved, and edge intelligence and cloud-edge collaboration have emerged, attracting considerable attention from academia and industry. The cloud-edge collaboration mechanism is shown in Fig. 3.14:

As various industries in smart cities increase their requirements for high-performance integrated storage and computation of big data, cloud computing and edge computing are developing toward a collaborative system. The computing resources of the cloud are used for big data and large model training and inference, whereas the computing resources of the edge are used for big data analysis and model inference. Therefore, in response to the trend of domestication of computing platforms and collaboration of cloud-edge computing resources, we propose building a common technical system for cloud-edge collaboration based on a domestic intelligent computing platform (as shown in Fig. 3.15). This system will include key software and hardware collaborative technologies based on the domestic intelligent computing platform, open neural network exchange technology, high-performance custom operator adaptation technology, collaborative training and inference technology between the central cloud and edge nodes, and "end, edge, cloud" collaboration and unified scheduling technology based on the domestic intelligent computing

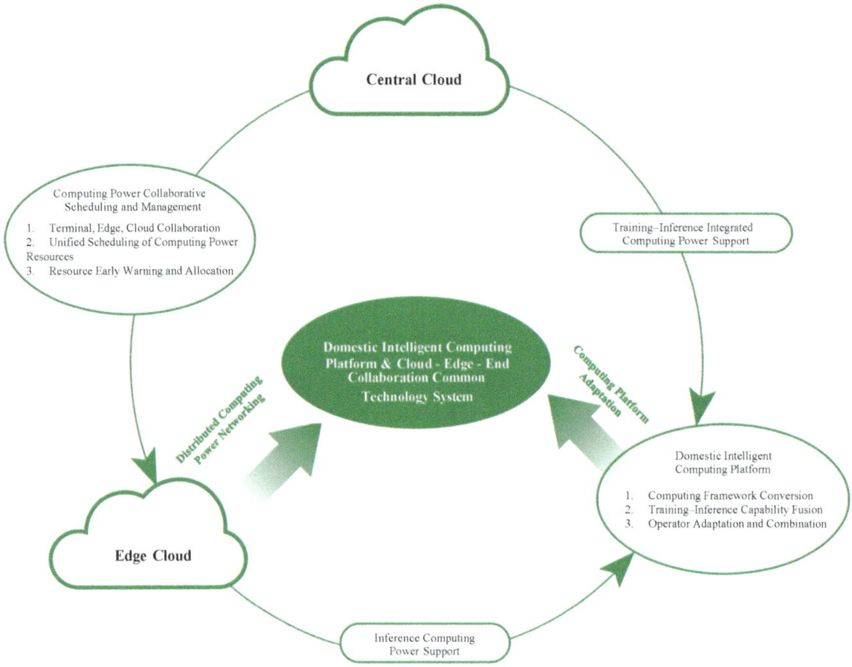

Fig. 3.15 Common technical system for cloud-edge collaboration based on the domestic intelligent computing platform

platform, all of which will meet the requirements of the domestic computing platform for cloud-edge computing resource collaboration and unified scheduling.

The key software and hardware collaborative technologies based on the domestic intelligent computing platform include model conversion and heterogeneous parallel computing technology compatible with mainstream frameworks such as Caffe, TensorFlow, PyTorch, open neural network exchange technology, high-performance custom operator adaptation technology, and AI process orchestration technology based on graph optimization. These technologies can enhance the AI access capability and computational efficiency of the domestic intelligent computing platform, collaborative inference, incremental learning, federated learning, model segmentation and pruning methods between the central cloud and edge nodes. We construct "end, edge, cloud" collaboration and unified scheduling technology based on the domestic intelligent computing platform, achieving resource collaboration, application collaboration, data collaboration, and intelligence collaboration, and deriving a product solution based on this platform. This approach is being promoted and applied in industries such as smart parks, smart energy, smart health care, and smart agriculture to achieve the goal of building smart and beautiful cities.

3.3.6 *Innovative Application Framework Based on a Cloud-Edge-End Collaboration Technical System*

Integration and Application Framework

In response to the new demands for digital, intelligent, and integrated innovative applications from thousands of industries in smart cities, we propose an integration and application framework for cloud-edge-end collaboration technology for smart cities (as shown in Fig. 3.16) based on the constructed smart city cloud-edge collaboration technical system, which integrates multisource heterogeneous information perception and intelligent Internet of Things technology, cross-platform edge computing and artificial intelligence based on national AI chips, intelligent cloud computing and big data analysis technology for smart cities, and cloud-edge-end collaboration technology based on the domestic intelligent computing platform. As shown in the figure, the focus is on solving the "island" problem that still exists in the context of technical combinations and synergy, such as device interconnections, data interconnections, business interconnections, and industrial chain interconnections. The integration of technology and the application framework promotes the formation of a technological ecosystem, greatly facilitating innovative application research and development and implementation in fields such as the smart city industry, urban governance, and the integration of industries and cities. This development will lead to the formation of an innovative application ecosystem, within which data exchange and model synergy can be realized.

Smart Park Application Architecture Based on Cloud-Edge Synergy Technology Integration and an Application Framework

To construct an IoT base cloud platform in smart parks, we leverage new generation information technologies such as the IoT, edge computing, cloud computing, and artificial intelligence. Through sensing, the IoT, and intelligence, we connect the physical infrastructure, information infrastructure, social infrastructure, and commercial infrastructure in the park. We solve the data sharing and linkage problems between different IoT application systems in various smart parks, forming core functional modules for IoT data collection, aggregation, analysis, and display. In addition, we configure functions according to the needs of different parks, creating an IoT information ecosystem for the park. In response to the research and development and upgrade needs of the IoT base cloud platform for smart parks, we also build an IoT base cloud platform technology integration and application research and development framework based on the cloud-edge synergy technology system. This framework includes a cloud base platform based on a high degree of resource-sharing cloud computing SaaS container, middleware for accessing massive heterogeneous IoT sensing data in smart parks, and a diversified application system for smart park construction.

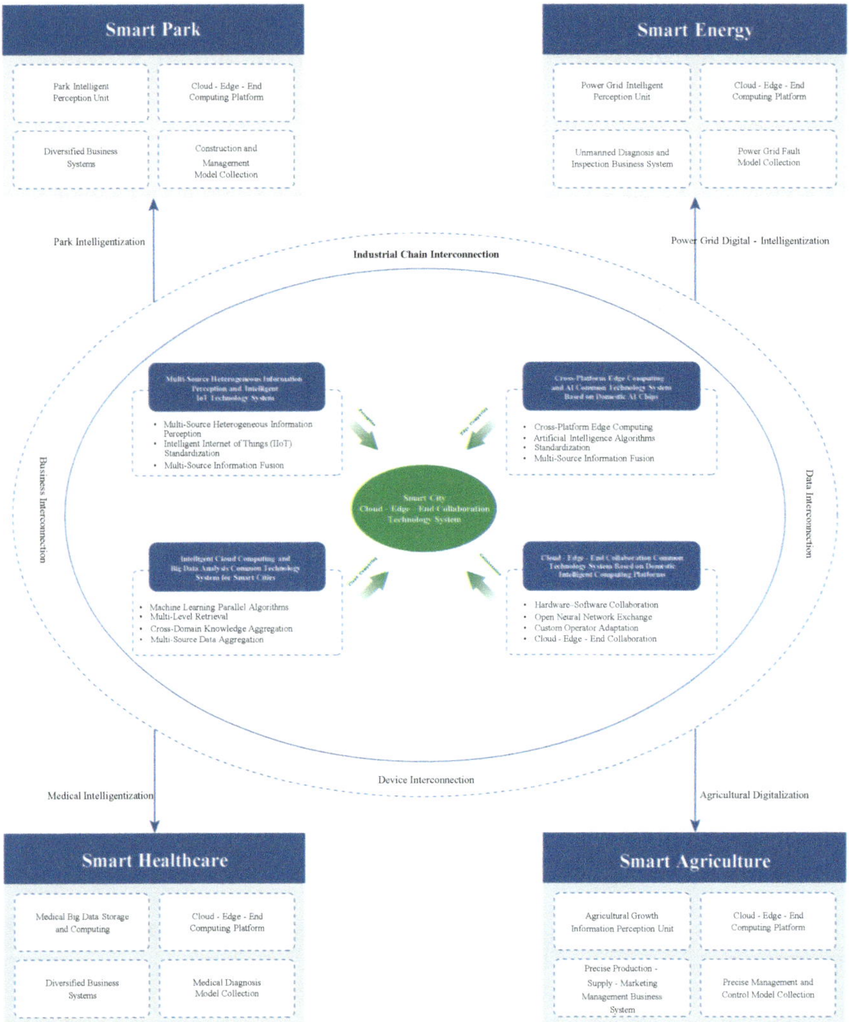

Fig. 3.16 Integration and application framework

The overall framework of the smart park technology integration and application scheme is shown in Fig. 3.17:

Cloud base platform based on a high-resource-sharing cloud computing SaaS container: The cloud base platform based on a high-resource-sharing cloud computing SaaS container enables cloud computing SaaS isolation mode shaping, platform multitenant management, physical database resource isolation, tenant basic information management, performance isolation and request scheduling, unified authority management, personalized customization, and system architecture design.

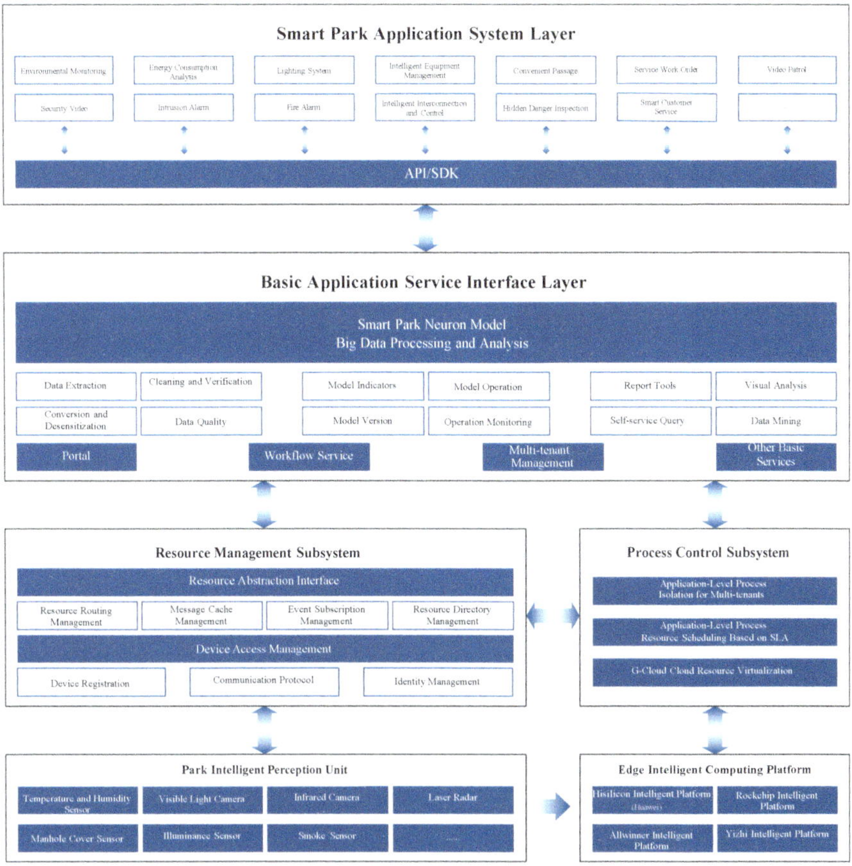

Fig. 3.17 Overall framework of smart park technology integration and application scheme

Middleware for accessing massive heterogeneous IoT sensing data in smart parks: This component provides unified access and management for massive heterogeneous IoT devices, establishing a bridge between IoT hardware and its existing information systems and various smart park applications. Through IoT devices, it collects, parses, stores, analyzes, mines, and displays big data of park signs, enabling the sharing and linkage of various data to meet the requirements of smart park management. Therefore, to shield the concurrency and heterogeneity problems of massive device access at the bottom layer, we build IoT middleware in adapter mode based on the multi-source heterogeneous information sensing and intelligent IoT technology systems and establish a virtual bridge between IoT devices and upper-layer applications to solve problems such as heterogeneous interface adaptation to facilitate the unified access of IP cameras.

Diversified application system for smart park construction: A diversified application system framework for smart park construction is built based on the cloud-edge synergy technology system constructed for different scenario requirements in smart parks. This framework supports functions such as data management, data monitoring, report analysis, environmental monitoring, and basic verification platforms. Users can build diversified thematic application systems for smart parks based on standard data interfaces and business interfaces, as well as test and verify IoT-related businesses in the park.

Smart Energy Application Architecture Based on Cloud-Edge Synergy Technology Integration and an Application Framework

The smart energy system is an important part of the construction of smart cities, and the digitization of grid operation and maintenance and collaborative scheduling is an important part of this system. The digital grid is a technological ecosystem formed by the fusion and innovation of artificial intelligence theory, technology, and methods with the physical laws, technology, and knowledge of the power system. In response to the development needs of the smart grid and energy internet, we use the massive amount of multisource data generated by the aerial inspection of transmission and distribution equipment, fixed video monitoring of substation equipment, and remote sensing monitoring of transmission and distribution lines and substations as data sources. We build a grid operation and maintenance and collaborative scheduling digitization technology integration and application framework based on the cloud-edge synergy technology system. This framework serves as the basis for innovative applications, business collaboration, and grid management mode upgrades of grid digitization. We carry out innovative application research and development of patrol image and video data processing, analysis, and interpretation methods, perform intelligent detection of power equipment defects, ensure the safe and reliable operation of the power grid system, and promote smart city construction.

The integrated and applied framework of smart grid operation and maintenance and collaborative scheduling technology includes transmission line video anti-external force damage warning technology and applications based on edge computing and artificial intelligence, grid wire tracking and tree obstacle analysis technology and applications based on edge computing, grid patrol fault diagnosis technology and applications based on edge computing and domestic smart computing platforms, intelligent patrol platforms of substations based on ground patrol robots, and deep processing and visualization analysis technology and the application of multisource information under the full-space system framework. With the help of the smart city cloud-edge-end collaborative technology system, equipment networking, data interconnections, computing resource collaborative networking, business collaboration, etc., in the grid operation and maintenance and scheduling applications can be realized. Thus, a cloud-edge collaborative grid patrol intelligence can be built, with refined operation and maintenance and unmanned fault monitoring, greatly advancing China's energy-saving and emission-reduction "carbon neutrality" strategy.

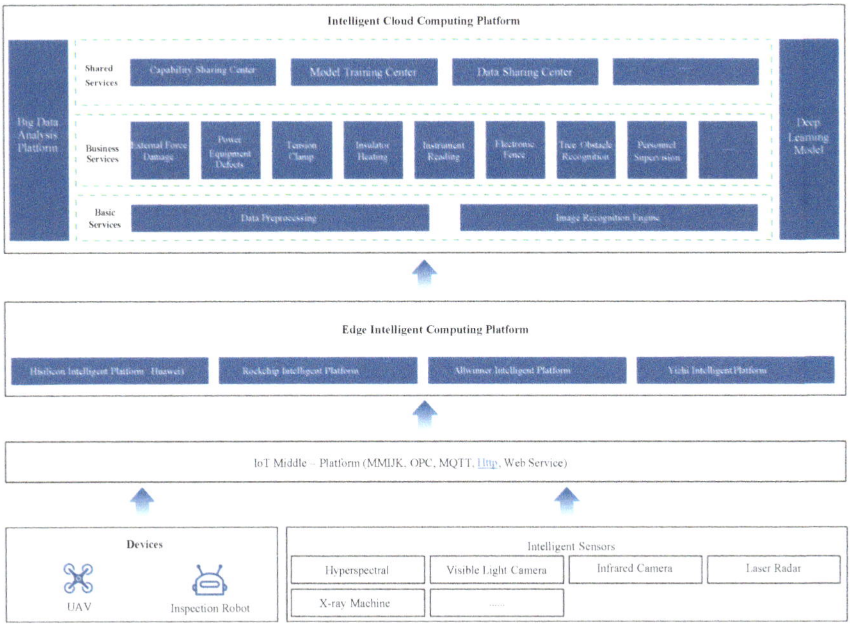

Fig. 3.18 Architecture of smart energy applications based on cloud-edge-end collaborative technology integration and the application framework

The architecture of smart energy applications based on the integration and application framework of cloud-edge-end collaborative technology is shown in Fig. 3.18:

Based on edge computing and artificial intelligence, video warnings of transmission lines against external force damage can be aimed at real-time detection, tracking, and identification of large mechanical moving targets. The development and application of intelligent recognition algorithms based on edge computing and artificial intelligence can perform image capture and warning of large cranes entering line protection areas, monitor black spot construction sites, and prevent accidents.

Based on edge devices and drones equipped with miniature airborne lidar, visible light, infrared, and other devices, real-time wire tracking and tree obstacle analysis technology can be developed. Combining the miniature airborne lidar system with the drone autopilot system, real-time tracking and autonomous inspection of transmission wires can be achieved, and real-time tree obstacle hidden danger analysis and warning can be achieved based on the power defect hidden danger level standard library.

Based on drones and edge platforms, drone front-end real-time target recognition and power equipment defect analysis algorithms, onboard AI terminal software and hardware systems, and intelligent scene perception and recognition can be developed.

Based on infrared and visible light, the detection of insulator damage and heating defects can be performed, and intelligent algorithm for transmission line insulator fault monitoring can be constructed, which include intelligent detection and segmentation algorithms for insulators in drone inspection images, intelligent fusion algorithms for infrared and visible light images, and detection algorithms for composite insulator overheating defects and glass insulator self-explosion defects.

Intelligent patrols of substations based on ground patrol robots, using visible light images, infrared images, sound, and other information collected by patrol tasks, and research on abnormal intelligent detection algorithms for substation power equipment, power meters, etc., can be conducted.

Deep processing and visualization analysis technology for multisource information under the full-space system framework can be achieved. Breakthroughs in automatic processing, seamless fusion, and efficient modeling technology of massive transmission line drone inspection data from multiple sensors, multiple scales, visible light, lidar, infrared, etc., under the full-space information system framework can be promoted. A transmission line defect hidden danger model library and intelligent analysis system for intelligent inspection applications can be built, and automatic processing and visualization analysis of drone inspection data can be performed.

Architecture of Smart Medical Applications Based on Cloud-Edge-End Collaborative Technology Integration and the Application Framework

Smart health care refers to the integration of artificial intelligence, cloud computing technology, and IoT sensing technology into the current public health care management system to make health care services more intelligent and digital. The smart health care system is also a large database that stores a large amount of medical data. With the continuous improvement in modern medical technology, the amount of medical data has increased substantially. How to apply artificial intelligence technology to analyze, process, and mine massive amounts of data is an important challenge. To address the demands of smart health care for massive data storage, model training, model inference, and technical and business collaboration, a smart health care technology integration and application framework is built with the help of the cloud-edge-end collaborative technology system. Within this framework, cloud-edge-end computing resource collaborative scheduling technology can facilitate the storage and calculation of massive amounts of medical data, and cross-platform edge computing and artificial intelligence common technology systems based on domestic AI chips can enable the training and inference of large medical models. Finally, the framework enables medical data collection, data annotation, intelligent algorithm training and inference, business operations and collaboration. The smart health care technology integration and application framework includes components such as medical signal data processing, medical image data analysis, medical information system visualization technology, etc.

The architecture of smart medical applications based on the integration and application framework of cloud-edge-end collaborative technology is shown in Fig. 3.19:

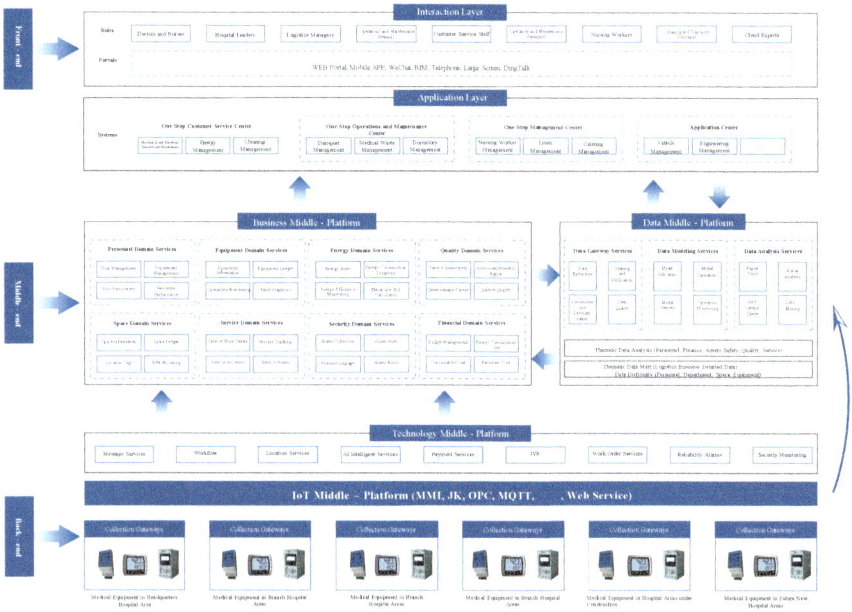

Fig. 3.19 Architecture of smart health care applications based on the cloud-edge-end collaboration technology integration and application framework

Based on unsupervised learning and semisupervised learning data efficient labeling and governance methods, an automatic model for multimodal data collection, cleaning, labeling, backup, and export can be established.

Based on deep learning, medical imaging multisource modality information fusion processing methods, cross-modality, cross-medical discipline data comprehensive analysis modeling technology, reinforcement learning, grid search medical model training and data dimension expansion technology can be established.

For different electromyography methods, body sounds, electroencephalograms, heart sounds, respiration and other electrical signal time–frequency distribution rules, which are based on the electrical signal time and frequency distribution characteristics of the filtering noise reduction method, a relevant distribution model can be established. An intelligent signal recognition method based on machine learning and deep learning can be constructed, and then, the corresponding software and hardware systems can be developed.

Multidimensional medical imaging visualization technology can be developed to provide multivisual, multilevel, multiattribute, multifunctional information to meet clinical applications and achieve fast, accurate, and convenient operation functions.

Smart Agriculture Application Architecture Based on a Cloud-Edge-End Collaboration Technology Integration and Application Framework

China is a major country in terms of agricultural production and agricultural product consumption. With the development of advanced technologies such as the Internet of Things, big data, artificial intelligence, and cloud computing, an innovative application ecosystem of smart agriculture, such as the agricultural IoT, agricultural big data platforms, and precision control platforms for agricultural production, has been formed. Moreover, with the digital upgrade of the entire chain of agricultural production, agricultural product sales, agricultural input production and input of agricultural products, new demands have emerged for the collection, storage, and operation of data resource systems and the coordination between information systems and agricultural production theory. Given these new demands and challenges, we propose a smart agriculture innovative application framework based on the cloud-edge-end collaboration technology system, which includes an upgrade of the two major sectors of planting and breeding. Digitalization can significantly improve the standardization and scale of the breeding process; use sensors, wireless communication, big data, cloud computing, IoT, artificial intelligence and other technologies for data collection and analysis; establish visualization models; and achieve precise management of breeding. With the help of digital technology, the developed breeding management system can work normally in a dynamic environment for a long time and identify problems in a timely manner. New technologies such as big data, artificial intelligence, and cloud computing can standardize and scale the breeding environment, reduce the risks in the breeding process, and thereby improve the breeding efficiency, technical level, and economic benefits.

The architecture of the smart agriculture cloud-edge system is shown in Fig. 3.20:

In terms of planting

Crop production environment and growth information perception heterogeneous IoT technology includes the following: regional main crop growth and production environment information access standards, precision control technology standards, data collection specifications and big data platform access specifications; the crop production environment, growth and ecological information collection perception technology and system; the integrated development of crop environment, growth and ecology and other new sensor components of the multifunctional interface; the multidimensional integration of satellite, remote sensing and sensors; and other multiple perceptions obtained by crop information perception technology. Based on agricultural IoT technology, a multidimensional perception big data platform for the main crop production information is constructed.

The main basic crop production information database, intelligent perception and precision analysis system: uses artificial intelligence, cloud computing and big data analysis technology to construct a precision control model that integrates crop planting theory and a collaborative information system, including the following: precise control of crop growth, yield, quality and resistance by the environment; climate, farming time, farming mode and pest and weed stress methods and models;

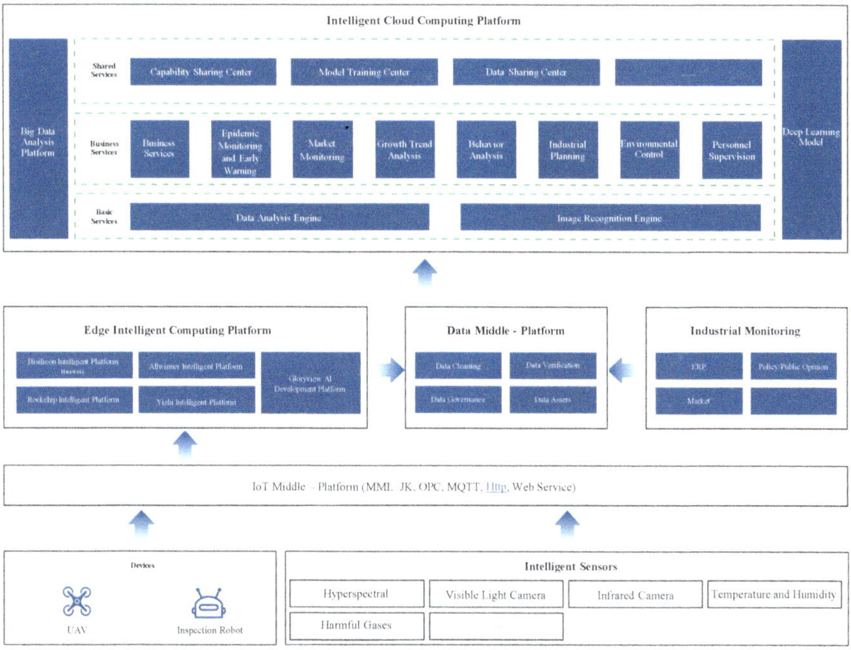

Fig. 3.20 Smart agriculture cloud-edge system architecture diagram

the big data deep mining model of the interaction mechanism of crop soil environment factors and the corresponding growth, yield and quality, with the help of multisource heterogeneous information perception; and an intelligent IoT technology system to build a main crop production information and multidimensional environmental information perception shared service big data platform.

In terms of breeding

Using artificial intelligence and big data analysis technology, digital characterization methods and acquisition technologies for the body shape, vital signs, health status, inventory count and other breeding process information of different physiological stage breeding objects are constructed. Precision control and precision feeding technologies for the breeding environment are studied.

Real-time online key technologies such as biomass estimation, behavior quantification analysis, and the disease prediction of breeding objects are developed. A pig management analysis platform based on big data from the breeding process is developed, and a green, efficient, and intelligent breeding factory is built to achieve efficient breeding.

Part II
Smart City Applications

Chapter 4
Application of a Digital Foundation

On March 12, 2014, the Central Committee of the Communist Party of China and the State Council issued the "National New Urbanization Plan (2014–2020)", which proposed the goals and key tasks of digital city construction and encouraged cities to promote the construction of digital infrastructure. On December 27, 2021, the Central Cyberspace Affairs Commission of the Communist Party of China issued the "14th Five-Year National Informatization Plan", which proposed the strategic choice of accelerating digital development and building a digital China, constructing a ubiquitous intelligent digital infrastructure system, and supporting the construction and application of the digital foundation. The digital foundation refers to a type of infrastructure used in the construction of smart cities that combines physical facilities and digital technology and aims to provide a platform to support and facilitate city management, services, and development. **The digital foundation has five technical features**:

Interconnectivity: Using internet and IoT technology, it connects various city facilities and equipment, thus achieving data intercommunication and sharing.

Intelligent Perception: Equipped with sensors, monitoring devices, etc., it can perceive real-time information about the city environment, traffic conditions, energy use, etc., and convert it into digital data.

Big Data Analysis: Through data collection and analysis, a large amount of city operation and management data can be obtained, and technologies such as artificial intelligence and data mining can be used for in-depth analysis and prediction, providing a basis for decision-making.

Interaction and Service: It provides interaction and service interfaces for residents and city managers, realizing information inquiry, public service application, city navigation, etc., through mobile applications, smart terminals, and other means.

Openness and Scalability: The platform architecture is open and can integrate and expand various applications and services to meet different city management needs.

F. Liu, *Smart and Digital Cities*, https://doi.org/10.1007/978-981-95-2243-9_4

The digital foundation helps city management departments solve the problem of "data islands" and collect more city operations data. In addition, all people, objects, and spaces can be maximally synchronized in the platform to restore a complete city operational status. Moreover, as a result of information connectivity, it forms a series of public service "assets" such as data, IoT, and AI, supports the rapid reuse of application scenarios such as smart governance, smart health care, smart education, smart tourism, smart energy, etc., and reduces redundant construction and investment.

The IoT platform, which has a comprehensive perception ability characteristic of the "IoT", collects the city's vital signs every moment, converts them into data streams, and connects every level of city operation. The data platform, with a full-process thinking ability characterized by "data connection", evolves from the current initial stage of data collection, storage, and sharing to the stage of intelligent judgment and independent analysis based on data; the AI platform, with a global collaborative ability characterized by "intelligence connection", is a process of continuous empowerment and self-evolution, which not only transmits the thinking judgment results of the "data connection" stage to specific scenario applications but also links other digital foundations in various fields, absorbing and integrating new capabilities and achieving continuous evolution. The digital foundation plays the roles of data interface, information processor, and service platform in the construction of smart cities, and the IoT platform, data platform, and AI platform are interrelated, together building a complete digital infrastructure to support the construction and development of smart cities.

4.1 IoT Platform

4.1.1 IoT Platform Overview

An Internet of Things (IoT) platform is an integrated platform that connects and manages IoT devices, collects and analyzes device data, and integrates capabilities such as device management, secure data communication, and message subscription. It can connect a massive number of devices and the collection of device data, and it provides APIs to applications, which can send commands to the device end, enabling remote control.

The system architecture of an IoT platform usually includes the following components:

Perception Layer: This layer is the bottom layer of the IoT system and includes various sensors, actuators, smart devices, etc. These devices can perceive the surrounding environment, collect data, and send data to the IoT platform.

Network Layer: The network layer is the middle layer of the IoT system and is used for managing and connecting perception layer devices and application layer systems. It can include various network technologies, such as Wi-Fi, Bluetooth,

ZigBee, and LoRa. In this layer, communication between devices and the platform can be achieved through various protocols.

Application Layer: The application layer is the top layer of the IoT platform and is used to process data from perception layer devices and provide corresponding services. This layer usually includes applications, analysis engines, artificial intelligence algorithms, etc. In this layer, the IoT platform can provide users with various services, such as data analysis, device management, and real-time monitoring.

Data Storage Layer: This layer is used by the IoT platform to store and manage data collected from perception layer devices. The data in this layer can be stored and managed in various ways, such as through relational databases, nonrelational databases, and distributed file systems.

The IoT platform has become an important part of the IoT ecosystem. As middleware connecting IoT terminal devices and applications, the IoT platform has the following core capabilities:

Device Connection and Management Capability: The IoT platform can support the connection and management of various terminal devices, such as sensors, smart home devices, and industrial automation devices. It can connect devices to the IoT platform through various communication protocols, such as MQTT, CoAP, and HTTP, and implement functions such as device registration, configuration, monitoring, maintenance, and upgrades. At the same time, it can support large-scale device management to ensure the stability and reliability of the IoT system.

Data Collection and Processing Capability: The IoT platform can collect data from various devices and process and analyze them to provide more accurate information and more refined services. It can use various data collection technologies, such as sensors, RFID, GPS, etc., to collect various data generated by devices, such as temperature, humidity, light intensity, and location information. At the same time, it can use various data processing and analysis technologies, such as big data analysis and machine learning, to process and analyze the collected data to discover potential value and insights in the data.

Application Development and Deployment Capability: The IoT platform can provide various applications and development tools to help enterprises and individuals quickly develop and deploy various IoT applications.

Data Security and Privacy Protection Capability: The IoT platform can provide various security measures to ensure the data security and privacy protection of the IoT system. It can use various security protocols and encryption technologies, such as SSL/TLS, AES, and RSA, to protect the communication security between devices and the platform. Moreover, it can provide various identity authentication and authorization mechanisms, such as OAuth and OpenID Connect, to protect the identity security of devices and users. In addition, it can provide various data privacy protection technologies, such as data desensitization and data encryption, to protect users' privacy data.

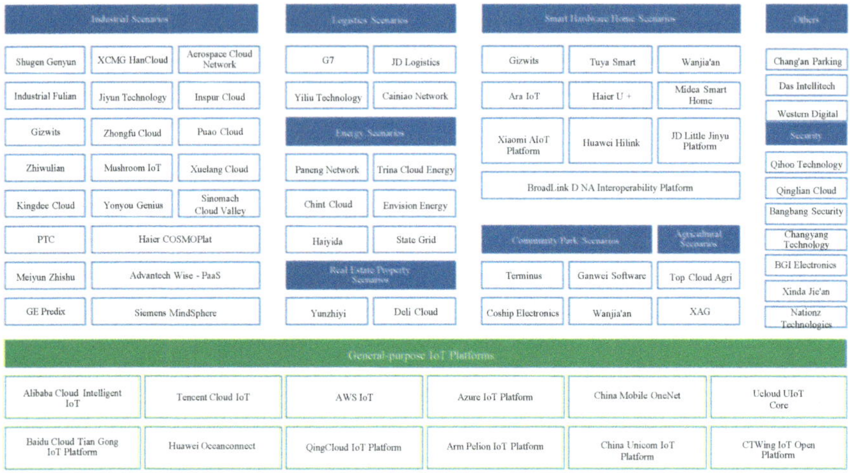

Fig. 4.1 IoT platform player map[1]

Data Visualization and Analysis Capability: The IoT platform can provide various data visualization and analysis functions to help enterprises and individuals better understand and utilize the collected data. It can use various visualization tools and technologies, such as charts, maps, and dashboards, to present data to users in an intuitive and easy-to-understand way. Moreover, it can provide various data analysis and mining technologies, such as data mining and predictive analysis, to help users discover patterns and trends in the data, thereby enabling more informed decision-making.

4.1.2 Platform Industry Products

The IoT platform is a complex system involving multiple fields, so many companies and organizations are involved in this field. These participants include software developers, cloud computing service providers, and data analysis companies. Their cooperation and competition drive the continuous innovation and application of IoT technology, providing important support for the construction of an intelligent society. The main participants are shown in Fig. 4.1:

We introduce and analyze the general-purpose IoT platform products of several mainstream platforms as follows:

Alibaba: Alibaba Cloud's Internet of Things (IoT) platform integrates a wealth of cloud computing and IoT technologies, providing global IoT services. It offers a complete IoT solution, including device access, data collection, cloud processing, and

[1] Source: IoT Media.

application development. It supports various communication protocols and device access methods and can flexibly adapt to various scenarios. It provides a visual console and API for easy management of devices and data and supports large-scale device management and upgrades. Owing to its strong technical strength and global deployment capabilities, the Alibaba Cloud can provide users with highly reliable and highly available IoT services. Alibaba Cloud's IoT platform provides a complete solution, and users can quickly build an IoT system with a simple configuration. Alibaba Cloud's IoT platform provides a variety of expansion capabilities, such as edge computing and AI, to meet the needs of different scenarios.

Tencent: Tencent's IoT development platform, IoT Explorer, integrates Tencent Cloud's powerful computing and IoT technologies and supports global deployment. It provides various device access methods and communication protocols and can flexibly adapt to various scenarios. It supports large-scale device management and upgrades and provides a visual console and API for easy management of devices and data. It provides a complete IoT solution, including device access, data collection, cloud processing, and application development.

Huawei: Huawei Cloud's IoT platform, OceanConnect, integrates Huawei Cloud's computing, storage, and networking technologies, providing a comprehensive IoT solution. It supports various device access methods and communication protocols and can flexibly adapt to various scenarios. It provides secure and reliable IoT services, including device authentication, data encryption, data backup, etc. In addition, it provides a visual console and API for easy management of devices and data.

Baidu: Baidu's Intelligent Cloud IoT platform (Ti Gong IoT platform) integrates Baidu Cloud's artificial intelligence, big data, and other technologies, providing users with intelligent IoT solutions. It provides various device access methods and communication protocols and can flexibly adapt to various scenarios. It also provides a visual console and API for easy management of devices and data. Moreover, it provides a wealth of templates and components, and users can quickly build applications with simple drag and drop.

4.1.3 Internet of Things Platform *(*ProBase*)*

Unlike the general-purpose IoT platforms of the above mainstream manufacturers, the IoT platform (ProBase) (as shown in Fig. 4.2) is designed for intelligent buildings, smart parks, smart agriculture, smart campuses, intelligent medical and other fields; provides comprehensive intelligent device unified support management; and is committed to providing comprehensive solutions of "full connection, deep integration, and panintelligence" for users in various industries.

Fig. 4.2 Internet of things platform (ProBase)

ProBase's IoT and data management platform. The platform adopts a completely modular design concept, advanced IoT technology architecture, overcomes the traditional siloed system construction and management model, converges various intelligent subsystems and sensor devices downward, and includes intelligent gateways, sensor monitoring devices, video surveillance devices, intelligent control and other devices. It provides an efficient, stable, and secure application platform between various services, application systems, and intelligent devices, facilitating device data collection, command issuance, and other business scenario construction. The intelligent devices managed by the platform include but are not limited to intelligent gateways, sensor monitoring devices, video surveillance devices, intelligent control devices, etc. The platform supports other service platforms, application platforms, and data platforms upward through a unified northbound standard OpenAPI interface, fulfilling the intelligent management requirements of various application scenarios, including "full data integration, a fully visible status, complete business management, and full event control".

The platform can adapt to various network environments and protocol types and can provide rapid access to various sensors and intelligent hardware. It provides rich application interfaces to support various industry applications, effectively reduces the cost of IoT application development and deployment, and meets the platform-level service requirements of IoT field device connections, protocol adaptation, data storage, data security, and big data analysis. The platform builds the IoT capabilities of the "cloud-edge-end" overall architecture, extends terminal adaptation access capabilities downward, integrates segmented industry applications upward, and can provide device access, device management and other basic device management capabilities, as well as device positioning, spatial mapping, data visualization, remote upgrade OTA, message queue MQ and other capabilities. The platform has the following features:

Unified Device Access

Provides MQTT, CoAP, and panprotocol standard access SDK and supports direct connection, gateway, cloud-to-cloud docking, and other access methods.

Unified Physical Model

Provides flexible physical model definitions and standard templates for all industries and categories, builds standardized data models, solves data fragmentation, and simplifies device and application development.

Unified Operation and Maintenance Monitoring

Provides device lifecycle management, business monitoring, and link monitoring capabilities, reflects the device access situation in real time, and realizes automated operation and maintenance of the platform business and unified alarm management.

Unified Data Interface

Provides a unified northbound standard OpenAPI interface, assisting various application software and display systems in quickly and seamlessly interfacing with massive data, achieving resource integration across systems and products, and accelerating project integration.

Location Space Services

Provides 2.5D spatial display and device location services, efficiently and accurately locating various smart hardware, solving difficulties in device operation, maintenance addressing and point finding, and performing device monitoring and operation and maintenance visualization.

Multi-scenario Support

Flexible rule configurations, supporting scenario linkages, data forwarding, etc., are all managed by a unified rule engine.

4.2 Data Platform

4.2.1 Data Platform Overview

A data platform is a system for managing, storing, processing, and analyzing data. It can integrate various data sources, including structured data (such as data in relational databases) and unstructured data (such as text, images, audio, etc.). The data platform can also provide data cleaning, transformation, standardization, and visualization functions to help users better understand and utilize data. Moreover, the data platform can analyze and model data through data mining, machine learning, and artificial intelligence technologies to provide deeper insights and predictions. The characteristics of the data platform include the following:

Data Integration: The data platform can integrate various data sources, including structured and unstructured data, to help users better manage and utilize data.

Data Processing: The data platform can perform data cleaning, transformation, standardization, etc., to help users better understand and utilize data.

Data Analysis: The data platform can analyze and model data through data mining, machine learning, and artificial intelligence technologies to provide deeper insights and predictions.

Visualization: The data platform can provide data visualization functions to help users better understand and present data.

Security: The data platform can provide data security guarantees, including data encryption and permission management, to ensure data security and privacy.

Real-time: The data platform can provide real-time data processing and analysis functions to help users quickly obtain data and make decisions.

The data platform construction path can be approximately divided into five steps:

Clarify the construction idea. Based on the current business situation, identify the core business domain all business modules, such as those related to research and development, manufacturing logistics, marketing, finance, and human resources, design standard data management, data quality management, quality evaluation and other related management processes, and assign responsibilities to individuals.

Data asset management. Support asset ownership, asset classification, asset overview, asset search, statistical analysis, lineage analysis, etc., and provide imports/exports of multiple format files.

Data standard management. Support information architecture management, template management, logical modeling, dimension modeling, model materialization, standard verification, and synchronization functions.

Metadata management. Metadata management is the basis for eliminating data islands and achieving unified data governance. It needs to include metadata collection, parsing, management of metamodels, support for metadata display and search, etc.

Data quality management. The data platform should have quality rules, rule verification, quality monitoring, rule association, and evaluation release functions.

4.2.2 Data Platform Industry

The main participants in the data platform industry are suppliers who help downstream enterprises build data platforms and provide services. Data platform suppliers are composed of five main types of manufacturers: leading internet companies, digital solution providers, big data companies, independent platform developers, and artificial intelligence manufacturers (as shown in Table 4.1). New players continue to enter the market. In addition, manufacturers of all types have different competitive advantages and are in the expansion stage of capturing market share and building entry barriers in their respective fields.

4.2.3 **Data Platform** *(***ProDate***)*

ProDate covers data storage, data fusion, data processing, and data sharing; integrates the two technical evolution directions of data lakes and data warehouses; and provides enterprise users with a cloud-native warehouse-lake integration solution. As a new data application platform based on big data technology, it provides a series of key technologies, such as cloud-native features, support for storage–computing separation architecture, inheritance of RDBMS database functions and ACID features, support for industrial-grade SQL standards, and a hybrid MPP + DAG computing engine. It helps enterprises effectively cope with large-scale, diversified, high-timeliness, and intelligent data application needs, providing a new data infrastructure for enterprises from BI to AI.

Data Integration

Efficient data lake entry from multiple heterogeneous sources is achieved, batch/stream/real-time data access can be supported in various ways, and the data foundation required by enterprises and user applications can be quickly realized.

Data Assets

Data asset management includes metadata inventory, metadata maintenance, a data asset classification framework, and data asset catalog construction. By taking stock of data assets, an enterprise-level data asset catalog is formed, laying a solid foundation

Table 4.1 Data platform product analysis table

Serial number	Product type	Features and advantages	Representative products
1	Public cloud manufacturers	They implement platform strategy internally, provide services externally after verification, and have the first-mover advantage. They possess full-stack technical capabilities at the bottom layer and are able to provide native, standardized packaged products and solutions. In addition, they output resources, core technologies, methodologies, and tool systems for industry development	Alibaba Cloud, Tencent Cloud, AWS
2	Data and intelligence companies	Possessing data resources, they can help enterprise customers address some data gaps and quickly implement applications. Their artificial intelligence technology and algorithm capabilities are strong, with advantages in data collection, cleaning, and computational inference. Additionally, they have a customer resource base and have accumulated a large amount of knowledge and insights into various vertical industries	Minglu Technology, Xinghuan Technology, Sensors Data
3	Digital solution providers	With experience in ToB services and a customer resource base, they have advantages in customer acquisition and expansion; they have accumulated a large amount of knowledge and insights into various vertical industries; they can quickly sort out enterprise business and processes to accurately identify customer needs; and they can output comprehensive digital transformation services based on the platform architecture	Yuanian Technology, Yonyou, Kingdee
4	Independent platform developers	Data platform-related technical tools constitute their core service. They often act as ecosystem partners for leading companies, and they specialize in business, deeply cultivating scenarios and vertical industries	Shulan Technology, Yunxi Technology, Kangaroo Cloud, Gloryview Technology

for enterprise data innovation applications. The metadata inventory collects metadata by connecting to enterprise business systems, data lakes, or data warehouses, automatically obtaining the original enterprise data dictionary and the relationships among the data and thereby forming an enterprise metadata map. Effective resource labeling establishes effective resource judgment rules based on the collected metadata; automatically identifies all data resources; filters out invalid resources such as empty tables, backup tables, and temporary tables; and supplements information to help understand the business meaning of data, thus forming more complete and effective resource metadata and paving the way for data asset classification. Data asset cataloging builds a data asset classification framework based on business lines, organizational structure, data characteristics, and other dimensions; improves data asset catalog information based on metadata information; supplements business and ownership information related to assets; and forms a data asset portal for data consumers. Data asset services establish a mapping relationship between the data asset catalog and physical resources, develop data service types for different consumers, and meet the diverse data needs of business systems, data analysts, and front-line business personnel.

Data Indicator System Construction

The indicator management solution is designed to manage indicators uniformly, can be used by both business personnel and technical personnel, and can provide users with standard indicator management and indicator analysis services. Various operational indicator data ensure that key data are well understood and that business trends are under control.

Data Exchange

The integrated information resources at the bottom layer are utilized to achieve interconnection and information sharing of business data in various departments at the application level, providing incremental identification, data cleaning, data transmission, data compression, consistency verification, and other means to ensure data access, standardize data circulation, and improve the level of data sharing.

Data Development

A visual data development platform, rich data development components, fully automatic job scheduling and monitoring capabilities, and support for multiperson online collaborative development are provided, greatly reducing the threshold for users to use big data and helping users quickly perform data processing. Through various analysis methods (including statistical chart display, report query, ad hoc analysis, report analysis, leadership cockpit, map analysis, etc.), various analysis scenarios (including comprehensive business queries, KPI indicator displays, quarterly reports, data self-service exploration, command center big screens, etc.) can be addressed.

4.3 AI Platform

4.3.1 AI Platform Overview

The integration of artificial intelligence technology with vertical industry applications poses different demands in multiple aspects, such as software platforms, intelligent technology, and the underlying hardware. A full-stack platform architecture oriented toward key industries is therefore crucial. By continuously improving artificial intelligence development frameworks, data processing model construction, deployment monitoring, and other research and development tool chains, we can accelerate the establishment of a full-stack intelligent technology service system. This system provides full-stack AI technology support that synergizes software and hardware, from basic computing power, basic system frameworks and platforms to intelligent applications. It explores the capabilities of basic and vertical industry artificial intelligence technology platforms, provides overall solution selection and design services, exploits the intensive production and management of AI capabilities, and continuously lowers the application threshold through automation capabilities.

As a modular and standardized platform tool, the AI platform overcomes the bottlenecks caused by traditional "chimney-style" business architectures, such as redundant resource construction and data flow barriers. Through efficient production, flexible organization, and convenient acquisition of full-chain service capabilities, it provides artificial intelligence development frameworks, data processing model construction, deployment monitoring, and other research and development tool chains. It provides a full-stack intelligent technology service system offering full-stack AI technology support that synergizes software and hardware, from basic computing power, basic system frameworks and platforms to intelligent applications. It provides a one-stop solution covering the entire lifecycle of research and development, deployment, and operation, helping to improve efficiency, enhance decision-making intelligence, and promote the rapid integration of intelligent technology with vertical industry scenarios. The AI platform includes three core layers: AI technology services, AI research and development platform, and AI management and operation:

AI Technology Service Layer: The AI technology service layer includes two major modules: general AI technical capabilities and industry-specific AI technical capabilities. It provides AI general capabilities, including basic technology functions such as computer vision, intelligent voice, natural language processing, knowledge graphs, and scenario-based AI technology services oriented toward industry applications, thus enabling enterprises to quickly produce diversified application scenario innovations.

AI Research and Development Platform Layer: As the core capability output of the AI platform, the AI research and development platform layer has large-scale deep learning computing support capabilities. It includes two major capability modules, a data service system and AI model development, which form the foundation of enterprise innovation capabilities. By establishing a data service system, it provides data cleaning, data labeling, data enhancement, and feature engineering capabilities for AI model development, ensuring a high-quality, efficient data supply for AI research and development. The AI model development module provides enterprises with machine learning and deep learning capabilities, including AI model construction, training, tuning, and evaluation, and lowers the model development threshold through automatic machine learning technology, accelerating intelligent enterprise application innovation.

AI Management and Operation Layer: The AI management and operation layer includes two major modules, basic resource management and AI core asset management, which support the entire process of AI capability production, service, and operation. Basic resource management consists of account management, permission management, storage management, computing resource management, monitoring and alerting capabilities to fulfill enterprise account permission review, resource management scheduling, and other internal management needs; improves AI resource utilization; reduces operations and maintenance management costs; and optimizes AI capability usage experience. AI core asset management centers on AI models, samples, algorithms, and other core assets; provides management, publishing, sharing, operation, and transaction capabilities, achieves the cross-organization, cross-platform management and application of AI core assets; helps enterprises achieve core asset accumulation and sharing; and promotes a collaborative innovation ecosystem within the enterprise.

4.3.2 AI Platform Industry

In recent years, the maturity of artificial intelligence technology in China has continued to improve, and the types of services have continued to be enriched. As an intelligent base for the rapid research and development, sharing, reuse, and efficient deployment management of AI technology in various industries, the AI platform is an essential infrastructure for ensuring the universal benefits of intelligent capabilities. With the rapid increase in the number of AI manufacturers in China, China has gradually eliminated its dependence on overseas manufacturers in the field of technology platforms. Tech companies such as Baidu, Alibaba, Tencent, Huawei, JD.com, and Kingsoft have successively released their AI platforms or built AI technology services via cloud services. The Chinese AI platform market is highly concentrated and competitive and is dominated by large manufacturers. There are three main reasons for this situation:

AI technology development requires data accumulation: Large manufacturers have an unmatched advantage over start-ups in data accumulation. Massive, low-cost data accumulation helps large manufacturers improve algorithm accuracy through training and optimization, while small manufacturers incur high costs to purchase a large amount of data to achieve training purposes.

A full-stack AI ecosystem enhances user stickiness: Large internet or tech manufacturers have a complete AI layout, providing AI capabilities to enterprises or developers while offering more additional services, such as cloud servers, cloud computing, incubators, and entrepreneurship classes.

Pricing advantage: Large manufacturers can attract developers to join by offering low or even free pricing advantages, obtaining data while profiting from additional services.

The analysis of the AI platform products of various manufacturers is shown in Table 4.2:

In recent years, CPUs, GPUs, DPUs, and other computing chips suitable for large-scale data cluster computing have been continuously developed and improved. Memory, virtualization, networking, and other IT infrastructure technologies are evolving toward large-scale concurrency and ultrahigh-speed transmission and are being optimized in conjunction with mainstream AI frameworks. The deep integration of future AI development with the above IT infrastructure will become a major trend, which may start with the cooperation of AI technology and hardware manufacturers and gradually deepen. The future development trend of China's AI platforms will include ease of use, strengthening of vertical scene specialization, comprehensiveness, industrial collaboration, and the participation of leading manufacturers.

Ease of use is receiving increasing attention: With the improvement in model libraries and the increase in acquisition channels, enterprise users will give more attention to the convenience and practicality of AI at the application level and place higher importance on product design and packaging.

Strengthening of vertical scene specialization: At this stage, AI algorithms and applications have strong universality. Owing to the lack of industry experience and professionals, most industry applications involve only fine-tuning of general models. This situation is expected to improve in the future.

Comprehensive AI service platforms: The AI services that enterprises need will become more closely integrated with big data services, cloud computing services, etc. Comprehensive AI service platforms will become the mainstream in the industry.

Industry ecology of production and use collaboration: The progress of AI will increasingly rely on autonomous model optimization and adaptive adjustment on the user side. Open source communities and open AI exchange platforms will play important roles in this process.

Leading manufacturers will aggregate AI: Considering factors such as talent and experience accumulation, IT infrastructure resources, R&D investment, and community interaction, combined with global AI industry experience, leading AI platforms will provide greater assistance to enterprises in achieving scale effects in the AI field.

Table 4.2 AI platform product analysis table

Serial number	Manufacturer name	AI platform overview
1	Amazon web services	The Amazon Web Services AI development platform boasts high compatibility and a rich array of application modules. The AWS Marketplace supports hundreds of algorithms and models. Its frameworks and toolkits are optimized for AWS services, helping data scientists to increase their productivity nearly tenfold. AWS application modules include automated data extraction and analysis, business metrics analysis, voice AI, machine vision, code and development operations, and other functionalities
2	Huawei ModelArts	Huawei ModelArts is a one-stop AI development platform for developers, providing massive data preprocessing and semiautomatic annotation for machine learning and deep learning, large-scale distributed training, automated model generation, and on-demand deployment capabilities for edge-cloud models. It helps users quickly create and deploy models, managing the full AI workflow
3	Baidu PaddlePaddle	PaddlePaddle is based on Baidu's years of deep learning technology research and business applications. It integrates a deep learning core framework, basic model library, end-to-end development suite, tool components, and service platform. Officially open-sourced in 2016, it is a comprehensive, technologically advanced, and fully featured industrial-grade deep learning platform. PaddlePaddle, which originates from industrial practice, is committed to deep integration with the industry. It is now widely used in industries such as manufacturing, agriculture, and services, serving more than 2.65 million developers and empowering an increasing number of industries to harness AI via its partners
4	Tencent cloud	Tencent provides AI development solutions for various business scenarios, continuously reducing the cost of building and applying AI capabilities. The Tencent Cloud TI platform can integrate the entire process chain of industry + AI implementation. It helps users quickly create and deploy AI applications, and its platform function modules can be precisely deployed on demand. Tencent Cloud TI platform includes the AI application service platform TI-Matrix, machine learning platform TI-ONE, and data annotation platform T1-DataTruth

4.3.3 AI Platform *(*Pro AI*)*

Pro AI is an integrated workspace that includes AI algorithms, computing power, and development tools and provides support functions for assisting development, invocation, research, and management. Through API calls, embedded development, and other forms, it provides efficient AI implementation capabilities for information systems, devices, analysis platforms, etc., and achieves low-cost development of AI products or flexible use of devices. From a construction framework perspective, there are four main aspects of construction:

Data Services—Building a data service module that integrates data access, management, and annotation.

Model Training—Building a model training platform that includes preset multi-type pretraining models, code compilation in conjunction with basic frameworks, and model training and correction. The training platform integrates various AI technologies, such as computer vision, NLP, video processing, and machine learning.

Technical Services—Building a technical service platform that includes model output invocation, management, and encapsulation.

Support Conditions—Building a support platform that includes computational resource allocation, device resource management, personnel collaboration, and model invocation management. It flexibly adjusts the resources needed for model training and computation and personnel collaboration. When users need a large number of resources, it expands capacity, improves computational power and stability, and reduces model training and computation time. When the user demand is small, it reduces the underlying resource configuration, reducing the system redundancy caused by resource occupation.

ProAI provides integrated artificial intelligence services for the construction and operational needs of smart cities; provides managers, operators, and decision-makers with integrated management of data resources; centralized management of high-quality data resources; and integrated services for artificial intelligence algorithm training, deployment, and computational output.

Coordinated Management of Data Resources

The management of various types of production data from the base, various information systems, and industry-public data, including images, videos, audio, and text, are coordinated. At the same time, data are divided according to their attributes, establishing deep learning data resources, including but not limited to object detection, object tracking, instance segmentation, public sentiment text, audio, and structured data, thereby facilitating data extraction and input to training models or analysis.

Establishing a High-Quality Data Resource Library

To achieve artificial intelligence, it is necessary to teach computers how to understand and judge things as humans would, and data annotation is the first step. Data annotation labels the images that need to be recognized and distinguished by the computer in advance, allowing the computer to continuously identify these image features, ultimately achieving autonomous recognition. Data annotation provides a large amount of labeled data for AI companies for machine training and learning, ensuring the effectiveness of the algorithm model. The AI platform will establish a data annotation tool with multiperson annotation, intelligent annotation, and open invocation, thereby establishing high-quality data resources. Within these data resources, data analysis can be performed directly by invoking the annotated data without repeating the annotation, thereby reducing the difficulty of data acquisition, annotation, and invocation.

Model Training, Deployment, and Computational Output

The AI platform will integrate AI algorithms, computational power, and development tools; open machine learning, deep learning, and training model development frameworks; and provide the computational power support needed for development. The development platform architecture can be divided from bottom to top into the infrastructure, framework tools, training platforms, and technical services. It provides two methods in the development architecture: one is the basic model architecture commonly used in computer vision (CV), natural language processing (NLP), machine learning, statistical analysis, etc., and the other provides basic frameworks (such as PyTorch, TensorFlow, Keras, Paddle, etc.), combined with notebook-like editing tools, to quickly and flexibly build models. On the one hand, it meets the convenient invocation requirements of basic models. On the other hand, it can build more accurate, small custom models. In terms of technical services, it provides elastic resource configuration, algorithm upgrades, and collaborative operation services, providing an integrated model service platform with stable computational power and short training time.

Chapter 5
Smart City Industry Applications

5.1 Smart Governance: Promoting the Efficiency and Intelligence of Government Work

Smart governance involves the use of technologies such as cloud computing, big data, the Internet of Things, artificial intelligence, and blockchain by the government to change traditional government work modes; to build an integrated online government service system; and to achieve cross-level, cross-regional, cross-system, cross-department, and cross-business collaborative management and services to improve the intelligence, efficiency, and transparency of office work, supervision, services, and decision-making.

The core goals of smart governance are to improve the operational efficiency of the government, improve the quality of public services, enhance the transparency of the government and public participation, and thereby increase the governance capacity of the government. It includes smart offices, smart supervision, smart services, and smart decision-making, covering government departments and administrative units at all levels from central to local. The vertical business scope is divided into smart city services, smart gardens, smart taxation, and other fields, providing efficient and convenient services for enterprises and individuals through multiple channels, all business operations, and the entire process of governance.

The informatization of government departments in China began in the late 1980s and early 1990s and has undergone the stages of "office automation", "e-government", and "smart governance" (as shown in Fig. 5.1). At present, China is in the construction stage of "smart governance".

The first phase includes the construction of a high-speed information communication network infrastructure and online office construction, which address the issue of "office automation". It transitions government document approval from offline to online, shifts government office work from paper-based to paperless, and changes information sharing from paper file borrowing to network information

F. Liu, *Smart and Digital Cities*, https://doi.org/10.1007/978-981-95-2243-9_5

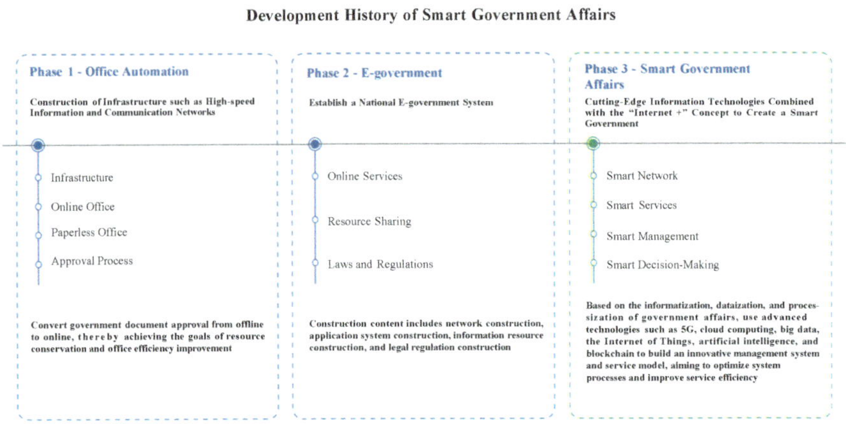

Fig. 5.1 Development history of smart governance

sharing, thereby achieving the goals of resource conservation and office efficiency enhancement.

The construction goal of the second phase is to establish a national-level e-government system, connecting the existing government networks of all levels and various commissions and bureaus, establishing an online service system, and forming unified technical standards and management norms. The construction tasks of this phase include network construction, application system construction, information resource construction, and legal and regulatory construction. The construction of e-government has promoted the transformation of government functions, improved the level of government management, strengthened the capacity of government public services, and is an important measure for the government to transition from management-oriented to service-oriented.

The third phase began in 2015, when the government work report at the third session of the 12th National People's Congress first proposed the concept of "Internet+", which can be regarded as the beginning of China's "smart government" construction. In July 2016, the Office of the Central Committee of the Communist Party of China (abbreviated as the Central Office) and the State Council Office issued the "National Informatization Development Strategy Outline", which proposed the strategic deployment of "establishing a national governance big data center". In March 2021, the "Fourteenth Five-Year" plan proposed to promote the coconstruction and sharing of government informatization, improve the national e-government network, strengthen the rapid iteration of government informatization construction, improve the efficiency of digital government services, and comprehensively promote the digital and intelligent goals of government operation methods, business processes, and service models. These top-level strategic designs are driving the steady progress of China's smart government construction.

Smart government is an upgraded form of e-government and is a necessary stage in the government's shift toward being service oriented and intelligent. It is an innovative management system and service model built on the basis of government informatization, datafication, and processization, using advanced technologies such as 5G, cloud computing, big data, Internet of Things, artificial intelligence, blockchain, etc., aiming to optimize process systems and improve service efficiency. Smart government is an important component of smart city construction, including smart networks, smart decision-making, smart management, smart services, and other aspects.

Since 2012, an increasing number of cities have started planning and preparing for smart government and smart governance. Beijing, Shanghai, Guangzhou, Shenzhen, and a batch of first-tier large and medium–sized cities have entered the implementation phase of smart governance, and smart governance has become a key component in the planning and construction of smart cities.

The market size of the smart government industry is also constantly expanding. According to the data, the market size of China's smart government industry has risen from 184.5 billion yuan in 2014 to 426.2 billion yuan in 2022, a year-over-year increase of 7.43% (as shown in Fig. 5.2). China's smart government market has always maintained a high growth trend, and it is expected that with the acceleration of smart city promotion in the future, the market size of the smart government will continue to grow.

In summary, smart government has become a long-term strategy and development goal of government departments and will continue to maintain a high growth trend

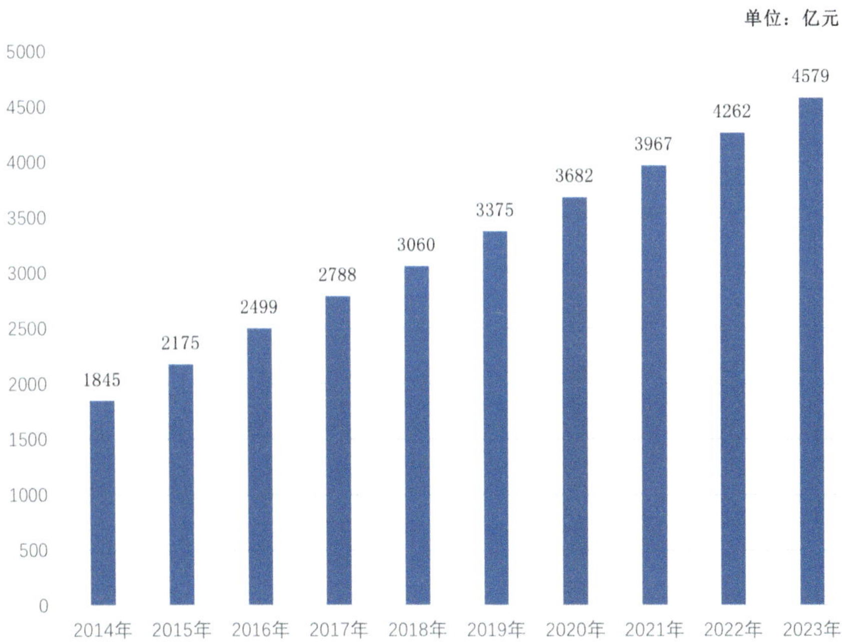

Fig. 5.2 Growth trend of smart government market size

in the future, gradually expanding from first- and second-tier cities to third- and fourth-tier cities. This market has enormous potential and a vast development space.

Gloryview Technology has been working in the field of smart government for many years. After years of accumulation, its government informatization products cover all areas of daily office work of party and government organs. The main products include a party and government comprehensive office platform, a supervision and inspection system, a public security micro alarm platform, an office affairs logistics management system, a grain digital management system, etc. In addition to products, there are solutions for applications in segmented industries. Below, we share our special applications and practice in "Xinchuang", landscaping, and taxation.

5.1.1 Special One: "Xinchuang" Practice, Breaking Through Technical Blockades, and Building Core Competitiveness

Xinchuang, or the information technology application innovation industry, is essentially based on Xinchuang software and hardware for informatization construction. It strives to break from foreign dependence in the field of computer information technology and gradually achieve domestic substitution and an independent supply. In general, the core content of Xinchuang covers four major sectors: basic hardware, basic software, application software, and information security. Among them, basic hardware includes chips, servers, PCs, storage, etc.; basic software includes databases, operating systems, middleware, etc.; application software includes office software, ERP and other software; and information security includes hardware security, software security, security services and other products.

Xinchuang Development History

In recent years, the United States has frequently imposed sanctions on China's technology companies, essentially suppressing potential competitors in the technology race, such as telecommunications, artificial intelligence, computers, and chips. Given the current situation, the United States will continue to use its advantages in the foundational areas of its technology industry chain to crowd out and suppress the upgrade of China's technology industry. Since 2018, sanctions against companies such as Huawei and ZTE have become commonplace. To counter this extremely passive situation, China has incorporated the information and communication technology (ICT) industry into its national strategy, essentially building an underlying software and hardware architecture system through ICT industry applications, with the aim of breaking from dependence on foreign information and communication technology industries. The development of China's ICT industry began in the 1990s and has generally gone through four stages of development.

Preliminary research and development stage: In 1993, China Computer Software and Technology Services Corporation launched the first generation of the domestic Linux operating system "COSIX1.0" based on UNIX, marking the birth of a domestic operating system.

Accelerated development stage: In 2008, Alibaba upgraded its internal IT information technology, fully carrying out independent and controllable research and development. In 2010, the civilian "Zhongbiao Linux" and military "Galaxy Kylin" merged. At the end of 2013, the Banking Regulatory Commission explicitly proposed ICT security requirements.

Pilot practice stage: In 2017, the second batch of major projects of the Nuclear High Base was launched. In 2019, the domestic CPU entered a period of harvest, with the appearance of the Zhaoxin KX6000 greatly improving performance. Feiteng released a new generation of desktop processors FT-2000/4. In 2019, several key special pilot projects were completed.

Application landing stage: In 2020, ICT projects in various provinces gradually started. In September 2020, four departments, including the National Development and Reform Commission, the Ministry of Science and Technology, the Ministry of Industry and Information Technology, and the Ministry of Finance of the People's Republic of China, jointly issued the "Guiding Opinions on Expanding Investment in Strategic Emerging Industries to Cultivate and Strengthen New Growth Points", proposing to accelerate the breakthrough of core technologies such as key chips and key software, vigorously promoting the construction of key projects and projects, and actively expanding reasonable and effective investments.

As a strategic emerging industry, the ICT industry has seen an increasing number of policies issued by central and local governments, and the importance attached to ICT by local governments is also increasing annually. ICT has become key to the digital transformation of the national economy and the promotion of industrial chain upgrades. The domestic ICT industry has entered a period of rapid development, and this momentum will continue in the coming years.

Issues Encountered in the Development of ICT

Imbalance of Supply and Demand in the ICT Ecosystem

In recent years, facing new situations, ICT has received more attention from China and industry and has been entrusted with many expectations. However, a complete industry ecosystem requires the integration and development of both supply and demand, which cannot be achieved overnight. The ICT ecosystem involves the organic combination of networks; security, hardware, software, and third-party suites; and talent. According to relevant statistics, the balance of supply and demand in the ICT ecosystem has a gap of up to 11%, indicating a serious imbalance in the ICT ecosystem. The supply cannot meet the demand, and the problems at the security, infrastructure, and underlying hardware levels are particularly serious.

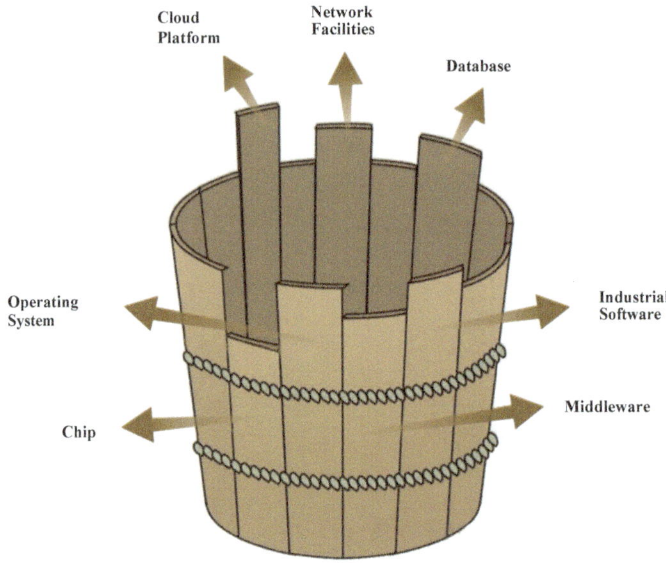

Fig. 5.3 The barrel effect

The Barrel Effect[1] Affects the Quality Improvement of the ICT Industry

With the continuous advancement of ICT work, China's independently innovative ICT industry has undergone rapid development. Considering the gap in technological maturity compared with the global level, China's ICT market is leading the world in terms of network infrastructure, databases, and cloud service platforms and is worth promoting and implementing. On the other hand, there are obvious technical shortcomings in areas such as chips, operating systems, middleware, industrial software, and IAAS (as shown in Fig. 5.3), and there is still a large gap in relation to the advanced technology level of developed countries worldwide. This situation introduces uncertainty to the steady development of China's ICT industry and requires increased efforts in technological breakthroughs.

The road to ICT is long, and there are many opportunities. However, it still requires all ICT market innovators to resonate at the same frequency, work together, create a closed-loop ecosystem where all parties cooperate and succeed, drive the development of the ICT industry with technological innovation, and avoid the situation where a short board in the ICT industry leads to resource waste and starting over.

[1] The barrel effect, also known as the "short plank theory," refers to the idea that the capacity of a barrel is determined by its shortest plank. This concept is often used as a metaphor to emphasize the significance of individual or organizational weaknesses, suggesting that one's shortcomings can limit one's overall performance. It also serves as a reminder to strengthen any weak areas without being entirely constrained by them.

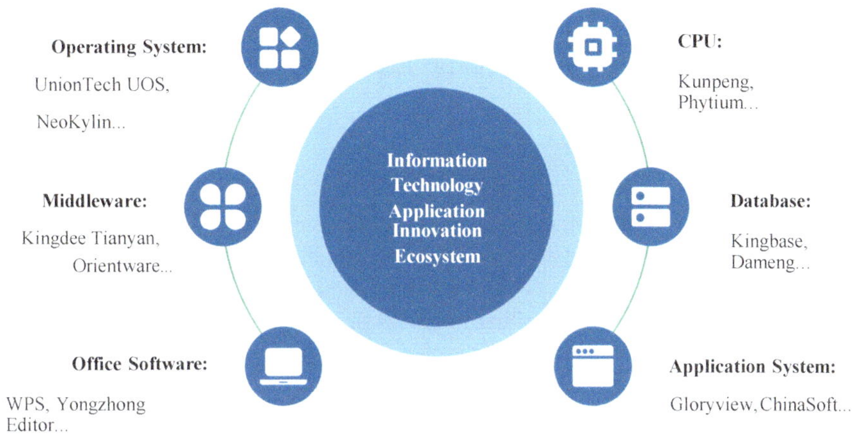

Fig. 5.4 XinChuang ecosystem

Gradual Improvement in ICT Applications and Ecosystems

The ecosystem of the ICT industry includes underlying hardware, basic software, platform software, application software, information security, etc. With respect to policy support, the technological innovation capabilities of various fields have greatly improved, and substantial progress has been made in terms of structural optimization and upgrading. The construction of the ICT ecosystem has gradually improved, resulting in full industry chain integration of the upstream, middle, and downstream regions of the ICT industry, as shown in Fig. 5.4. As shown in the figure, the mainstream products and representative manufacturers in the current XinChuang ecosystem include domestic CPUs such as Kunpeng and Feiteng; databases such as Dameng and RenDa JinCang; application systems such as Gloryview and ZhongRuan; operating systems such as Tongxin UOS and Zhongbiao Kylin; middleware such as Kingdee Tianyan and Dongfangtong; and office software such as WPS and Yongzhong Editor. In terms of products, solutions, and markets, they are all engaging in comprehensive cooperation, continuously adapting to each other, and gradually forming a complete, stable, and mature XinChuang ecosystem.

Sharing XinChuang Practice Cases

Since 2019, as a rapidly progressing field, a series of XinChuang projects have been implemented. Information technology companies continue to provide safe and reliable digital solutions for government agencies. Leveraging years of experience in the government sector, Gloryview Technology has developed XinChuang products with independent intellectual property rights, such as electronic document collaboration, paperless meetings, and work supervision, which are applied in different project scenarios. Below, we share three of our cases.

XinChuang Practice Case 1: Electronic Document System XinChuang Project in a City

Project Background: With the accelerated advancement of "e-governance", online and electronic methods have become important means for government and enterprise units to manage official documents. The electronic document system of a case city, as the central hub for handling internal office affairs of party and government organs, involves important processes such as document approval, document circulation, requests for instructions, and information retrieval. It is the internal system with the widest coverage, highest usage frequency, and largest peak usage among the city's party and government organs. In accordance with the unified work requirements of the higher-level departments, the electronic document XinChuang work is prioritized in party and government organs. Accordingly, the city proposed a comprehensive XinChuang project for the electronic document system.

Construction Content: The electronic document system XinChuang project includes software such as an integrated office portal, document scanning software, a document approval system, a supervision and supervision system, a document registration system, a meeting room reservation system, a dispatch system, an instant messaging platform, and distributed file synchronization. It also includes the following construction needs:

Procurement of domestic terminals, including domestic computers, servers, operating systems, printers, scanners, etc.

Adaptation of infrastructure, including domestic servers, domestic operating systems, middleware, databases, etc.

Supporting software, which refers mainly to flow, version, signature, and adaptation with the electronic document system and infrastructure.

Compliance with security requirements, including the use of national secret algorithms, etc., to optimize and transform the system.

Project Results: Overall integration is highly important to compensate for quality shortcomings and control. It contributes to the collection of terminals, software adaptation, product integration, and operation and maintenance services throughout the process. It has accomplished the full-stack XinChuang replacement of servers, chips, operating systems, middleware, and databases at the city, county, town, and village levels, achieving the goal of full-stack, single-track operation and successfully completing the full implementation of XinChuang adaptation of party and government electronic documents issued by higher-level departments.

To date, domestic terminal equipment has been fully deployed for use in the city's party and government organs, and the operation of the electronic document system in the XinChuang environment has been stable, which has been highly praised by the city's leaders on multiple occasions.

XinChuang Practice Case 2: Integrated Meeting System Information Project in a City

Project Background: Due to the needs of meeting work, it is necessary to improve the efficiency and quality of meeting management by the Municipal Party Committee Office and further enhance the support level and management efficiency. The Municipal Party Committee Office of the case city plans to migrate the city's meeting system to the XinChuang environment through cloud migration, upgrades, and transformation, thereby ensuring data security, operation security, and business security and achieving business collaboration between the meeting and the meeting units under the XinChuang platform.

Construction Tasks: The upgrade and transformation include improving the overall function of the current meeting system, enhancing system operations and performance, and improving system functions while adding channels based on mobile office meetings. XinChuang adaptation: the project is designed according to the standard of XinChuang replacement to ensure the full adaptation of servers and terminals. To improve work efficiency, the XinChuang version of seat arrangement software should be provided, seamless docking with integrated meeting information should be achieved, participants should be automatically collected, and personnel collection, seat arrangement, and seat ticket printing full process service should be realized.

Project Results: The XinChuang Integrated Conference Management Platform is compatible with mainstream domestic chips (Loongson, Kunpeng, and Feiteng) and operating systems (Tongxin UOS and Kylin), supports SM2, SM3 and other national secret algorithms. Combined with intelligent seating arrangement software, it has been widely used in various large, medium and small meetings at all levels of government, political consultative conferences, and people's congresses in the city. At this stage, the platform has officially been launched and used by various government departments in the city and has successfully assisted the city in carrying out the work of the two sessions multiple times, including conference seating arrangements, notification management, and leadership rankings. At the same time, it automatically identifies multiple abnormal situations, such as incorrect leadership rankings, duplicate names, incorrect seating arrangements, and incorrect housing arrangements, greatly reducing the workload of conference staff and receiving high praise from the city's conference management department.

XinChuang Case Study 3: Intelligent Seating Arrangement Software XinChuang Edition

Project Background: In the daily work of provincial and municipal parties and government agencies, there are many important meetings. For meeting organizers, tasks such as seating arrangement, effect diagram production, and seat information distribution are always complicated.

First, the work is mainly manual, making it difficult to improve efficiency. In terms of seating arrangement, effect diagram production, and seat information distribution, manual operation is the main method, which lacks effective system software support.

Second, there are many errors due to human operation. All aspects of conference affairs are controlled by staff, leading to operational errors and a lack of experience.

To address these difficulties, a provincial People's Congress Standing Committee purchased a set of Intelligent Seating Arrangement Software XinChuang Edition.

Construction Tasks: The intelligent seating arrangement software required for this project specifically includes the following five major functional modules:

Venue Management: Provides more than three initial templates for the venue, including settings for common venue templates such as round, oval, and rectangular.

Attendance Management: Maintains the list of people who need to attend and arrange seats for a certain meeting; and supports the creation, modification, import, deletion, manual dragging and adjustment of the order, etc.

Personnel Division: Facilitates the addition of unit lists, the creation of unit personnel data, imports, reference sorting, manual sorting, duplication checking, statistics, the creation of areas, personnel allocation to areas, and other management functions.

Area Seating: Includes eight mainstream seating methods: middle-left–right, middle-right-left, from left to right, from right to left, fixed middle-left–right, fixed middle-right-left, Z-shaped arrangements, and S-shaped arrangements.

Venue Production: Displays the created venue diagram effect and seating arrangement effect through a graphical interface; adjusts seats through graphical dragging, moving, swapping, etc.; and supports text notes, printing, exporting, etc.

The software is deployed on the operating system of a domestic server, with Kingdee Tianyan as the middleware and KingbaseESV8.0 as the database management system, running on the e-government external network.

Project Results: The project has been successfully launched and has successfully assisted the province in holding a series of meetings, such as the plenary session of the People's Congress, the presidium meeting, the convener meeting, and the group meeting. The official application of the software has reduced the time required to arrange seating from several days to just a few minutes, greatly shortening the working time of conference organizers. Hence, conference seating will be easy, worry-free, and more intelligent, and conference organization work will usher in a new era of digital and XinChuang-supported office environments. In the future, intelligent seating will continue to be optimized and iterated, with the goal of providing the best experience to users.

Gloryview closely follows the key points of national, provincial, and municipal digital government reform work and, through the mode of "joint innovation and production-construction integration", provides comprehensive services for party and government agencies at all levels. Moreover, using hosts and servers with XinChuang mainstream core chips, operating systems, and XinChuang third-party software plugins, it has built a self-developed laboratory characterized by self-developed XinChuang technology, XinChuang project low-code development technology, etc.

Through XinChuang practice in the party and government industry, a comprehensive office platform with controllable XinChuang code has been formed, mainly for party and government agencies. On the one hand, it is used to solve the problems of difficulty in cross-basis environments and application integration when various office systems of party and government agencies are migrated and transformed in batches. In addition, it integrates the domestic environment and X86 environment application systems; quickly achieves unified organization, unified users, unified authentication, unified to-dos, unified messages, cross-business process integration, etc.; reduces the impact of the basic environment and complex system architecture on application system business integration; and achieves the goal of comprehensive office and business collaboration. On the other hand, the controllable code comprehensive office platform provides functions such as document receipt and dispatch, personal file cabinets, internal exchange, data statistics, meeting management, instant messaging, work groups, address book, notices and announcements in the application scenarios of "document handling", "affairs handling", "meeting handling", "information disclosure", etc., within party and government agencies.

5.1.2 Special Project Two: Smart Gardens, Building Green Cities

Smart gardens are an integral part of smart cities and fall under the categories of urban greening and livability. Urban green spaces, street trees, greenways, and ancient and famous trees are important natural and environmental resources of a city. The area, quantity, and distribution of these resources are fundamental indicators of the livability of a city's natural environment. Smart gardens not only provide city managers with intelligent management tools but also serve as ways for residents to connect with nature, participate in greening, and protect the environment.

Current Status of Garden Greening Management

With the construction and development of urban garden greening, new opportunities and challenges have emerged in terms of its management. Owing to the complexity of garden greening classification and challenges in management, it is becoming increasingly difficult to manage garden greening through traditional methods. It is difficult to update basic data in a timely manner, let alone provide data support for planning and decision-making.

Current urban garden greening management still faces difficulties in accurately capturing information about various green spaces, ancient and famous trees, parks, scenic spots, etc. Greening planning and evaluation still rely on manual work and experience. The collected data cannot be scientifically and effectively utilized, the efficiency of daily management work is low, and resource allocation is unreasonable. It is impossible to obtain the current status and changes in urban green space resources in a timely, accurate, and dynamic manner. There is a lack of real-time data updates, and it is difficult to conduct reasonable spatial analysis, which in turn make scientific and rational planning of garden greening highly difficult.

As an important aspect of smart cities, garden greening management is a key management task in building national ecological garden cities and an important indicator in the pilot index system of national smart cities (districts, towns). With the successive introduction of the national "Fourteenth Five-Year" plan and the "National Garden City Standards" compiled by the Ministry of Housing and Urban–Rural Development, local governments have begun to pay attention to the planning and layout of garden greening in urban construction. The "National New Urbanization Plan (2014–2020)" and the "Guiding Opinions on Promoting the Healthy Development of Smart Cities" (Development and Reform High Tech [2014] No. 1770), which were jointly issued by the National Development and Reform Commission and eight other departments in 2014, require efforts to solve real problems restricting urban development and build green, low-carbon, intelligent cities with the support of technological innovation. The "Urban Garden Greening Evaluation Standards" (GB/T 50563-2010) and the "National Garden City Series Standards" also clearly propose the establishment of a digital information library for urban garden greening, information release and social service information sharing platforms; the establishment of a dynamic supervision mechanism for the construction and management of urban garden greening; and the guarantee of public participation and social supervision. Therefore, cities across the country urgently need to establish digital information management systems for urban garden greening, improve the level of business and management efficiency through intelligent information systems, and carry out comprehensive management of urban garden greening.

Problems in Garden Greening Management

As an important aspect of smart city management, garden greening management is a key management task in building national ecological garden cities and an important indicator in the pilot index system of national smart cities (districts, towns). However, many problems remain in garden greening management.

Garden greening classification is complex, data statistics and updates are slow, and management is difficult.

Garden greening planning, construction, and management are operated independently, forming their own systems, which are homogeneous and fail to meet individual needs. There is a lack of a global, systematic, and standardized management system.

The pattern of the green space system lacks systematic consideration, and the phenomenon of "building first, filling in green spaces later" is serious.

The awareness of urban garden greening is generally low, and there is a phenomenon of "government enthusiasm, citizen indifference."

The garden management in each district and the comprising subsystems are relatively independent, creating "information islands" with a low level of intelligence.

Data collection is isolated, and system linkage is difficult to achieve.

The application has poor scalability and high expansion cost.

It is impossible to achieve efficient and convenient centralized management, and the operation cost is high.

The operation status of important equipment cannot be monitored in real time, and accident warning is difficult to implement.

Application of New Technologies in Smart Gardens

Smart gardens rely on a new generation of information networks, integrating cutting-edge technologies such as artificial intelligence, big data, cloud computing, the Internet of Things, GIS, etc. (as shown in Fig. 5.5), combined with "Internet+" thinking and ecological garden management concepts, to achieve fast and accurate garden information collection, calculation, and processing. Using various sensing devices (environmental temperature and humidity, soil moisture, carbon dioxide, images, etc.), intelligent terminals, and automated equipment to intelligently link people with nature, smart gardens can achieve intelligent perception, intelligent warning, intelligent analysis, intelligent irrigation, and online guidance in garden management. They provide technical support in areas such as refined cultivation, visual management, and intelligent decision-making, making cities more livable and making the environment more harmonious through intelligent management and services.

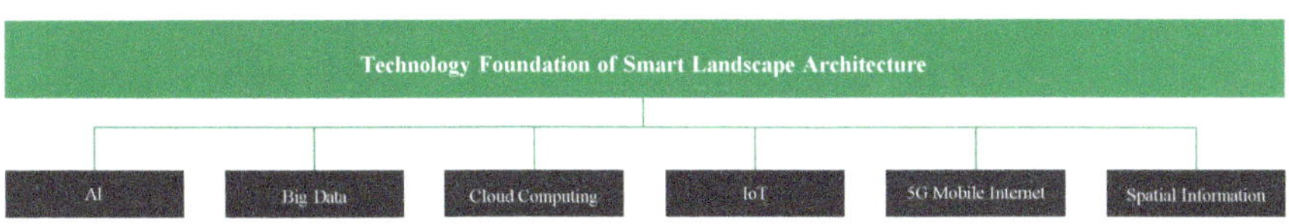

Fig. 5.5 Foundation of smart garden technology

Application of Artificial Intelligence Technology in Smart Gardens

Artificial intelligence technology is widely used in the construction of smart gardens. For example, image recognition technology can automatically identify and classify plants in the garden, provide information on plant morphology and habits, and offer a convenient and precise method for garden design and management. It can also be combined with technologies such as robots and drones to automate maintenance tasks such as cleaning, pruning, and weeding in the garden, thus improving work efficiency and quality.

Application of Big Data Technology in Smart Gardens

The application of big data in smart gardens includes dozens of data types, such as geographic data, spatial data, business data, carbon sink data, thematic data, impact data, and three-dimensional data. The platform needs to process and mine TB-level data. The use of big data technology to store, manage, mine, and analyze these data provides a basis for decision-making for management units, improves decision-making efficiency, and achieves scientific and intelligent garden maintenance. The use of massive amounts of data to train AI models establishes standards and process management for the entire process of inspection, management, and maintenance. Through data sharing technology, it breaks data silos, completes business system data linkages, reduces the manual data entry workload, improves office efficiency, and accelerates approval speed.

Application of Cloud Computing Technology in Smart Gardens

Cloud computing technology refers to the provision of services by aggregating a large amount of computing resources and data storage resources through the network. Users can use these resources as needed and interact through the network. In the field of smart gardens, cloud computing technology has a wide range of applications, with the following main aspects:

Intelligent management: Using cloud computing technology to build a smart garden management platform for comprehensive, real-time monitoring and management of garden resources. This platform can obtain real-time environmental data collected by various sensors and provide reasonable garden resource management schemes through cloud-based algorithm analysis.

Data analysis and decision support: Through big data analysis technology in cloud computing, a comprehensive analysis of garden resources can be conducted to provide scientific decision references for decision-makers. For example, by combining historical data statistics, weather change trends can be predicted, and corresponding irrigation and fertilization plans can be formulated.

Information sharing and exchange: Cloud computing technology can be used to build a smart garden platform, establish online communities, bulletin boards, etc., to facilitate information sharing and exchange between garden practitioners and promote the accumulation and sharing of experience.

System upgrade and maintenance: Cloud computing technology can achieve the scalability and upgradability of software and hardware resources, ensuring the efficient and stable operation of the garden management system. Moreover, through remote maintenance and monitoring in the cloud, various fault problems can be solved in a timely manner, thereby improving the quality of garden services.

Application of Internet of Things Technology in Smart Gardens

Using RFID, sensors, video recognition, global positioning systems, and other devices and technologies, various necessary information, such as sound, light, heat, electricity, chemistry, biology, and location, is collected. Through various forms of network access, connections between things and people are established, achieving intelligent perception, recognition, and management of garden greening-related information and processes. This achievement reduces human resource input, improves equipment operation efficiency, and enhances data continuity.

To achieve precise and intelligent maintenance management of garden greening, front-end IoT intelligent sensing devices are indispensable. The platform uses advanced intelligent devices and various sensors to monitor soil, air temperature and humidity, soil temperature and humidity, CO_2, PM2.5, pH, diseases, pests, etc., in the park. Through the platform's image analysis algorithm, the AI algorithm identifies diseases and pests and other related monitoring content and reports the monitoring data to the platform in real time.

Application of GIS Technology in Smart Gardens

GIS, or geographic information systems, is an interdisciplinary discipline that includes geography, cartography, remote sensing, and computer science. GIS is also a computer-based tool that can analyze and process geographic information, spatial information, and remote sensing images, integrating the unique visual effects and geographic analysis functions of maps with general database operations (such as queries and statistical analysis). The basic data of the smart garden system are related to geographic and spatial information; thus, the GIS system is the foundation and support for the creation of smart gardens, and all upper-level businesses must perform corresponding operations based on the GIS system.

Sharing of Smart Garden Practice Cases

At present, the construction of smart gardens in China is still in its infancy. Some city garden work authorities have not yet realized the importance of the smart garden system platform. There are problems such as insufficient investment in informatization construction, repeated construction of business systems, inability to eliminate data islands, large investments in manual data collection, slow data update speeds, and the inability to aggregate and analyze various data.

Therefore, to build a green city, a garden city, and an ecological city, it is necessary to create an advanced smart garden management platform. This platform lays the foundation for smart gardens and empowers smart garden management. It gradually achieves the unified acquisition, updating, and management of garden information, data, and materials; creates a comprehensive process and facilitates detailed analysis and management of garden greening business, providing information and data support for leadership decision-making; and enables public participation, thus improving citizen satisfaction with urban greening. Below, we present a case study of the construction of a smart garden by a city's garden bureau:

The Forestry and Garden Bureau of a city in Guangdong Province aims to standardize garden business management processes and upgrade intelligent management methods, ultimately aiming to reduce management costs, improve management efficiency, and build a smart garden management system. With the goals of smart greening, smart management, smart ecology and the use of information technology as a means, this bureau initiated the construction of a comprehensive smart garden management platform to solve the current problems in garden management and achieve the intelligentization of garden management work.

The project lasted a year and a half. It involved summarizing and sorting existing data, conducting a baseline survey of the city's gardens, green spaces, greenways, and famous old trees, and forming databases of garden green spaces, city parks, famous old trees, and other resources. On this basis, the smart garden comprehensive management platform has built more than a dozen business subsystems, such as for city green space management, park landscape management, and mobile maintenance, and includes dozens of forestry garden thematic databases for the city's gardens, green spaces, greenways, street trees, parks, famous old trees, etc. The application involves city-level, district (county)-level forestry garden management departments and subordinate units, including resource management, administrative approval, administrative law enforcement, emergency command, and other aspects, and covers nearly ten thousand users, forming a "big data, big service" comprehensive management platform for the city's garden industry.

With a "graphic integration" data system at its core, the platform uses the approval business collaborative office as a breakthrough, the ecological garden big data base platform as support, intelligent and refined garden management as a driver, decision-making support as an enhancement, and information resource industrialization as a long-term mechanism is realized. Through these resources, it establishes a unified

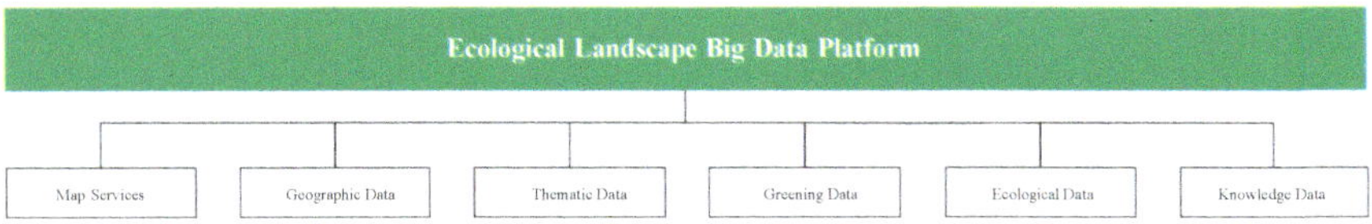

Fig. 5.6 Composition of the big data platform

business platform covering the city's garden greening, forest resources, and ecological protection business and assists in achieving the intelligentization of city garden decision-making, standardization of office work, transparency of supervision, and convenience of service.

The ecological garden big data platform is the basic platform for the creation of smart gardens. It includes basic maps, data storage, data analysis, data sharing, data interfaces, and other basic services.

The platform is based on "Internet+", combined with advanced technologies such as the Internet of Things, big data cloud computing, the mobile internet, and intelligent information terminals. It integrates the core business of forestry gardens and assists in the sustainable green development of ecological gardens. The composition of the ecological garden big data platform is shown in Fig. 5.6.

Map Service

The map service uses GIS technology to provide access to electronic maps. Through GIS software, vector and raster data sources are specified, thematic maps are created, annotations are set, and then these processes are published as map services. Map services include dynamic map services and tile map services. Tile map services can precreate map caches to improve map display and access efficiency. Map services are the most common GIS services. In addition to providing basic map browsing functions, basic map service functions can be used to implement map queries, map searches, map editing, and other services.

Geographic Data

The construction of a basic database that can be shared by multiple garden greening application systems and used by all relevant garden greening departments is planned. Relevant data standards are formulated, and the provided urban 1:10,000, 1:2000 full-element digital topographic maps, satellite remote sensing images, digital elevation models (DEMs), etc., are sorted into the database. Unified geographic coordinates, unified geographic coding, and data format consistency with urban digital city planning are ensured.

Basic geographic data also include multiyear 1:500 electronic maps of the city, city topographic maps, exquisite garden 3D models, pedestrian overpass greening 3D models, common city tree species tree models, key ancient trees, 720 tourist attractions, 360-degree panoramic data, etc.

Thematic Data

The construction of the garden greening thematic database includes the sorting and storage of four types of data: spatial data, attribute data, document data, and metadata. The thematic database includes garden greening data (including parks, urban green spaces, urban green lines, overpass greening, and famous trees) and the database of urban garden greening practitioners.

Greening Data

The platform conducts a baseline survey of the green spaces, street trees, famous old trees, park scenic areas, greenways, overpass greening, and other garden greening data in the built-up area, collecting green space patches and street tree data, with an accuracy of up to 0.25 m.

Ecological Data

Ecological protection data include city-wide wild animal and plant distribution data, annual forest pest distribution data, forest fire prevention equipment and facility data, carbon storage data, video surveillance data, drone data, and other data related to ecological protection.

Knowledge Data

These data include basic plant and animal databases, tree classification databases, forest pest and disease knowledge databases, various standards and norms related to forestry and gardening, and shared data from various units. The ecological big data platform, driven by forestry and gardening operations, establishes standards for garden greening spatial data, business data, and data update guidelines, standardizing the orderly update of various types of data and ensuring the timeliness, standardization, and accuracy of the data.

Based on the practical cases of this project, we believe that the key aspects of the construction of smart gardens are as follows:

Building an Integrated Smart Garden Management Platform

The Smart Garden Management Platform relies on new generation information technologies such as spatial information technology and big data analysis technology to tightly integrate and manage the city's basic spatial database, professional gardening database, and nonspatial database. This design forms a specialized digital information management system for urban gardens that integrates information perception, planning assistance, comprehensive management, and public participation. It achieves dynamic monitoring, deep management, and effective sharing and use of garden resources. By integrating data resources, breaking down data silos, improving office efficiency, reducing labor inputs, ensuring timely data updates, and fully utilizing existing data and resources, it realizes the sustainable development of gardening work, accelerates the intelligent process of garden management, and promotes the construction of smart ecological cities.

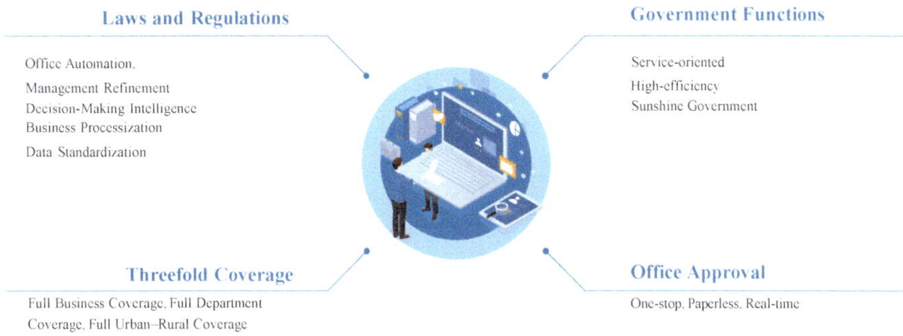

Fig. 5.7 Goals of the integrated management of smart gardens

The construction of the Smart Garden Management Platform involves topographic map information, garden planning and greening support, garden green space management, real scene image information, greening project engineering supervision and management, garden greening spatial data, famous historical sites, parks, famous old trees, permission allocation, fine processing, business expansion construction management, etc. This platform aims to comprehensively enhance the digital, informational, and intelligent level of garden greening; promote the professionalization of the fine management of garden greening; and greatly improve the macrolevel management and service capabilities of the entire industry. The goal of the integrated management of smart gardens is shown in Fig. 5.7.

Improving Management Systems, Enriching Management Means, and Innovating Management Methods

Gardening has entered the era of smart ecological gardens. Reasonable garden planning and construction can not only create an environmentally friendly and a natural ecological environment and drive the development of related industries such as tourism and leisure but also constitute an important way to comprehensively improve the management and service level of gardens and contribute to a city's image.

The Smart Garden Management Platform innovates management systems, means, and methods; relies on a new generation of information networks, which are based on smart ecological garden big data; integrates spatial information technology, cloud computing, big data analysis, the Internet of Things, the mobile internet and other technologies; and builds a standardized, systematic, intelligent, and public service-oriented urban "smart garden" comprehensive management platform. This design allows for fine, intelligent management of the entire process of garden greening before, during, and after an event, improves citizens' understanding and participation in urban greening work, comprehensively improves the management of garden greening, and achieves a new urban "smart garden" management model of "one network for the whole city, one line for supervision, one platform for display", thus

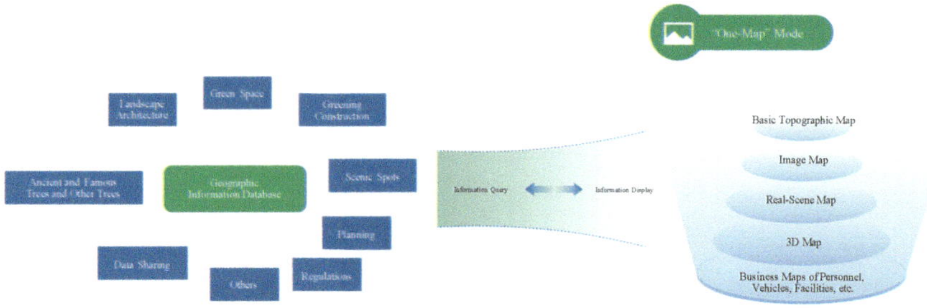

Fig. 5.8 Use of the platform to achieve one library, one map

achieving the management goals of data digitization, resource perception, information sharing, intelligent decision-making, system integration, business collaboration, and humanized service.

Big Data, Decision Support, Data Sharing, Integrated Office, Multiplatform Integration

The Smart Garden Management Platform includes several parts, such as a data storage computing platform, decision support platform, big data mining analysis platform, data sharing platform, and integrated collaborative office platform. The platform integrates and analyzes various garden greening resources and data to construct a "one map, one library" for garden greening, that is, one spatial data map and one garden resource library (as shown in Fig. 5.8).

By building a powerful data storage computing platform, a large amount of basic garden data is saved and managed, providing data support for garden planning and reducing management costs and complexity.

By constructing a decision-making platform for auxiliary planning based on mathematical models, data and information are quickly transformed into a decision-making basis, assisting the garden department in quickly and accurately completing relevant planning and decision-making.

Through a big data mining analysis platform, the effective management of risk prevention, early warning, and contingency measures can be realized, to focus on innovative work.

A data sharing platform is established to facilitate the docking and interaction of garden data with other platform data and achieve the open sharing of platform data information through the API interface.

Through an integrated collaborative office platform, we aim to improve and enhance services, increase the level of plant protection, optimize work processes, and increase management efficiency.

5.1.3 *Special Topic Three: Smart Taxation, Empowering Tax Business Upgrades*

Smart taxation is a new tax management model that uses information technology to digitize, intellectualize, and data-fy tax management, collection, and supervision. It can increase tax management efficiency, improve tax collection and supervision, optimize tax policy formulation, enhance tax services, and promote the development of the digital economy. It is one of the important developments in modern tax management and is highly important for promoting scientific, standardized, and convenient tax work.

Problems and Challenges Encountered in the Development of Tax Digitalization

Although China has achieved certain achievements in tax digitalization, it still faces some challenges and difficulties.

In terms of information security, the security of tax data is very important. If the data are leaked or attacked by hackers, tax collection and management work may be severely affected. Therefore, the government needs to strengthen data protection and security control, enhance the research and application of security technology, and improve information security capabilities.

In terms of technological innovation and application, with the continuous emergence of new technologies, tax digitalization transformation needs to continuously explore innovation, actively apply new technologies, and improve tax collection efficiency and service levels. However, these tasks require the government and enterprises to have sufficient technical strength and talent reserves to meet the needs of digital transformation.

In terms of the user experience, the goal of government digital services is to better serve the people and improve their satisfaction. However, if digital services cannot meet the needs of the people or if the user experience is not good, then the promotion and application of digital services will be restricted. Therefore, the government needs to pay more attention to the user experience, actively collect user feedback and needs, continuously optimize digital services, and improve user satisfaction.

In terms of data standardization, tax digitalization transformation requires considerable data support. However, due to the different formats and standards of data between different departments and regions, there are many difficulties to data collection, integration, analysis, and sharing. Therefore, the government needs to unify data standards, formats, exchange protocols, etc., establish data sharing mechanisms, and achieve data intercommunication, mutual use, and interconnection.

The problems and challenges of the development of tax digitalization are shown in Fig. 5.9.

These challenges and difficulties exist not only in tax digitalization transformation but also in other areas of digital government services. The government needs to

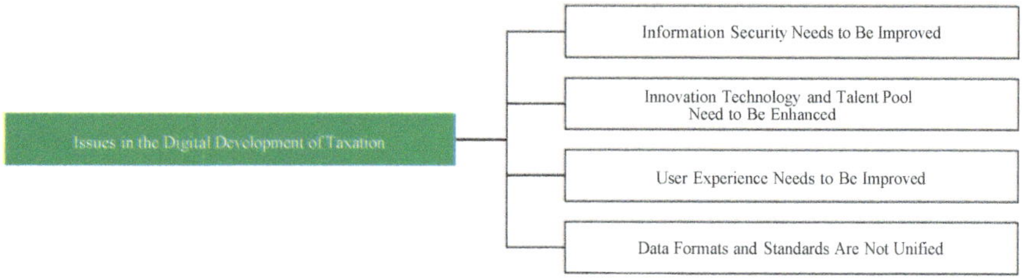

Fig. 5.9 Problems and challenges of tax digitalization development

formulate scientific policy planning; strengthen cross-departmental, cross-regional, and cross-field collaboration; continuously innovate and improve digital services; and promote improvements in digital transformation.

New Technology Empowers Innovation in Smart Taxation

In response to the challenges and problems faced by tax digitalization transformation, smart taxation has generated new solutions and innovative applications with new technologies such as artificial intelligence, big data, cloud computing, blockchain, and the mobile internet.

Innovative application of artificial intelligence in smart taxation: Artificial intelligence technology can be used to automate some repetitive and regular work, such as tax data analysis and processing and taxpayer information verification. This application can improve the efficiency and quality of tax work and reduce error rates and omission rates. For example, artificial intelligence technology can be used to develop intelligent tax systems, achieve automated tax collection and services, and improve efficiency and quality. In addition, artificial intelligence technology can be used to develop intelligent customer service systems; provide more convenient, efficient, and personalized services; and meet different user needs.

Innovative application of big data in smart taxation: The tax system needs to process a large amount of data. Big data technology can help tax agencies quickly and accurately analyze and apply these data and thereby improve tax collection efficiency. Tax authorities can strengthen the verification and supervision of taxpayers through big data analysis methods, thus reducing illegal tax behaviors. Using big data technology to analyze and mine massive amounts of tax data and extract useful information and patterns provides a scientific basis for tax decision-making.

Innovative application of cloud computing in smart taxation: Smart taxation can use cloud computing technology to migrate the tax management system to the cloud, achieving more secure, stable, and convenient information services. Tax authorities can improve their ability to respond to emergencies through cloud computing, thus ensuring the continuity and stability of tax collection and management work.

Innovative application of blockchain in smart taxation: Blockchain technology can assist tax authorities in establishing a more transparent and trustworthy tax collection and management system. By using blockchain technology, the security and integrity of tax data can be ensured, preventing data tampering and leakage and enhancing trust between taxpayers and tax authorities. Blockchain can also be used to address information security issues by storing tax data on a distributed ledger and encrypting, verifying, and controlling access to data through mechanisms such as smart contracts, thereby ensuring data security and privacy protection.

Innovative applications of mobile internet technology in smart taxation: Mobile internet technology can provide more convenient and faster services for smart taxation, such as mobile apps, WeChat official accounts, etc. This feature allows taxpayers to conveniently handle declarations, inquiries, payments, etc., at any time and place, improving the level of tax services.

In summary, the integration and innovative applications of new technologies in smart taxation can help tax authorities improve tax collection efficiency, establish a trustworthy tax collection system, automate repetitive tasks, and increase the level of tax services. The introduction and application of these new technologies will further promote the development of smart taxation and advance the modernization of tax management.

Case Study of Smart Taxation Practice: Empowering Digital Innovation in Smart Taxation with Intelligent Customer Service Products

Project Background: With the continuous advancement of tax collection and management system reform, tax and fee reductions, support for enterprises to resume work and production, and the final settlement and payment of personal income taxes, the consultation needs of taxpayers and payers have exploded, and the volume of calls, number of seat agents, and volume of voice calls are all in a state of continuous growth. The high rate of manual consultation seriously affects the quality and efficiency of tax services. Owing to its complicated menu and deep levels, the button-style IVR[2] system of the original Dalian Tax Service Customer Service Center has reduced its scalability amid an increasing number of new businesses, which has severely affected the user experience. Even with the continuous increase in the number of human agents employed in the customer service center, it still cannot meet the demand during peak call volume periods. In addition, rising labor costs have had a negative impact, posing an increasingly large challenge for the customer service center.

Construction Functions: To accelerate the intelligent and informatized construction of the 12,366, framework while considering actual local conditions, we actively explore the development of intelligent voice consultation functions to provide

[2] IVR denotes interactive voice response system. It is an automated voice response technology used in telephone or other communication systems that enables interaction with users through voice prompts and keypad inputs.

taxpayers with more efficient and convenient consultation services. The main construction functions are divided into two parts:

First, a bureau enterprise number precision push system is constructed to proactively deliver the consultation needed by taxpayers, resolving the contradiction between resource shortages and the growth of tax consultation demand.

Second, by introducing an intelligent voice IVR system, the menu is flattened to solve the problem of too many horizontal nodes and overly deep vertical levels in the button-style IVR service while achieving arbitrary jumps between voice node levels, intelligent interruption, and return. Based on the existing intelligent voice service mode of the Dalian Tax Bureau, it is expanded to self-service voice service. It provides interactive self-service voice service, preprocessing voice for some important issues and common issues, using an intelligent voice recognition engine and semantic analysis engine, and docking with the semantic understanding module to facilitate intelligent voice interaction, intelligent diversion, menu navigation, and other functions.

The architecture of intelligent customer service, which is divided into user, customer service system, backend service, and other modules, is shown in Fig. 5.10.

The intelligent customer service system has facilitated the application of functions such as data integration and cleaning, natural language processing, machine learning and decision engines, and artificial intelligence algorithms.

Using big data technology, the taxpayer data are integrated and cleaned to ensure the accuracy and completeness of the data.

Natural language processing (NLP) technology is used to process the questions raised by taxpayers, converting them into a form that computers can recognize and interpret. Natural language processing technology includes word segmentation,

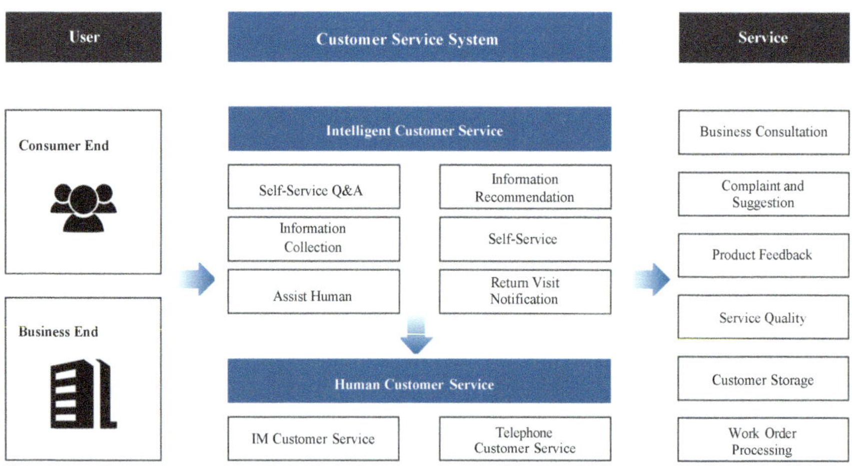

Fig. 5.10 Intelligent customer service architecture diagram

semantic analysis, text classification, keyword extraction, etc., which can help the system better understand the needs of taxpayers.

A machine learning and decision engine is established. The intelligent customer service system integrates machine learning and decision engine technology to analyze and process large amounts of data, providing personalized services for taxpayers. Machine learning technology can automatically optimize system performance and improve system accuracy and efficiency, and decision engines can automatically classify and answer questions, thus improving service quality.

Artificial intelligence algorithms, including neural networks, genetic algorithms, fuzzy logic, etc., are used to analyze taxpayer historical behavior and trends, improving the intelligence level of the system. Moreover, a training and testing system for artificial intelligence algorithms is established to improve the intelligence level and accuracy of the system.

Project Effectiveness: Currently, the intelligent voice customer service system is online and operational at the Dalian Municipal Taxation Bureau of the State Taxation Administration. The data show that the robot's effective response rate has reached over 85%; the accuracy rate of voice intelligent consultation is higher than 80%; and the volume of manual calls has significantly decreased, reducing the number of seats. Through continuous education using process data, related indicators are steadily rising. It plays a crucial role in reducing costs and increasing the efficiency of tax services by resolving the contradiction between the shortage of tax consultation resources and the gradually increasing demand for tax consultation.

Compared with traditional manual customer service, intelligent customer service improves user satisfaction, reducing the operating costs of the tax bureau. Moreover, the use and extension of accumulated data are more conducive to the support system of big data operation architecture for tax business development, creating a virtuous cycle based on business development and user data and further improving the business operation service matching capability.

5.2 Smart Health Care: A Key Driver of the "Healthy China" Strategy

In 2022, at a cardiovascular- and cerebrovascular-themed exchange meeting in Guangzhou, a female patient from Zhanjiang said that she had to pull many strings to make an appointment at a well-known tertiary hospital, yet in the end, the consultation with the doctor lasted less than 3 min. This topic resonated with many patients present at the meeting, many of whom had similar experiences. These issues have caused dissatisfaction among many patients who have spent energy, time, and money seeking medical treatment in Guangdong. For many patients, it is common to experience difficulty making appointments and experience short consultation times in tertiary hospitals.

According to data released by the National Bureau of Statistics on January 17, 2023, China's total population has decreased. The population at the end of 2022 was 850,000 less than that at the end of the previous year, and the natural growth rate of the population was -0.60%. This is the first time that China's population has experienced negative growth over 60 years, and the problem of aging is becoming increasingly severe. With the increase in the aging population, there may be a continuously increasing demand for medical services. Medical insurance and social security funds may increasingly shift toward the retired population, so medical expenses are also increasing annually, and it is difficult to further alleviate the problem of "expensive medical treatment" for residents.

The difficulty and high cost of seeking medical treatment have become among the most prominent livelihood issues in China. Livelihood issues are not only social management issues but also institutional issues, that is, directly related to a country's and the government's institutional arrangements during a certain period. How can these problems be solved? The construction of smart health care is arguably one important means. Smart health care has a promoting effect on various links, such as registration, diagnosis, treatment, rehabilitation, payment, hospital management, and regional medical resource allocation. Using technologies such as artificial intelligence, big data, cloud computing, the Internet of Things, and the mobile internet, a patient-centered medical information management and service system can be built to provide services such as medical information interconnection, remote diagnosis and treatment, shared collaboration, clinical innovation, and scientific diagnosis. The rapid development of various technologies has promoted the formation of a patient-centered medical data network, ushering in an era of smart health care for the medical industry, which is also an important driving force in implementing the "Healthy China" strategy.

5.2.1 Problems in Medical Services and Hospital Management

According to data from the National Bureau of Statistics, by the end of 2022, there were a total of 1.033 million medical and health institutions nationwide, including 37,000 hospitals, of which 12,000 were public hospitals and 25,000 were private hospitals; there were 980,000 primary medical and health institutions, including 34,000 township health centers, 36,000 community health service centers (stations), 321,000 outpatient departments (clinics), and 588,000 village health rooms; and there were 13,000 specialized public health institutions, including 3385 disease prevention and control centers and 2796 health supervision offices (centers). China has a total of 3275 tertiary hospitals, accounting for approximately 12% of the total number of first- to third-level hospitals in the country, but the number of consultations in tertiary hospitals accounts for more than 60%. Moreover, tertiary hospitals in China are concentrated mainly in large cities such as Beijing, Shanghai, and

Guangzhou, and medical resources in small and medium–sized cities are relatively insufficient. By the end of 2022, China had a total of 11.55 million health technicians, including 4.4 million practicing physicians and assistant practicing physicians, and 5.2 million registered nurses, while the total number of consultations throughout the year was 8.4 billion, placing considerable pressure on the medical supply. Under such circumstances, the following problems may arise in medical services and hospital management:

First, how can medical resources be allocated reasonably to meet the medical needs of the people?

In April 2017, the General Office of the State Council issued the "Guidance on Promoting the Construction and Development of Medical Consortia", which explicitly required the full initiation of pilot projects for the construction of medical consortia in various forms, with all tertiary public hospitals participating and playing a leading role. The proposal of medical consortia actually aims to solve the problem of medical resource allocation. However, after many years of development, the allocation of medical resources between large urban hospitals and small hospitals is still unbalanced, which is more prominent against the backdrop of the COVID-19 pandemic. Large hospitals are crowded, while grassroots hospitals or community hospitals are deserted. Currently, 80% of the country's medical resources are concentrated in large cities, with 30% concentrated in large hospitals. Approximately 80% of the people who visit large hospitals each year have common and prevalent diseases that can be treated at grassroots hospitals. The hierarchical diagnosis and treatment model of the medical consortium is shown in Fig. 5.11.

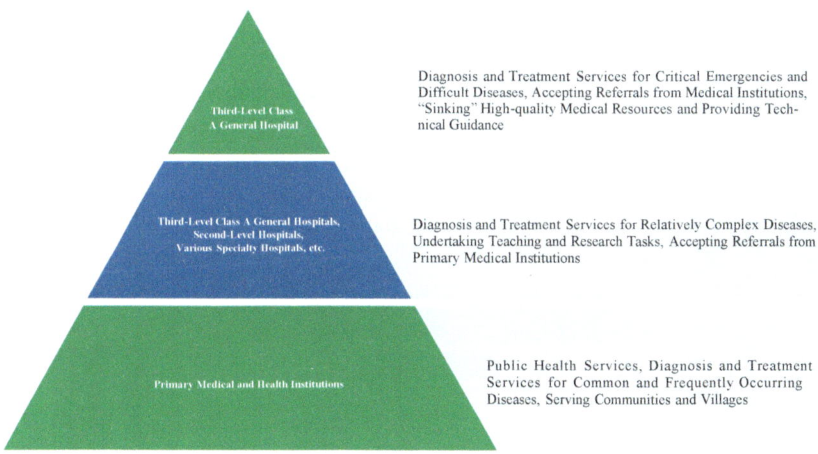

Fig. 5.11 Hierarchical diagnosis and treatment model of the medical consortium

Second, the "three longs and one short" problem has been alleviated, but "large hospitals" are still overcrowded.

In November 2022, the National Health Commission of the People's Republic of China (abbreviated as the National Health Commission), the National Administration of Traditional Chinese Medicine (abbreviated as the National Administration of Traditional Chinese Medicine), and the National Disease Prevention and Control Bureau (abbreviated as the National Disease Control Bureau) issued the "14th Five-Year Plan for National Health Informatization". It reviewed the main work during the "13th Five-Year Plan" period. During the "13th Five-Year Plan" period, the health industry vigorously promoted the integration and implementation of the two major strategies of Healthy China and Digital China, deeply implemented the "13th Five-Year Plan" for national health informatization development, and accelerated the standardized application of big health and medical data and the innovative development of "Internet + Medical Health". Hospitals at the second level and above nationwide have fully implemented "Internet + Medical Health" with 10 services and 30 measures and deepened the "five ones" service action for the convenience and benefit of the people. In addition, hospitals at all levels have generally begun to offer online services such as internet health consultations, appointment registration, in-clinic settlement, medical insurance networking, examination and test results inquiries, mobile payment, etc.; optimized and reformed the medical process; and the problem of "three longs and one short" in medical treatment has been somewhat alleviated. However, owing to the lack of talent, poor equipment, and low medical level in third- and fourth-tier cities, these problems still exist and are difficult to improve significantly in a short period of time. Patients' first choice is still top-tier hospitals in large cities, resulting in "large hospitals" still being overcrowded. The optimized medical process model is shown in Fig. 5.12.

Third, the level of informatization in logistics and operations management is insufficient, making it difficult to meet the overall development requirements of hospitals.

Hospital logistics and operations management are important links in the medical and health service system. As it aims to integrate all business processes within the hospital and provide patients with better medical services, the internal processes of

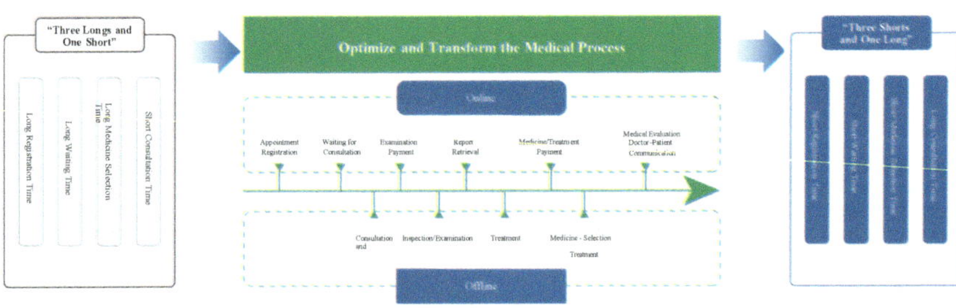

Fig. 5.12 Optimized medical process model

hospital logistics management have a significant impact on patients' experience in the hospital. With the increasing awareness of service experiences and consumer rights and under current medical policy support, there are clear development guidelines for doctor–patient relationships, medical quality, and the work efficiency of medical staff, which also increase hospitals' attention to logistics management. However, at this stage, the level of informatization in logistics and operations management is insufficient to meet the overall development requirements of hospitals.

The serious imbalance of medical resource supply and demand, the uneven regional distribution, the aging population, the rapid growth of chronic diseases, and underperforming logistics and operation management in hospitals have created an enormous demand for smart health care. The current social development situation has led to new requirements for medical and health services, making it urgent for hospitals to change their operation mode and continuously explore the digital transformation of health care.

5.2.2 The Development History of Smart Health Care

China introduced medical informatization technology as early as the 1980s and 1990s. At that time, a few large hospitals with conditions purchased computer hardware and software and established IT systems and databases to improve hospital management workflows and promote efficiency. The initial results of medical informatization were concentrated on the financial settlement system of the hospital and the management system of medical supplies such as drugs. This endeavor was the earliest form of smart health care, which centered on identifying and resolving issues in nondiagnostic services.

In the twenty-first century, with the further maturity of IT technology and the explosion of the internet, medical informatization entered the second stage. An increasing number of hospitals began to build their own computer rooms and networks and purchase and use hospital information systems, electronic medical records, self-service equipment and other information management systems. At this stage, smart health care began to focus on business, which is dedicated to patient-centered services and serves doctors, nurses, and patients to improve the efficiency of medical resource utilization.

After 2010, with the rapid development of smartphones and the mobile internet, the technology of medical informatization was stimulated to upgrade and evolve again, entering the third stage. An increasing number of hospitals have launched convenient medical services based on the mobile internet, such as remote mobile app registration and app report viewing. The smart health care of this stage focuses on interconnection to create an overall data integration platform, aiming to eliminate the "information islands" within the medical system and integrate data from different systems.

Today, with the promotion of new technologies such as 5G, artificial intelligence, cloud computing, the Internet of Things, and big data, medical informatization has

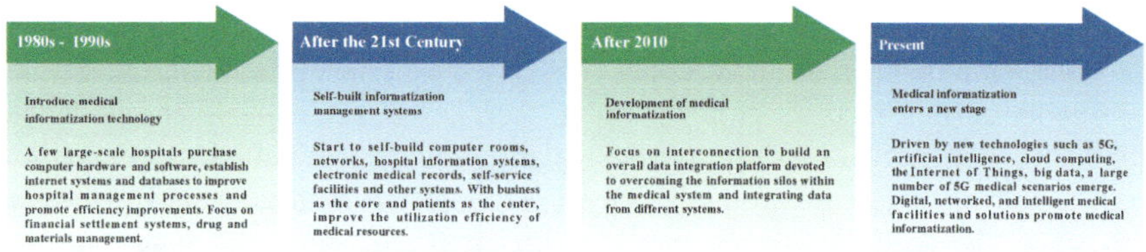

Fig. 5.13 Development process of smart health care

entered a new stage. A large number of 5G medical scenarios have emerged, and digital, networked, and intelligent medical facilities and solutions have truly come to the forefront. Especially during the COVID-19 pandemic, various IT and communication technologies have been integrated into traditional medical scenarios, empowering medical institutions and medical staff and providing great help for overcoming the pandemic. The development process of smart health care is shown in Fig. 5.13.

5.2.3 *The Upgrade of Smart Health Care is the Key to High-Quality Development in Health Care*

Smart health care is an important application in terms of people's livelihoods. With the introduction of various digital medical policies and the deepening of new medical reforms, relevant government departments in China have actively responded to the construction and investment of smart health care, improved intelligent medical technology in multiple dimensions, such as new technology application and policy encouragement, and promoted medical informatization. The development of technologies such as artificial intelligence, big data, cloud computing, the Internet of Things, and sensors has made it possible to assist in medical decision-making and medical methods. At the same time, hospitals cooperate with medical insurance and social service departments to simplify processes in all links, such as prediagnosis, diagnosis, postdiagnosis, and medical support, allowing medical information to flow and share among patients, medical equipment, hospital information systems, and medical staff, greatly improving the efficiency and the quality of medical services.

National Policies Accelerate the Rapid Advancement of Smart Health Care

The "Overall Layout Plan for Digital China Construction" issued by the Central Committee of the Communist Party of China and the State Council on February 17, 2023, noted that it is necessary to consolidate the foundation of Digital China construction, where one strategy can be to open up the major arteries of digital infrastructure. This first strategy calls for accelerating the coordinated construction of 5G networks and gigabit optical networks; promoting the full development and application of the mobile Internet of Things; improving the overall application infrastructure; and strengthening the digital and intelligent transformation of traditional infrastructure. The second strategy is to smooth the large circulation of data resources, build a national data management system and mechanism, and improve data coordination management institutions at all levels. In addition, it calls for promoting the convergence and utilization of public data and building national data resource libraries in important fields such as public health, science and technology, and education. Smart health care is an important part of the construction of Digital China, and China has launched a number of policies to promote the development of smart health care. The policies devised for smart health care over the years are shown in Table 5.1.

The "Three-in-One" Construction Principle Clarifies the Direction of Smart Hospital Construction

Smart health care is the result of the integration of life sciences and information technology and is an important part of modern medicine and communication technology. In general, smart health care services include smart hospital services, regional medical services, and home health services. Smart hospitals are an important part of smart health care and an important focus of smart health care development.

In 2019, leaders from the National Health Commission proposed that the scope of smart hospitals includes three main areas: "smart health care" for medical staff (centered on internal hospital operations), "smart management" for hospitals, and "smart services" for patients.

In the field of smart health care, in December 2018, the Office of the National Health Commission issued the "Electronic Medical Record System Application Level Grading Evaluation Standards (Trial)", which aims to establish an electronic medical record system application level evaluation and continuous improvement system suitable for China's national conditions to increase the level of clinical diagnosis and treatment with electronic medical records as the focus. In the field of smart services, in March 2019, the Office of the National Health Commission issued the "Hospital Smart Service Grading Evaluation Standard System (Trial)", which guided hospitals in building functional, information-sharing, and intelligent smart service information systems to improve the patient experience. In the field of smart management, in March 2021, the Office of the National Health Commission issued the "Hospital Smart Management Grading Evaluation Standard System (Trial)", which guided medical institutions to scientifically and standardly carry out smart hospital construction and improve the level of refined and intelligent hospital management. The evaluation standards for smart services and smart management in smart hospitals are shown in Fig. 5.14.

The construction policy of "three-in-one" medical care, service, and management has clarified the construction direction of smart hospitals, enabling the hospital's business to gradually upgrade from the traditional independent operation of medical and logistical services to an integrated online management model.

Table 5.1 Policies devised for smart health care over the years

Time	Issuing unit	Meeting or policy document	Main content
October 2016	Central Committee of the Communist Party of China, State Council	"Healthy China 2030" Planning Outline	Promoting the construction of a healthy China is an important foundation for comprehensively building a moderately prosperous society and basically accomplishing socialist modernization. It is a national strategy to comprehensively improve the health quality of the Chinese nation and achieve coordinated development of people's health, the economy and society. It is a major measure to actively participate in global health governance and fulfill the international commitment to the 2030 Sustainable Development Agenda
April 2018	General Office of the State Council	Opinions on Promoting the Development of "Internet + Medical Health"	Promote the implementation of the Healthy China strategy, improve the modernization level of medical and health management, optimize resource allocation, innovate service models, improve service efficiency, reduce service costs, and meet the growing medical and health needs of the people
December 2018	Office of the National Health Commission	"Notice on the Issuance of the Graded Evaluation Management Method for the Application Level of Electronic Medical Record Systems (Trial) and Evaluation Standards (Trial)"	Continuously promote the informatization construction of medical institutions with electronic medical records as the core
March 2019	Office of the National Health Commission	"Notice of the Office of the National Health Commission on the Issuance of the Graded Evaluation Standard System for Hospital Smart Services (Trial)"	Hospital smart services refer to the application of information technology to improve the patient medical experience, strengthen patient information interconnection and sharing, and enhance the level of intelligent medical services

(continued)

Table 5.1 (continued)

Time	Issuing unit	Meeting or policy document	Main content
March 2021	Office of the National Health Commission	"Notice of the Office of the National Health Commission on the Issuance of the Graded Evaluation Standard System for Hospital Smart Management (Trial)"	The evaluation targets are hospitals that use information and intelligent means to carry out management. It also clarifies the purpose of establishing a graded evaluation standard system: to define the functions achieved at each level of smart hospital management, to provide a reference for hospitals to strengthen related work; to guide the evaluation of the current status of smart hospital management construction and establish a continuous improvement system; and to perfect the top-level design of the "three-in-one" smart hospital construction, making it an effective tool for improving the modern management level of hospitals
November 2022	Office of the Ministry of Industry and Information Technology, etc	"Notice on Organizing the Application for the 2022 Smart Health and Elderly Care Products and Services Promotion Catalog"	Smart health and elderly care services include two main types of smart health services: personalized health management, and internet + health consultation/science popularization; and four types of smart elderly care services: internet + home-based elderly care life care, mutual aid elderly care, elderly ability assessment, and online elderly education/shopping
March 2022	Office of the State Council	"14th Five-Year Plan for the Development of Traditional Chinese Medicine"	Promote the construction of smart traditional Chinese medicine hospitals integrating smart medical treatment, smart services, and smart management
December 2022	National Development and Reform Commission	"14th Five-Year Plan for Expanding Domestic Demand Strategy Implementation"	Carry out digital pilot demonstrations in key areas, such as intelligent transportation, smart logistics, smart energy, smart medical care, smart health and elderly care, and vigorously develop the third-party big data service industry

(continued)

Table 5.1 (continued)

Time	Issuing unit	Meeting or policy document	Main content
February 2023	Central Committee of the Communist Party of China, State Council	"Overall Layout Plan for Digital China Construction"	Promote the universalization of digital public services, vigorously implement the national education digitization strategy, improve the national smart education platform, develop digital health, and regulate the development of internet diagnosis and treatment and internet hospitals
February 2023	Office of the Ministry of Finance, Office of the National Health Commission	"Notice on Organizing the Application for Central Financial Support for Public Hospital Reform and High-Quality Development Demonstration Projects in 2023"	Focus on strengthening the construction of smart hospitals. Promote the construction of smart hospitals integrating electronic medical records, smart services, and smart management, and the standardization of hospital information, and support the establishment of a regional information sharing standard for test results

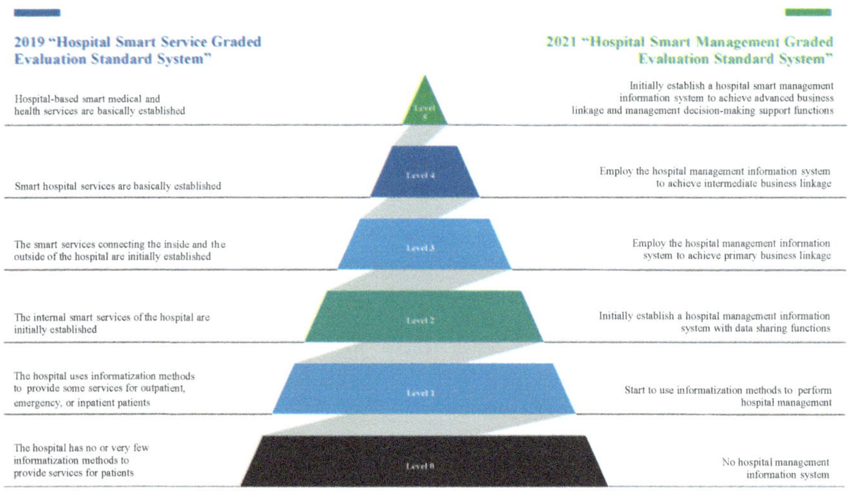

Fig. 5.14 Evaluation standards for smart services and smart management in smart hospitals

The Construction of Medical Consortia and Regional Medical Centers Promotes the Balanced Distribution of Medical Resources

Medical consortia integrate the medical resources of different types and levels of medical institutions through vertical or horizontal methods. They can solve problems such as the uneven distribution of regional medical resources, suboptimal patient medical treatment patterns, and weak service capabilities of primary medical institutions. They play an important role in promoting the vertical integration of high-quality medical resources, improving the service capabilities of primary medical care, and constructing a scientific and reasonable health service system. They are also an important means to promote graded diagnosis and treatment. On January 28, 2011, the first "Regional Medical Consortium" in Shanghai was launched in Luwan District. It was led by Ruijin Hospital, which is affiliated with Shanghai Jiaotong University School of Medicine, promoting and integrating six primary and secondary medical institutions in the region to serve the residents within the consortium.

The National Regional Medical Center is responsible for the diagnosis and treatment of difficult and critical diseases in the region; demonstrating and promoting appropriate and effective diagnostic and treatment technologies; developing medical science and improving medical service capabilities in the region; training the backbone of talent and discipline leaders; leading clinical research on major diseases in the region; producing timely clinical application of research results; integrating existing resources; promoting disease prevention and health care services, leading to the construction of medical service and disease prevention and control networks in the region; cooperating with the National Medical Center; strengthening academic exchanges and regional cooperation; improving China's medical service system; and improving regional medical service levels. On April 13, 2023, the National Health Commission held a press conference in Beijing to introduce the expansion and sinking of high-quality medical resources and the balanced layout of regions. The next step will be to study and formulate plans for the establishment of the National Medical Center and the National Regional Medical Center over the next five years to guide the directional flow of high-quality medical resources.

Medical consortia and regional medical centers are currently important measures for optimizing the allocation of medical resources and improving the level of medical services. The main roles of medical consortia and regional medical centers are shown in Fig. 5.15.

Achieving medical collaboration: Medical consortia and regional medical centers can achieve medical collaboration between different medical institutions; strengthen medical information sharing and exchange; promote the exchange of diagnostic and treatment experience, technology, and treatment plans; and improve medical quality and efficiency.

Enhancing medical service capabilities: Medical consortia and regional medical centers can improve medical service capabilities and levels by integrating medical

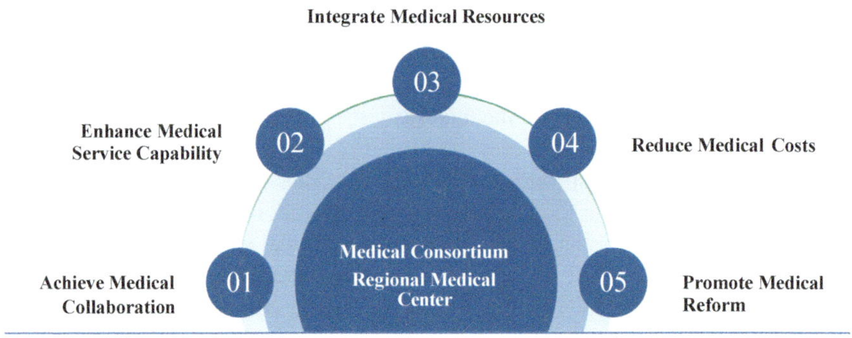

Fig. 5.15 Main roles of medical consortia and regional medical centers

resources, implementing graded diagnosis and treatment, and facilitating the professional division of labor, thereby providing patients with better medical services at their doorstep.

Integrating medical resources: Medical consortia and regional medical centers can integrate medical resources in the region, including talent, equipment, and drugs, to avoid the waste and duplication of medical resources and improve resource utilization.

Reducing medical costs: Medical alliances and regional medical centers can effectively reduce medical costs and alleviate the patient burden by optimizing the allocation and management of medical resources.

Promoting medical reform: Medical alliances and regional medical centers are important measures for medical reform and are capable of driving innovations and improvements in medical service models, management models, operational mechanisms, and medical insurance payment methods.

In March 2023, the General Office of the Communist Party of China Central Committee and the General Office of the State Council issued the "Opinions on Further Improving the Medical and Health Service System" (hereinafter referred to as the "Opinions"), proposing the midterm goal of building a "resilient integrated medical and health service system" by 2035 and emphasizing the "important supporting role of informatization in ensuring the efficient operation of the medical and health service system". The "Opinions" provide a series of policy guarantees for eliminating "information islands" and promoting the integrated development of regional medical informatization, including the construction of smart hospitals, medical alliances, and regional medical centers; strengthening the urban and rural primary medical and health service network; and building internet platforms in the medical field.

As an auxiliary means for the construction of medical alliances and regional medical centers, informatization is an important means of implementing resource integration and information sharing. The information platform of medical alliances and regional medical centers, supported by information technology and centered on resident health, explores system applications such as two-way referrals, telemedicine, mutual recognition of test results, chronic disease management, and remote medical treatment. Through online and offline business collaboration between medical institutions, the service chain within the medical alliance is extended and broadened, achieving continuous and integrated medical services within the region.

Scientific and Technological Development Empowers the Construction of Smart Health Care

The information infrastructure supports the rapid development of the digital economy. The state attaches great importance to the development of information infrastructure. It proposes to accelerate the construction of new infrastructure; strengthen the strategic layout; accelerate the construction of high-speed ubiquitous, integrated digital information infrastructure that is intelligent, agile, green, low-carbon, secure and controllable; and unblock the "main artery" of economic and social development.

As an important application in urban strategic planning with respect to people's livelihoods, smart health care integrates multiple new technologies, which are conducive to further enhancing medical technical capabilities and medical quality and effectively addressing current medical pain points and difficulties, and is a future development trend.

The intelligent upgrade of smart health care is inevitable for the digital transformation of health care. The beneficiaries of smart health care are not only patients but also hospitals and governments, which also benefit from digital transformation. New technologies can further penetrate the core of health care, have a profound impact on clinical and scientific research fields, and contribute value to hospital management, especially information management.

Application of Artificial Intelligence in Smart Health Care

In recent years, the development of artificial intelligence technology has made rapid progress, and artificial intelligence technology has entered people's lives with unstoppable momentum. With the elevation of the artificial intelligence industry to a national strategy, various ministries, provinces and cities have successively introduced related policies, and the overall artificial intelligence-related industry has shown a rapid growth trend, gradually expanding to smart health care and other application scenarios.

Application of Artificial Intelligence in Clinical Medical Diagnosis: Artificial intelligence is often used in clinical medical diagnosis for medical expert systems. It uses the design principles and methods of expert systems to simulate the thinking process of medical experts in diagnosing and treating diseases. It can help doctors solve complex medical problems, serve as an auxiliary tool for doctors' diagnosis, and carry forward valuable theories and rich clinical experience from medical experts. In general, artificial intelligence can fulfill the following roles in the medical field:

It can provide doctors with complete and effective information, thereby providing a scientific and reliable basis for the diagnosis and treatment of diseases.

It can greatly improve the automation of the measurement and analysis process of medical data, thereby greatly increasing work speed, reducing human work intensity, and reducing subjectivity.

It can concentrate expert knowledge and assist doctors in making more reliable and correct diagnoses. With increases in the number of cases, it can also enrich the system's knowledge, automatically or manually accumulate and analyze knowledge, and improve medical quality. It can discover patterns and knowledge from large-scale historical medical data, thereby providing decision support for future disease prevention and control.

Structured Video Analysis, Assisting the Construction of Safe Hospitals: In the context of significant imbalances in the supply of medical resources and personnel allocation, artificial intelligence, with its intelligent and automated features, has been applied in various medical scenarios, such as medical imaging, drug research and development, and hospital management. It can assist in improving the efficiency of hospital diagnosis and treatment and the level of operational management, thereby alleviating the problem of insufficient medical resources in China to some extent. Structured video analysis has a very wide range of applications in all aspects of hospital management. In the event of an abnormal incident, intelligent analysis can provide real-time, intelligent monitoring and alarm reminders, supporting users in timely decision-making and correct action. The principle of structured analysis technology is shown in Fig. 5.16.

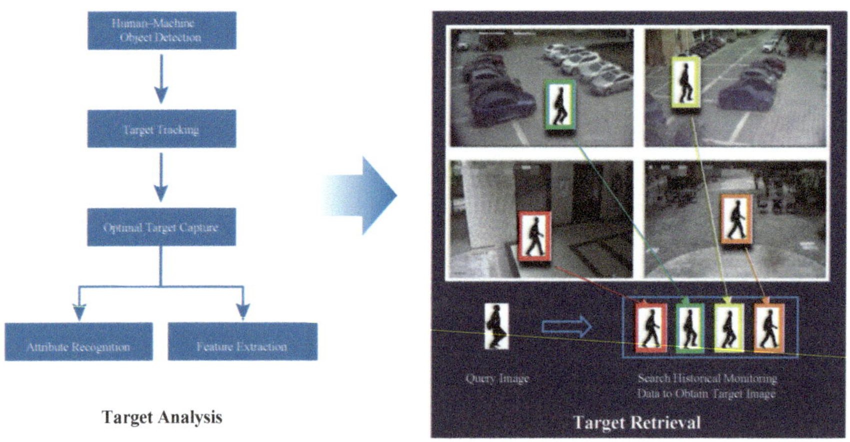

Fig. 5.16 Principles of structured analysis technology

In structured video analysis technology, edge intelligence analysis is an important component. Edge intelligence analysis refers mainly to endowing edge nodes with more perceptual computing capabilities, enabling edge nodes to collect and transmit scene content and generate structured data. For example, edge nodes can provide agile and timely warnings for high-risk population control; in complex scenes with large changes in light, edge nodes can autonomously adjust exposure, fill light, and other conditions to accurately detect targets, preserving more feature details for lossless front-end modeling. In addition to agile and precise data collection, edge intelligence can greatly reduce network bandwidth pressure and latency, reserving network bandwidth for large-scale networked intelligence.

Application of Big Data in Smart Health Care

With the rapid development and widespread application of big data technology in recent years, a variety of innovative medical solutions have emerged in the smart health care industry. Modern medicine, which relies on advanced scientific and technological advantages, has the basic characteristics of intelligence, interconnection, and convenience. The application of smart health care will facilitate functions such as telemedicine, platform appointments, and mobile medical information, bringing disruptive reforms to traditional medical service models. It will also provide more flexible and convenient channels for the functioning of medical services, ultimately providing an important impetus for the development of medical services in our country.

Digitalization of Medical Examination: Big data technology analysis of recent medical data shows that the clinical value of red cell distribution width (RDW) is very important in nonhematological diseases. Experts and scholars have conducted big data research on patients who die of cardiovascular diseases through years of follow-up surveys. The results show that the higher the RDW is, the higher the incidence and mortality of cardiovascular diseases in ordinary individuals, and the worse the prognosis of patients. Using big data analysis methods of smart health care to process biomarker data can effectively enhance the early warning and diagnostic functions of diseases, thereby achieving the goals of early diagnosis of diseases and improving long-term prognosis.

Application in Clinical Decision Support Systems: With the rapid development of internet health care and related industries, traditional medical information management systems can no longer meet the actual needs of refined medical management. How to build a medical cloud service platform for hospitals on the basis of existing traditional medical information systems and how to store, analyze, and mine medical big data to improve consultation efficiency and quality are the core goals of refined medical management at this stage. Traditional medical diagnosis in our country relies heavily on the individual professional level of medical staff, and it is difficult to avoid misdiagnosis because of human factors under this mode. Smart health care based on big data technology can assist in clinical decision-making, improve the accuracy of clinical decisions, reduce misdiagnosis rates, reduce medical costs, and provide better medical services to society in aspects such as preoperative diagnosis, effect evaluation, and prognosis prediction.

In addition, big data technology can be combined to perform precise analysis of physiological data and therapeutic effect data, construct personal health medical portraits, closely track the health status of citizens, and carry out more detailed health management, such as disease prevention and control and rational drug distribution.

Application of Cloud Computing in the Health Care Field

In traditional medical information systems, servers, networks, and storage are all maintained and used separately by different medical institutions. The services provided by medical information systems are also vastly different. These scattered systems cannot provide patients with effective medical guidance and information sharing, nor can they provide convenient statistical information. Cloud computing can integrate the information systems of different medical institutions to form a unified standard medical health information infrastructure. In addition to this infrastructure, integrated business services are provided, which will greatly improve the overall service level of the medical industry.

Computational resources can provide efficient support for pharmaceutical research and development. For example, the computation of targeted cancer drugs involves matching cancer cells with millions of proteins to identify those with high affinity. On the fastest single computer currently available, finding such a target could take several years or even decades of computational time, which is unacceptable for patients in urgent need of drug treatment. In the context of distributed computing in cloud computing, such tasks can be distributed to thousands or even hundreds of thousands of computers, each computing a portion of the protein matches. Thus, the computational results can be obtained within a day or even a few hours, meeting the needs of medical treatment.[3]

Application of the Internet of Things in Smart Health Care

In 2018, the Planning and Information Department and the Statistical Information Center of the National Health Commission released the "National Hospital Informatization Construction Standards and Specifications (Trial)", which clearly proposed the application of Internet of Things technology to support hospital management systems and clinical medical information systems. This issuance marked the beginning of the construction of IoT-enabled hospitals, with the ability to connect to the IoT being an important assessment and scoring criterion. The platform-based construction is the key evaluation object for application effects. It is clear that the adoption of IoT-connected devices has begun to impact the entire health care industry. The National Hospital Informatization Construction Standards and Specifications index system is shown in Fig. 5.17.

The deep application of IoT technology in the medical industry can effectively optimize the diagnosis and treatment process, refine asset management, conserve

[3] Bingyong Tang, Ting Xu, Rui Zhang. *Cloud Vision · Cloud Path: Evolution and Applications of Cloud Computing Technologies*. Beijing: China Machine Press, 2021: 119.

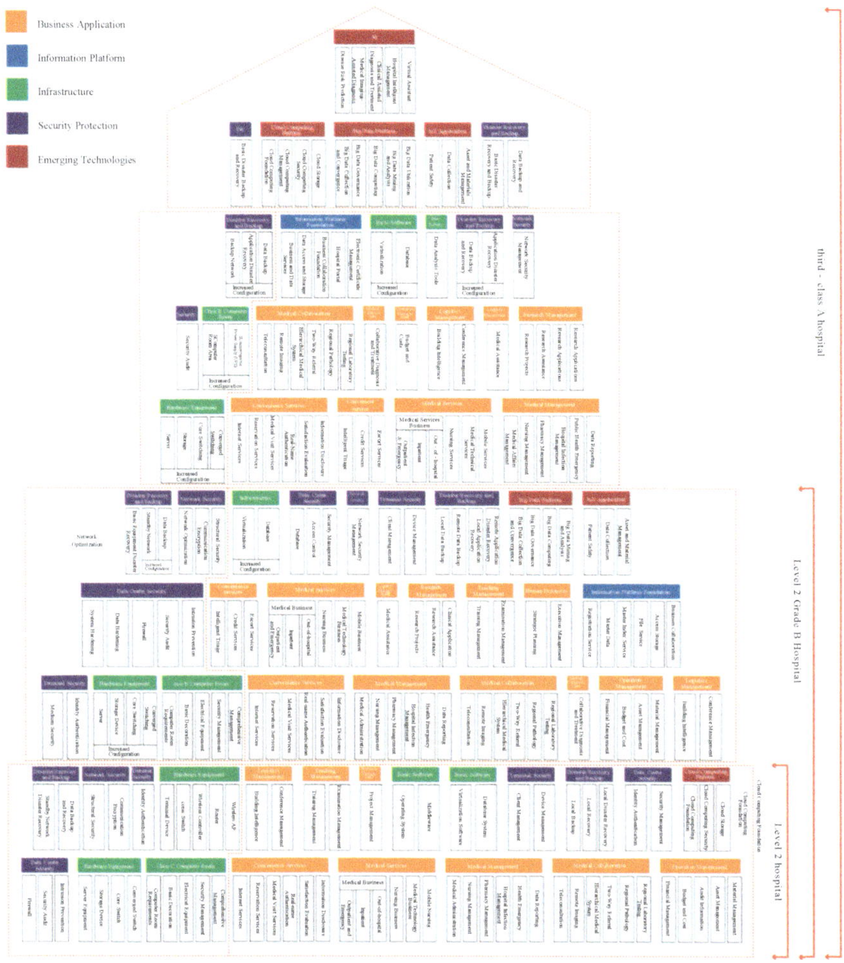

Fig. 5.17 National hospital informatization construction standards and specifications index system (Department of Planning and Information, National Health Commission; Center for Health Statistics and Information, National Health Commission. *National Standards and Guidelines for Hospital Informatization Construction (Trial)*, 2018.)

funds and reduce labor costs, help innovate diagnosis and treatment models, improve service efficiency, and better serve the health of the people. The application of IoT technology in the medical industry is shown in Fig. 5.18.

Application of the All-Optical Network in Smart Health Care

Requirements for "high-bandwidth" network construction: In recent years, the Ministry of Industry and Information Technology has deeply implemented the decision-making arrangements of the Central Committee of the Communist Party

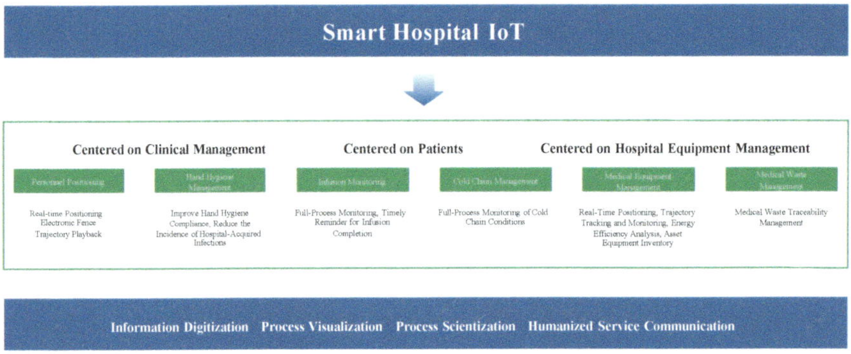

Fig. 5.18 Application of IoT technology in the medical industry

of China and the State Council. It has taken the construction of "Gigabit Cities" as an important focus, formed a joint effort to solve problems and promote development, and promoted the significant results in China's "Dual Gigabit" network development. The network coverage continues to expand, and 110 cities across the country have built "Gigabit Cities" that meet the construction standards of Gigabit Cities. In Gigabit Cities, the dual gigabit network composed of 5G and gigabit optical networks, owing to its characteristics of ultralarge bandwidth and ultralow latency, can drive the innovative development of various fields, such as the industrial internet, intelligent manufacturing, smart cities, rural revitalization, and cultural tourism. The stages of broadband development are shown in Fig. 5.19.

As an important part of smart cities, smart hospitals should continuously increase the depth and breadth of "dual gigabit" network coverage. Moreover, with the development needs of smart hospitals, various information management systems, such as hospital information management systems, electronic medical records, and medical imaging information systems, coexist, and each system network has high requirements. The all-optical network is an innovative all-optical network. Compared with

Fig. 5.19 Stages of broadband development

traditional LAN networks, it innovates system architecture and transmission media. It uses the key characteristics of all-optical network technology, such as large bandwidth, low latency, high security, and easy maintenance, fundamentally changes the hospital's network, and efficiently carries out the application construction of smart hospitals. With the trend of hospital intelligence and the development and promotion of all-optical network technology, all-optical network technology will have wider and deeper applications in the construction of smart hospitals and thus better improve the application level of smart hospitals and provide patients and medical staff with better services.

"Metaverse" Medical Application—New Diagnosis and Treatment, Medical Research

What will health care look like in the metaverse?

The future of health care may undergo an enormous transformation, but what will health care in the metaverse look like?

Patients do not need to register and queue in reality, and their virtual avatar in the metaverse can help with intelligent appointment and registration; some symptoms can be fed back to the doctor through the metaverse. If the symptoms are not serious, a diagnosis can be made directly and medication can be prescribed; if there are problems, an appointment can be made to visit the physical hospital in the real world.

Before surgery is performed on a patient, the patient's structure can be simulated from the inside out based on their data, clearly displaying each part and resolving blind spots in the field of vision. The entire surgical process can also be simulated preoperatively to avoid various problems and reduce surgical time, the incidence of complications, and radiation exposure. Before and after surgery, interactive 3D images can be used to explain issues to patients and their families, thus better alleviating their concerns.

Experts do not need to travel for consultations. They can consult with, examine and diagnose patients in the metaverse and guide local doctors in adjusting treatment plans, accelerating patient recovery.

The potential of health care in the metaverse may exceed our imagination. It will not only revolutionize current diagnosis and treatment but also promote improvements in medical education. All of these developments could happen in the future.

5.2.4 *Case Study 1 in Smart Health Care: Dongshan Hospital, Building a Trusted Smart Hospital for the Public*

Meizhou Dongshan Hospital is a large-scale comprehensive tertiary hospital designed and constructed by Gloryview Technology. The project is located in the

Fig. 5.20 Schematic of Dongshan hospital's smart hospital integrated management platform

central urban area of Meizhou. Dongshan Hospital is a large comprehensive hospital aimed at building a smart hospital and a tertiary hospital. Its digital construction started from scratch, from practical applications and economics, following the construction principle of "overall planning in one step, specific plans implemented in stages", combined with the expansion capability of Gloryview's digital base. Through continuous system upgrades and plan improvements, the development goal of the smart hospital was finally achieved. A schematic of Dongshan Hospital's smart hospital integrated management platform is shown in Fig. 5.20.

Establishing a Comprehensive Patient-Centered Medical Experience

Patients are always the service objects of the hospital, so the ultimate goal of smart health care is to provide better medical services and experiences for patients. Owing to the imbalance between the number of hospitals, the number of medical staff, and the total population, the number of patients visiting the hospital daily is large. In addition, the growth of the hospital area, the increase in departments, and the complexity of the layout lead to more prominent problems, such as the "three longs and one short" problem of long registration times, long waiting times, long medication collection times, and short consultation times. Dongshan Hospital provides a better medical experience to patients through the following means:

Optimization of the Medical Process

According to statistics, approximately one-third of daily outpatient patients visit the consultation desk or randomly find medical staff to consult about location issues, and over 90% of medical workers have encountered random patients asking for directions. Although most hospitals have set up multiple consultation points, the pressure load of the consultation desk is still overloaded, and medical staff waste considerable time and energy being repeatedly asked for directions, leading to the waste of medical resources. The traditional manual calling, guiding, and triage process is inefficient, and the fixed guidance signs in the hospital cannot meet the needs of different patient groups. If patients cannot be diverted and guided in a modern, informative, and effective way, it will be impossible to alleviate the inconvenience of the patient experience and reduce the unnecessary burden on medical workers.

Dongshan Hospital has built an intelligent navigation system within the hospital. Specifically, it has achieved the interconnection and linkage of medical data and map navigation data (as shown in Fig. 5.21), providing patients with online appointment registration, hospital navigation, self-service guidance, department navigation, self-service payments, examination/medication navigation, report printing, and other full-process intelligent guidance services for medical treatment. It has moreover developed a full-coverage navigation system for patients from home to hospital, from the parking lot to the outpatient clinic area, from the outpatient clinic area to the medical auxiliary area, from the outpatient area to the inpatient area, and from the medical area to the administrative service area, so that patients no longer need to ask for directions and no longer take the wrong path, thus truly making information more efficient and provide patients with more accessible information Moreover, when patients have multiple target locations, data analysis technology is used to plan a reasonable process and route for patients, reducing patient queues and wait times. It also extends the hospital service window, effectively diverts and distributes, improves the hospital's informatization image, improves service quality, reduces the consultation volume of medical guidance work, and reduces the rate of doctor–patient conflicts.

Optimization of Inpatient Experience

As a large-scale tertiary hospital, in addition to the large patient visit flow, the daily number of inpatients is another focus of hospital work.

The bedside care system provides a reliable and effective means of communication between medical staff and inpatients. Through the screen status of the device and the combination of sound and light, the patient's needs are accurately conveyed to the medical staff. This feature improves the work efficiency of medical staff, enabling them to better serve patients with the assistance of the system.

Inpatients can make calls and use intercoms between the bedside display screen and the call handle and the nurse station telephone, make emergency calls in the bathroom, and display prompts on the door and corridor equipment. In addition to

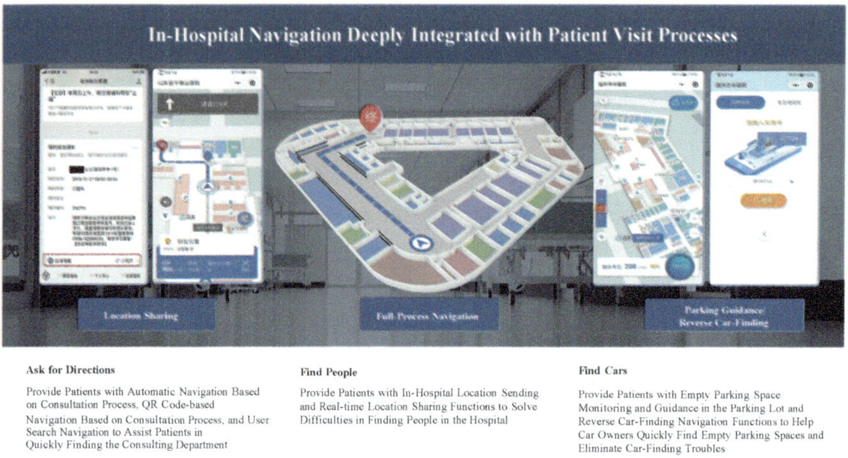

Fig. 5.21 Combination of medical process and in-hospital navigation

the telephone at the nurse station, there is a patient list, which can check the call reminder situation of each ward in real time.

In addition to basic medical intercom functions, bedside terminals can provide intelligent health education and facilitate medical arrangement inquiries, cost inquiries and settlements, bedside nutrition ordering, life entertainment services, remote visitations, orders, information inquiry, and inpatient payment functions.

The ward infusion monitoring system can automatically detect the infusion process, automatically change the liquid halfway, automatically close the infusion pipeline after the infusion is completed, and display and update the infusion status in real time. This system thus improves the patient's medical experience and prevents medical accidents and medical disputes during the infusion process.

Providing a Good Office Experience for Medical Staff

Improving the work efficiency of medical staff and ensuring their personal health are also important evaluation aspects of the smart medical mechanism. Dongshan Hospital provides medical staff with a highly efficient and stable network experience and an intelligent security system.

Efficient and Stable Network Experience

Within the Dongshan Hospital, an F5G all-optical network has been built, which provides a stable network foundation for the efficient medical diagnosis work of medical staff.

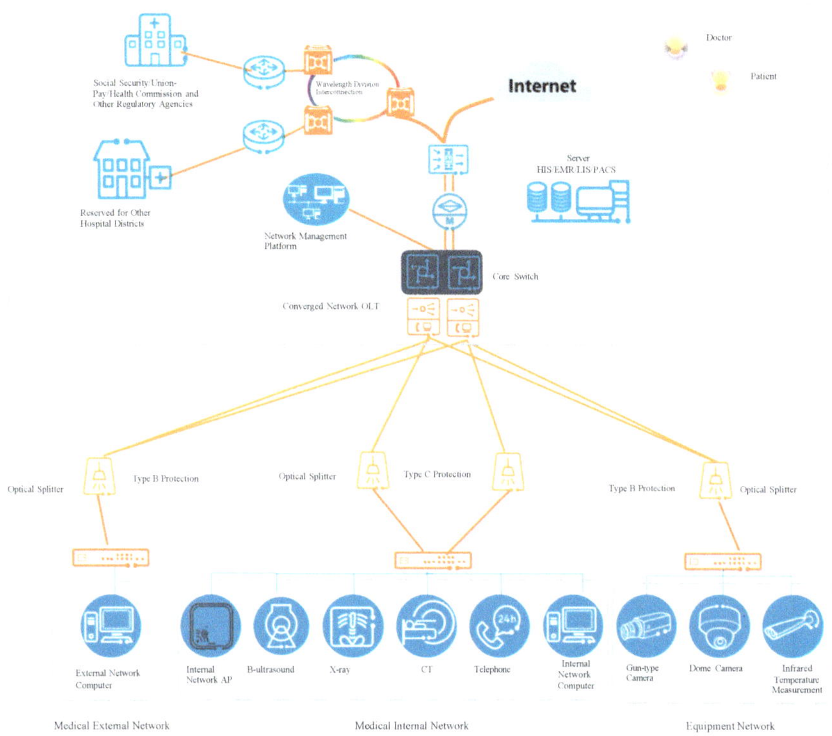

Fig. 5.22 Schematic diagram of the F5G all-optical hospital network logical architecture

F5G All-Optical Hospital Network Logical Architecture

Dongshan Hospital's all-optical network adopts a simple network architecture of IP + POL (passive optical local area network). The logical architecture is divided into four layers: the export layer, the core layer, the convergence layer, and the access layer, as shown in Fig. 5.22.

Providing Efficient Operation Management Methods for Hospital Managers

New situations and medical reforms force hospitals to move from extensive scale expansion development models to refined quality and beneficial development paths. Refined management for hospitals is a management concept that highly fits their work requirements. Refined hospital management refers to a management process that revolves around the core quality of people and runs through the entire medical system of the hospital. For hospital managers, meticulous management of people, vehicles, and objects in the hospital area; interconnection and data sharing between

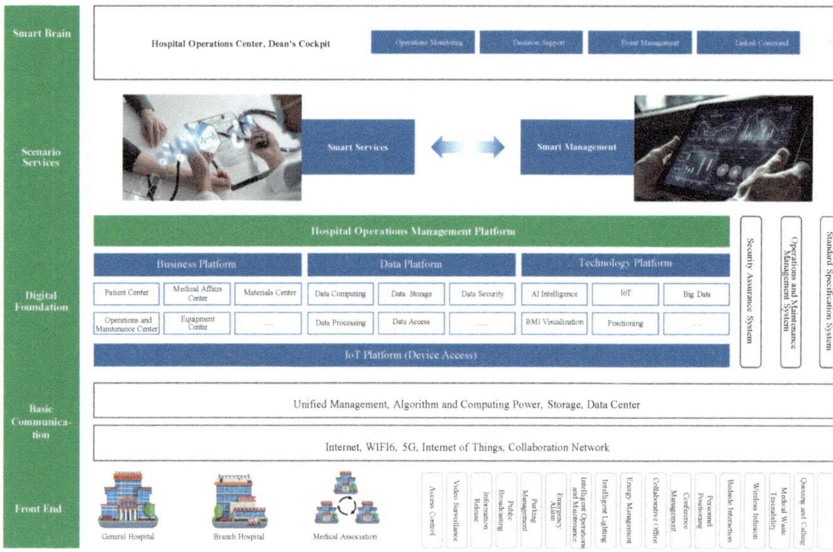

Fig. 5.23 Smart hospital operations management platform

various medical systems; and real-time control of various operational data in the hospital are prerequisites for achieving refined hospital management.

Dongshan Hospital has built a smart hospital operation management platform (as shown in Fig. 5.23) based on the Macroview IoT platform (ProBase) to achieve unified support management of front-end smart devices and meet the application requirements of various scenarios.

Unified Access Control of Front-End Devices Based on the IoT Platform

The IoT platform (ProBase) converges various intelligent subsystems and sensor devices of Dongshan Hospital and integrates intelligent gateways, sensor monitoring devices, video surveillance devices, intelligent control devices, medical special devices, etc., into platform management. The platform supports the smart hospital operation management platform through a unified OpenAPI interface, realizing the intelligent management requirements of "data full integration, status full visibility, business full management, event full control" for various application scenarios.

All-Around Management of Hospital Operations

The Smart Hospital Operation Management Platform is a comprehensive management platform based on information technology and hospital business processes. It is an important part of the construction of Dongshan Smart Hospital and the core of future hospital operations. This platform can help hospital managers achieve comprehensive control and management of all aspects of hospital operations, including medical services, hospital resources, personnel management, equipment management, etc. The Smart Hospital Operation Management Platform is highly important for improving hospital operation efficiency, optimizing resource allocation, and improving the quality of medical services, etc., and is an important part of the digital transformation of hospitals.

The Smart Hospital Operation Management Platform provides a unified management cockpit for managers at all levels of the hospital, intuitively displaying a multidimensional, multilevel overview of the hospital's situation. The operation management platform integrates the hospital's dispersed intelligent subsystems and important equipment, collects important operating parameters and alarm information in real time, links alarm handling with professional service management processes, and forms a closed-loop management process. Any alarm and handling of abnormal situations has a complete supervision process, with full records that can be traced back, and the handling process and results of important faults are reported to management. For example, hospital directors and logistics management personnel can learn the number of patients in the hospital, bed data, security alarms, energy consumption data, asset equipment distribution, equipment operation status, etc., all of which can be intuitively displayed on a large screen through a mobile app. When the equipment is running abnormally, the system sends an alarm message to the hospital director and logistics management personnel, automatically generates a work order, and distributes it to the operation and maintenance personnel to check the fault situation and complete the repair process. Once the repair is completed, the entire repair process is recorded. Through the Smart Hospital Operation Management Platform, the overall management efficiency of hospitals improves, and logistics management personnel can be streamlined and optimized, thereby reducing operating costs.

5.2.5 *Smart Health Care Practice Case 2: Bijie Third People's Hospital, the Keystone and Support in the Medical Consortium*

The new project of the Third People's Hospital of Bijie City, Guizhou Province, is located in Jinhai Lake New District of Bijie City. The construction project includes an outpatient medical technology building, an infectious disease inpatient building, a general inpatient building, and an administrative building, with a total of 1000 beds in the hospital.

Since June 2016, Guangzhou city has carried out assistance and cooperation work with Bijie, and with the Third Hospital of Bijie as the fulcrum, it has promoted the sinking of medical resources and expanded the scope of assistance services. With the assistance of the Guangdong Shengda Medical Group, the Cardiovascular Treatment Center of the Third Hospital of Bijie was established, encompassing 202 grassroots hospitals basically covering the grassroots medical institutions in Bijie City, forming a close medical consortium, implementing high-quality medical resources at the grassroots level, and establishing the "Bijie City Remote ECG Network System". With the Bijie City Cardiovascular Specialist Alliance as the center, a remote ECG network system connecting various county and town hospitals in Bijie city has been established, promoting the connection of 23 township hospitals (including village health rooms) to build a remote ECG system, forming a remote ECG emergency network system in Bijie city, and, with the help of the remote medical expert background of the Guangdong Shengda Medical Group, carrying out the diagnosis and treatment of cardiovascular diseases, team building and training of doctors, and "Internet + remote medical" work. The Third Hospital of Bijie plays an important role in the medical consortium and is the keystone and an important support for the construction of the medical consortium cooperation network. In addition, Gloryview participated in the construction of the new hospital. The main construction results of the project are as follows:

Provides good information technology support capabilities. The project builds a stable, secure, and high-speed internal and external network, enabling smooth interconnection and data sharing between various departments within the hospital and various basic hospitals while ensuring network security and preventing various network attacks and data leaks.

Establishes a highly reliable and secure primary and backup data center. The data center centrally stores various medical data and conducts deep mining and application through cloud computing, big data analysis, artificial intelligence, and other technical means. A schematic diagram of the primary and backup redundant computer rooms is shown in Fig. 5.24.

Data security and privacy protection. In the medical consortium, patients' personal identities and health information need to be strictly protected. Therefore, the hospital information system must have high security and privacy protection capabilities to prevent data leakage and information theft. This project adopts the Grade 2.0 Level 3 standard to build an information network security system to ensure the safe operation

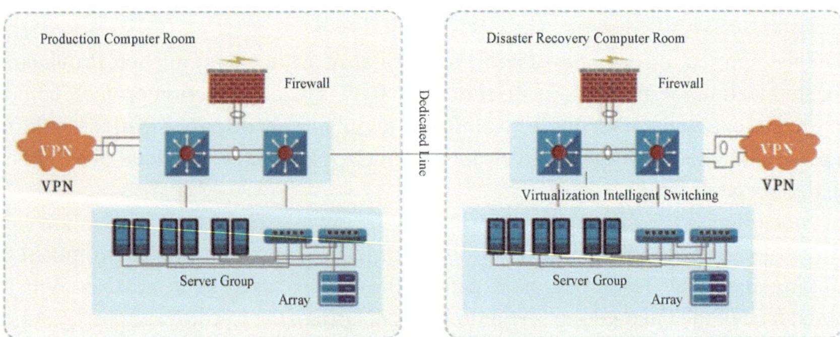

Fig. 5.24 Schematic diagram of the primary and backup redundant computer room

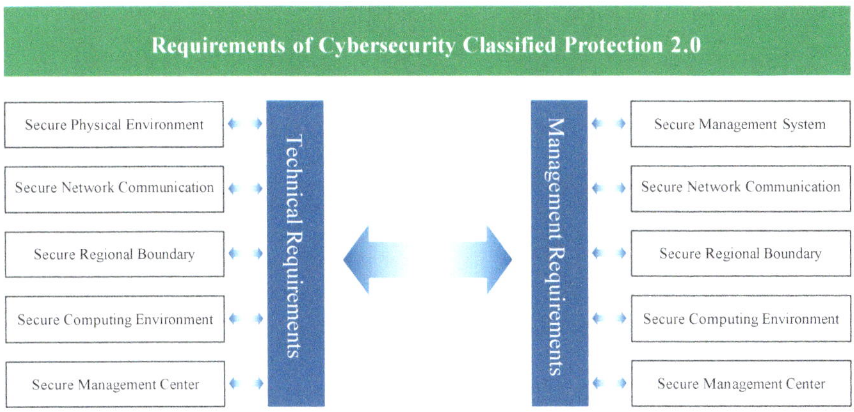

Fig. 5.25 Grade 2.0 technical management requirements

of the information system. The technical management requirements of Grade 2.0 are shown in Fig. 5.25.

Unified standards and interfaces: To ensure the correctness and reliability of information exchange and data sharing, the project follows unified information standards and interface specifications to ensure that data between different systems can be transmitted and interpreted.

Today, the digitalization and intelligentization process of the medical industry is accelerating, and the new generation of digital technology is driving the in-depth development of the medical industry, which is promoting the digital transformation and intelligent upgrades of the industry. Smart hospitals highly integrate the physical space and digital space of the hospital, use digitalization to reconstruct and empower the physical space, and make the hospital area "smart". According to the data in the "New Era Smart Hospital White Paper" by Johnson Controls, with the cooperation of systems such as online registration, automatic laboratory allocation, and electronic transmission of test results, the time for patients to obtain test reports can be shortened from 3 to 1 h. Dongshan Hospital and Bijie Third People's Hospital have effectively improved the medical experience and service level through digital and intelligent construction.

For patients, a smart hospital means a better medical experience; for medical staff, it means a safer and more comfortable working environment; for hospital administrators, it means managing the hospital with strategic planning and rational decision-making, thereby achieving refined management and precise services. With the rapid development of modern technology, hospitals are no longer synonymous with "disease" but are important places for physical and mental care. Gloryview Technology has consistently empowered the construction of smart hospitals, helping to build digital, green, and ecologically beautiful hospitals.

5.3 Smart Education: Promoting Digitalization and Supporting High-Quality Education Development

Education is a major focus of planning for the country and the party. The comprehensive advancement of the great rejuvenation of the Chinese nation cannot be separated from the strong support of education. With the unremitting efforts of the whole country, in the past decade, China's education has achieved great achievements. As of 2022, there were a total of 518,500 schools of all levels and types nationwide, with 293 million students and 18.8036 million full-time teachers, and the average number of years of education for the new labor force has reached 14 years. The level of education popularization has reached or exceeded the average level of middle- and high-income countries, among which the level of compulsory education has reached the average level of high-income countries worldwide, and higher education has achieved a historic leap from massification to universalization.

General Secretary Xi Jinping noted in the report of the 20th National Congress of the Party: "We must adhere to the principle that science and technology is the primary productive force, talent is the primary resource, and innovation is the primary driving force. We must deeply implement the strategies of rejuvenating the country through science and education, strengthening the country through talent, and driving development through innovation, opening up new fields and tracks for development, and continuously shaping new momentum and advantages for development." The report of the 20th National Congress of the Party clearly stated that "education, science and technology, and talent are the basic and strategic supports for the comprehensive construction of a modern socialist country."

5.3.1 Problems Encountered in High-Quality Education Development in the New Era

Under the leadership of the Party and the state, with the continuous advancement of national education reform and the education governance system, the level and quality of basic education in China have been continuously improved, the standardization of education management has been enhanced, the sense of gain and happiness of people's education has been further improved, and the fairness, quality and efficiency of basic education have been further guaranteed and improved. However, some problems have gradually emerged in the high-quality development of basic education in China, such as the dramatic increase in education anxiety, the uneven distribution of education resources, the imperfect education evaluation system, the need for improvements in education quality, and campus safety issues (as shown in Fig. 5.26), which have affected the progress and level of basic education development to an extent. In addition, these problems have profoundly impacted the comprehensive reform of education in China and improvements in the basic education governance system. To promote the high-quality development of basic education and enhance

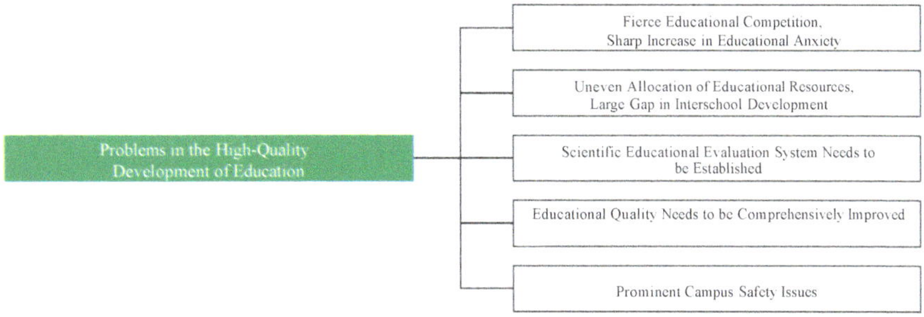

Fig. 5.26 Problems encountered in high-quality educational development

the function of basic education in serving national and social economic construction, it is necessary to pay attention to and solve prominent problems in the high-quality development of basic education in China.

Intense Educational Competition and Dramatic Increase in Education Anxiety

In recent years, many parents have complained about the heavy burden of education. They believe that the cost of education is high and often feel anxious about education. At the primary and junior high school stages, many parents have already set their sights on the entrance exams for junior high school. Although junior high school is part of compulsory education, the entrance exam after three years of junior high school is selective. As the ratio of general to vocational education is approximately equal, parents dare not relax their children's studies. Moreover, because most parents regard general high school as the goal of further education, the competition for entrance to high school is extremely fierce; parents' anxiety also leads children to enter the intensive practice stage prematurely, which adds extra homework, enrolls students in various tutoring classes, and causes them to develop education anxiety. Under the influence of parents' excessive anxiety and actions, students usually experience excessive homework loads and training, experience education anxiety prematurely, and even develop a dislike for learning, making it difficult to experience the joy of learning. In addition, the quality of students is difficult to improve comprehensively. Finally, teachers also develop a certain degree of education anxiety. Given the expectation of parents' academic achievements and the awareness of social education competition, excessive external utilitarian factors also intensify teachers' education anxiety, and teaching is disturbed to a certain extent.

Uneven Distribution of Educational Resources and Large Disparities in School Development

Owing to disparities in socioeconomic development, there are significant differences in the levels of economic development and the allocation of educational resources in the eastern, central, and western regions of China. Furthermore, there is a substantial educational gap between urban and rural areas. The majority of China's basic education funding is concentrated in urban areas, with high-quality educational resources, whether in terms of teaching staff or infrastructure investment, typically favoring large cities. The educational gap between urban and rural areas is very apparent, and even within the same city, imbalances in educational resources often occur. Moreover, with the continuous advancement of the new curriculum reform and the implementation of policies such as the reform of basic education evaluation, many rural schools not only lack hardware resources but also face an outflow of excellent teachers. The structural shortage of teachers in rural schools remains a significant issue. This issue has hindered the effective implementation of the new curriculum reform concept, leading to large disparities in school development and restricting the high-quality development and reform of the education model in basic education.

A Scientific Education Evaluation System Awaits Construction

Currently, China's basic education evaluation still overly emphasizes the selective function of education evaluation. To some extent, this orientation alienates the purpose of China's basic education, is not conducive to focusing on and promoting the development of all students, and it is not beneficial to the comprehensive and individual development of adolescents. Second, the effect of educational informatization in promoting the construction of the education evaluation system is not prominent enough. Advancing the construction of the education evaluation system through educational informatization and big data construction remains an important problem to be solved.

Educational Quality Requires Comprehensive Improvement

Educational quality is a decisive factor in school development, and current basic education still presents issues of low quality in its development process. To achieve high-quality development, basic education schools must earnestly work to improve education quality. However, in the process of education reform and development, the teaching quality of some basic education schools is not high, which has affected the school's reputation and student enrollment choices. There is a serious shortage of students in high-quality schools, whereas ordinary schools lack appeal. These issues have affected the school's educational benefits, and there is a significant gap in establishing a high-quality basic education system.

Campus Safety Issues

Students are the future of the nation, and campus safety is a barrier to protect student safety. Maintaining campus safety is a common consensus among all sectors of society. However, various types of campus safety accidents have occurred in recent years, such as food safety accidents, traffic safety accidents, teaching facility safety accidents, fights, deaths by drowning, campus violence, and suicides, which have sounded alarms indicating the need for campus safety improvements. Moreover, many loopholes and weak links exist in school safety management. Ensuring a safe and comfortable teaching environment for teachers and students in schools is also an important part of high-quality educational development.

5.3.2 The Transformation to Intelligent Education Is the Key to High-Quality Educational Development

Thus, many problems still need to be solved to achieve the high-quality development of China's education. However, in recent years, with the strong support of national policies and the rapid development of information technologies such as artificial intelligence, big data, the Internet of Things, and AR/VR, an increasing number of high-tech technologies have been applied to education. For example, through information network cloud technology, students can share high-quality education resources; schools can apply AR and VR teaching enhance classroom vitality and stimulate students' interest in learning; big data can support more targeted teaching; and artificial intelligence can ensure campus safety. The benefits of the transformation to intelligent education are obvious, and intelligent education has become a powerful tool to solve the problems encountered in high-quality educational development.

At present, China's education industry has ushered in a large-scale development period of intelligent transformation. It has also become an important goal and practical area of focus for the achievement of educational informatization in China. Under the support of national policies and technological development, intelligent education has constructed a new pattern for the high-quality development of China's education (as shown in Fig. 5.27).

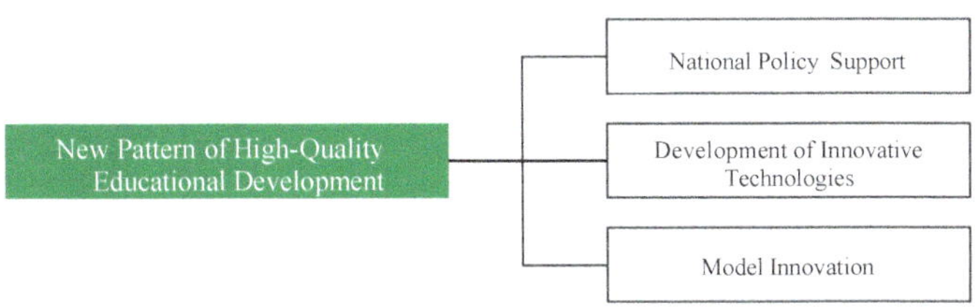

Fig. 5.27 New pattern of high-quality educational development

National Policies Promote Accelerated Transformation into Intelligent Education

To support the development of smart education, the government has continuously introduced a series of policies in recent years. In 2018, the launch of the Education Informatization 2.0 initiative marked a new stage in the practice of smart education, namely, the stage of smart education development promoted by the government. In April 2018, the Ministry of Education of the People's Republic of China (hereinafter referred to as the Ministry of Education) issued the "Education Informatization 2.0 Action Plan", which listed the "Smart Education Innovation Development Action" as one of the "Eight Major Actions" to promote education informatization. The Ministry of Education has successively identified 18 smart education demonstration zones, including Dongcheng District and Haidian District in Beijing, Yuncheng City in Shanxi Province, and Wuhou District and Chenghua District in Chengdu. With the support of local governments, these demonstration zones are coordinated by education administrative departments and related institutions, fully utilizing the market mechanism to provide personalized support and precise services for students, teachers, and parents using new generation information technology. They collect and utilize data on the status of participant groups and the process of education and teaching, promote learning at any time, any place, and any pace, provide a high-quality education supply with high learning experience, facilitate high content adaptation, and ensure high teaching efficiency to promote educational equity and improve educational quality.

In August 2022, to promote scientific literacy for all, the Ministry of Science and Technology, the Propaganda Department of the Central Committee of the Communist Party of China (hereinafter referred to as the Central Propaganda Department), and the China Association for Science and Technology (hereinafter referred to as the China Science Association) jointly compiled the "14th Five-Year Plan for the Popularization and Development of Science and Technology", which emphasizes the deep integration of science popularization and school education. It insists on initiating science popularization with children; building a progressive science education system for primary, junior high, and high school stages; and integrating in-school and out-of-school education. It systematically considers new scientific knowledge, new learning needs, and new teaching methods; increases the development of high-quality science education resources and excellent science popularization courses; effectively enhances scientific, systematic, appropriate, and interesting aspects; and enriches the content of science and technology education in primary and secondary schools. It strengthens science education and science popularization practices at the higher education stage and encourages and supports university teachers and students in performing science popularization social practice.

In February 2023, at the World Digital Education Conference cohosted by the Ministry of Education and the Chinese National Commission for UNESCO, the Chinese Academy of Educational Sciences officially released the "China Smart Education Blue Book (2022)" and the 2022 China Smart Education Development Index Report. The Blue Book, which highlighted the implications of smart education as the main line, proposes 16 specific features from four dimensions, namely, environment, teaching, governance, and talent; summarizes the experience of China's smart education development; and proposes seven topics and five initiatives that should be focused on in the future.

Innovative Technology Applications Scale Up and Open New Spaces for Smart Education

Smart education is a response to and concern with the educational transformation landscape in the digital age. The smart classroom, as a perceptible form of smart education enabled by various intelligent terminal technologies, has promoted changes in traditional teaching modes and transformed students from passive recipients to active learners and teachers from knowledge transmitters to learning organizers. It has accelerated the integration and application of intelligent terminal classroom teaching, and classroom research has begun to become systematic, gradually encompassing teaching objectives, teaching processes, teaching evaluations, etc. Under this driving force, smart education has gradually received attention in practice, and various innovative technology applications have scaled up, opening new spaces for smart education. Artificial intelligence can provide personalized learning plans and recommended courses for students through big data analysis, machine learning, and other technical means and can support automatic homework correction, answering questions, and perform other functions to improve learning efficiency. Virtual simulation technology can simulate complex experiments and scenes in the subject through computer technology, allowing students to practice in a virtual environment and improving learning quality and effects. Metaverse technology can combine virtual content with real scenes, providing students with a more intuitive and vivid learning experience, resulting in better educational effects.

Application of Artificial Intelligence in Smart Education

In July 2021, the General Office of the Communist Party of China Central Committee and the General Office of the State Council issued the "Opinions on Further Reducing the Homework Burden and Extracurricular Training Burden for Students in Compulsory Education" (hereinafter referred to as "Double Reduction"). This policy emphasizes that the quality and service level of school education and teaching should be further improved, homework assignments should be more scientific and reasonable, after-school services should meet student needs, extracurricular training institutions should be fully regulated, and various irregularities in subject-based extracurricular training should be eliminated. The new teaching model empowered by technology in smart education provides quality and efficiency improvements for the implementation of the "Double Reduction" policy:

Transforming digital resources into dynamic potential for precise teaching: In the smart education classroom, data technology can be used to capture and mine massive amounts of data in the teaching process in real time, providing teachers

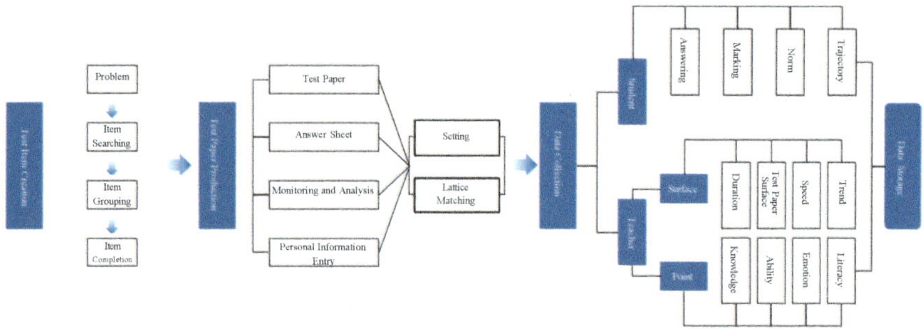

Fig. 5.28 Data-driven precision teaching process diagram

with insights into students' performance in various dimensions, such as cognition, behavior, knowledge mastery, and emotional motivation. Based on detailed research on learning situations, adjustments in teaching content, the reconstruction of teaching processes (as shown in Fig. 5.28), subject teaching is conducted in the best way to increase the quality of classroom teaching.

For example, by combining an intelligent dot matrix pen, MEC computing power, and AI intelligent analysis, process data such as answer notes, thinking trajectories, and exam data can be collected and integrated. This feature makes students' thinking visible, quickly identifies the reasons for students' loss of points, and achieves the goals of reducing the burden of teaching, improving quality, and increasing efficiency.

Empowering fine-grained, tiered homework to improve homework quality: Reducing the homework burden for students is an important manifestation of the "double reduction" policy. It requires schools not only to reduce the amount of homework but also to improve the quality and form of homework, truly performing the functions of homework assignment, diagnosis, consolidation, and learning situation analysis. AI technology and cloud computing can perform computational analysis of massive amounts of data and generate personalized homework and targeted exercises through the network platform system.

For example, a homework management system specifically set up for students' after-school homework. If the homework load exceeds the standard, a reminder is issued. After-school AI video teaching courses can be offered, exercises are pushed after the video ends. Once the student completes the exercises, the system pushes new wrong question resources, similar to a game, and the student progresses step by step. Each student answers the questions they can. Each question is "custom-made" for the student. Teachers can assign flexible and personalized homework to promote more precise and targeted homework to "control quantity and improve quality" and provide teaching guidance.

Application of Cloud Computing in Smart Education

The application of cloud computing technology in the field of smart city education can enrich educational resources, strengthen the management of student learning, and thus improve learning efficiency and teaching quality. Specifically, the application of cloud computing technology in the field of educational services can achieve the following goals:

First, a cloud education learning platform can be created, learning resources can be gathered, and teaching modes can be enriched. A city-wide education cloud learning platform can be built, and high-quality educational resources in city schools, such as lecture videos of famous teachers in various subjects, excellent teaching plans, and test question sets, can be collected to expand and share educational resources and improve educational fairness. Moreover, the cloud education learning platform can provide students with all-weather, various learning services, including audio and video on-demand, intelligent Q&A, and cloud live broadcast. With the cloud learning platform, students can completely overcome the limitations of traditional classrooms, no longer be restricted by space and time, and can arrange their learning time according to their own learning needs.

Second, student learning behavior can be analyzed, and targeted teaching services can be provided. The cloud learning platform can record and summarize students' learning data and exam data, mine the patterns and characteristics of student learning, and grasp the learning situation of students in a timely manner. As a result, the platform can then formulate personalized learning plans for students, thus enhancing the pertinence and effectiveness of student learning.

Third, the teaching ability and level of teachers can be improved based on student feedback. Using the cloud learning platform to collect and statistically analyze various learning information of students, including learning behavior and learning performance, each student's problem can be scientifically diagnosed in a timely manner, and teachers can identify their own shortcomings in combination with the diagnosis results, optimize teaching plans, improve teaching methods, enrich teaching methods, and thus achieve high-quality teaching goals with high efficiency.

In addition, the use of cloud computing technology to analyze various types of educational data comprehensively, including the distribution of educational funds, the distribution of educational institutions, and the situations of student enrollment, withdrawal, and transfer, can provide a scientific and reliable basis for educational management. The overall architecture of smart education cloud services is shown in Fig. 5.29.

Application of Digital Twins in Smart Education

Through the application of "cloud + AI + digital twin" and other technologies, a smart campus based on digital twins, can achieve a full view of the screen and manage the entire school with one network. The safety situation of the campus and dormitories can be viewed in real time; the usage of public classrooms, conference rooms, laboratories and other spaces is clearly visible; the statistics, abnormal situations and disposal of water and electricity use in various places are clear at a glance; and the construction of an academic style and scientific research achievements are fully apparent. The "people, places, things, and events" on campus are fully

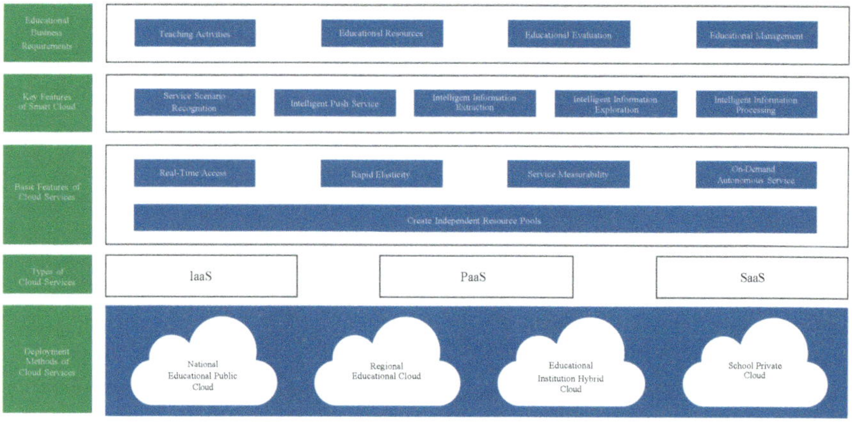

Fig. 5.29 Overall architecture of smart education cloud services

controlled, and the functions of routine operation monitoring, the analysis and judgment of various data indicators, and coordinated disposal in emergency situations can provide intelligent and efficient support for campus managers.

Application of the "Metaverse" in Smart Education

"Metaverse" Education—Immersive teaching refers to the use of virtual reality technology and other technical means to place students in a realistic virtual environment for comprehensive learning and experience. Compared with traditional teaching methods, immersive teaching is more intuitive and vivid, helping students better understand and grasp the learning content, thereby improving learning outcomes.

The impact of the metaverse on the education industry will be profound, which can be seen from the following aspects:

Providing a more vivid and intuitive learning experience: The metaverse can create various realistic virtual scenes through virtual reality, augmented reality, and other technical means, allowing students to immerse themselves in learning and experience, thereby enhancing the interest and appeal of learning.

Enhancing learning outcomes: The metaverse can provide various multimedia learning resources, such as 3D models, virtual laboratories, and gamified teaching, to help students better understand and grasp the learning content, thereby improving learning outcomes.

Realizing intelligent and personalized education: The metaverse can intelligently recommend learning resources and learning plans according to students' learning situations and needs through artificial intelligence and other technical means, thus achieving a more intelligent and personalized education model.

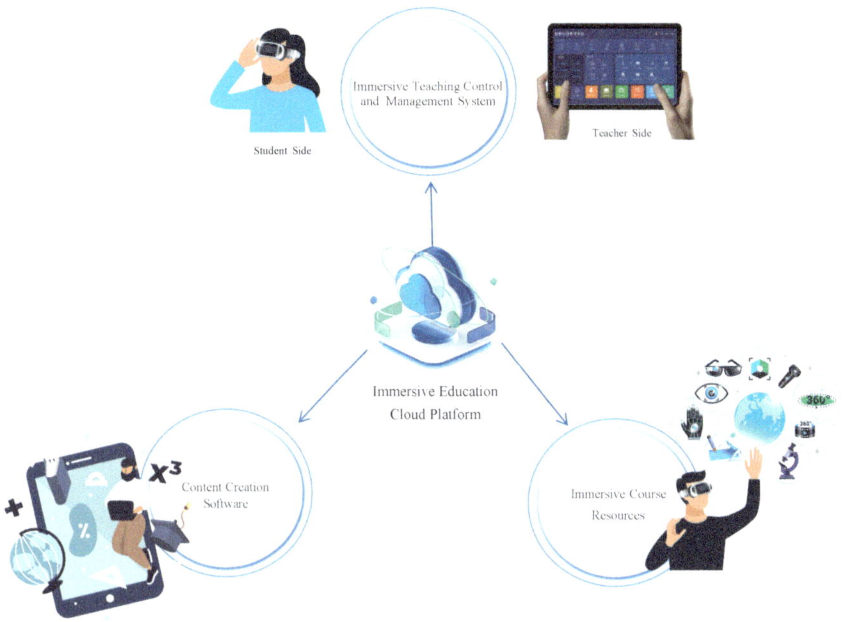

Fig. 5.30 Architecture of immersive teaching

Improving the teaching efficiency of teachers: The metaverse can provide various teaching tools and resources to help teachers improve their teaching, thus increasing teaching efficiency and teaching quality.

The architecture of immersive teaching is shown in Fig. 5.30.

In conclusion, smart education uses many innovative technologies that continuously drive the education industry forward. The application of smart education can not only improve education efficiency and quality but also meet the diverse learning needs of students, promoting the continuous upgrade and reform of the entire education field.

The Emergence of Innovative Models Sparks New Ideas in Smart Education

In 2022, the Shenzhen Municipal Government issued the "Implementation Plan for the Shenzhen Citywide Science Literacy Action Plan Outline (2022–2025)". The plan proposes promoting the deep integration of science popularization and school education; advocating for heuristic, inquiry-based, and open teaching; protecting students' curiosity; and stimulating their desire for knowledge and imagination. Additionally, it proposes strengthening the construction and equipment of primary and secondary school science education infrastructure; promoting the balanced development of science education activities and resources; promoting the deep integration of information technology and science education; and implementing scenario-based, experiential, and immersive learning. With the development of our country's education, teaching models are constantly changing, and various models have emerged, sparking new ideas in smart education.

Big Science Popularization Innovation Model: Led by the Big Science Plan, early scientific research practice content oriented toward future and unknown sciences can be established, with first-tier city main science museums as the core, county-level science museums as the main nodes, and schools as the network terminals of the teaching practice platform. By establishing standardized classrooms and curriculum systems, building diverse technology platforms, and supporting virtual and augmented reality technologies, the gap in school teaching resources can be filled, the early transformation of original creativity and imagination into scientific research capabilities can be realized, and national standards for youth science literacy can be established.

Personalized Teaching: Use artificial intelligence, big data, and other technical means to analyze and evaluate students' learning process and behavior, thereby implementing targeted personalized teaching plans and improving student performance.

Blended Teaching: Combine online and offline teaching activities organically, make full use of digital course resources and online learning platforms, and improve teaching outcomes through various methods such as lectures, practices, interactions, etc.

Cross-School Alliance Teaching: Use network technology to establish communication alliances between schools, facilitate resource sharing and communication cooperation between teachers and students from multiple schools, and promoting the integration and sharing of educational resources.

Practical Teaching: Through virtual simulation, experimental observation, social practice, metaverse, etc., combine knowledge and practice, thereby improving students' practical ability and application level.

5.3.3 Smart Education Practice Case: The First "Future School" in Eastern Guangdong—Hanwen Experimental School

The Hanwen Experimental School in Xiangqiao District, Chaozhou city (also known as Hanwen Academy), is a key construction project under the "Six Major Projects" for people's livelihoods in Chaozhou city. It was the first high-end private boarding school in the eastern Guangdong region to adopt the "academy system". It is also the first school in the eastern Guangdong region to be built according to the "Future Roadmap" Experimental School Development Guide 1.0, a standard jointly sponsored by the School Planning and Construction Development Center of the Ministry of Education and the Future School Research Institute, among others.

The Hanwen Experimental School aims to build a training system that embodies the characteristics of Chaozhou culture and the development of excellent traditional Chinese culture. It introduces the design concept of an open learning community, builds on the foundation of a "green campus, smart campus, and harmonious campus", and highlights the elements of Chaozhou "Han" culture and the academy system in campus design. It adopts a multidimensional integrated layout and meets the principles of internationalization, openness, compatibility, and diversity in functional design. It deeply applies artificial intelligence on campus. With the support of the Internet of Things platform and data platform, it provides efficient teaching, management, living services, and security guarantees. It aims to provide students with a high-quality and efficient learning experience, build a "future school" with more national characteristics in the eastern Guangdong region, inherit the culture of Chinese historical academies, operate schools with independent characteristics, innovate modern education mechanisms, and provide education that satisfies the people.

The construction of the smart campus of Hanwen Experimental School is based on data networks, with system integration and data analysis at its core. It uses regional networks, the internet, and Internet of Things technologies to build intelligent teaching, campus, and social environments. It exemplifies the deep integration of information technology and education; promotes system reconstruction and reengineering of teaching, teaching research, education management, and living services; improves the quality of education and teaching and the level of education management decision-making; and forms a new campus ecology of "full perception, full interconnection, and full intelligence". The main construction scope includes information infrastructure, smart campus information applications, and smart campus integrated management platforms.

The smart campus integrated management platform, supported by the self-developed Internet of Things platform and Macroview data platform, connects the data flow from north to south, creating a smart education environment, smart teaching resources, smart campus management, smart campus services, and other applications (as shown in Fig. 5.31).

As a newly established school, Hanwen Experimental School has advanced educational concepts and plans. It aims to fully integrate educational goals, teaching, research, management, and campus life through the use of information technologies such as the internet, cloud computing, and big data. It aims to realize intelligent campus management, integrated campus life, digital campus facilities, lively classroom teaching, seamless communication between home and school, and promote the integration of education and technology at Hanwen Academy, providing a high-quality, efficient living environment for every Hanwen student.

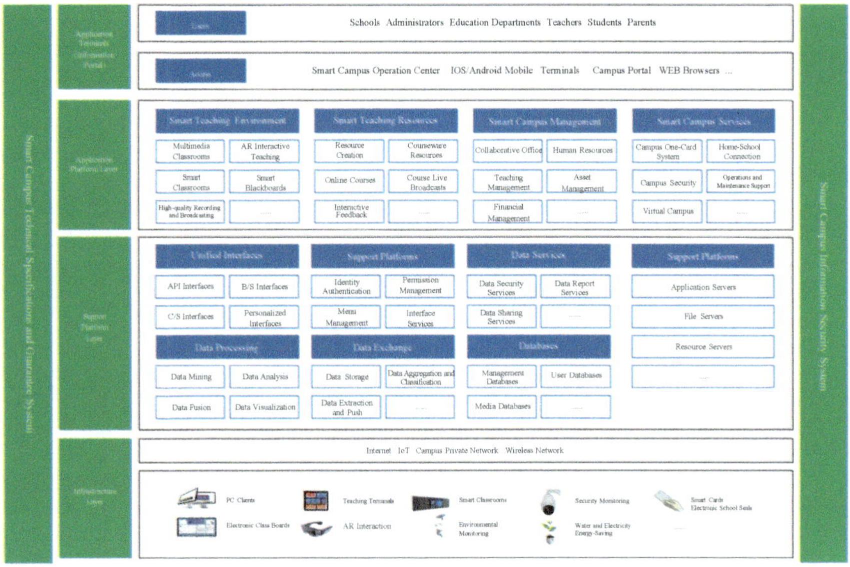

Fig. 5.31 Hanwen school smart campus integrated management platform

To promote the implementation of the school's planning, on the one hand, Hanwen Experimental School needs to establish a daily management mechanism for campus operations to ensure that the school's work can run efficiently, which is the foundation of the school's establishment. In addition, to achieve the goal of building a "future school", information services will be integrated into teaching, research, management, campus life, home–school communication, and other scenarios. Different segmented information systems will provide different services for the school. Therefore, it is necessary to further perform well in the data docking and unified management of various systems to ensure the sustainable development of information construction in schools. Focusing on the construction needs of the above two aspects, Gloryview Technology and Hanwen Experimental School have engaged in in-depth cooperation to jointly build the "future school".

Introducing a Smart Campus Integrated Management Platform Based on School Business Logic

A school is a system with various management departments and a clear division of labor. Each business link of each management department is related to the business links of other departments. The smart campus integrated management platform logically connects each business link of the school management department. The data generated at each link automatically flow to the next link that needs these data, which can greatly reduce the burden of repetitive work for administrative management teachers. Moreover, the introduction of the platform makes academic management, moral education management, daily office work, and home–school collaboration more relaxed and efficient.

For example, in academic administration, Hanwen Experimental School centers its approach on learners. In higher grades, it implements a system of flexible class scheduling for language, mathematics, and English classes. This means that while the subject classrooms and teachers remain consistent, students choose the class level that suits their own development based on their ability and interests. This design enhances students' initiative in learning, broadens their thinking, and gradually teaches them how to assess their own abilities accurately in the process of flexible class scheduling, thereby improving their core competencies. To carry out stratified education smoothly, the platform construction process involves one-on-one communication and in-depth discussions with school leaders to jointly develop a flexible class scheduling plan. This plan is then entered into the platform, and system algorithms ensure its feasibility. Moreover, the platform focuses on assisting school users in completing mixed scheduling of flexible classes and administrative classes, thereby optimizing the ratio of teacher resources and classroom resources to avoid waste. The platform system fully considers scheduling rules and flexible class scheduling needs. Using the platform's scheduling function, the academic affairs office can quickly generate a class schedule, teacher schedule, student schedule, and venue schedule, and the corresponding schedules can be synchronized in real time for teachers, students, and class signs.

Building a Broad Vision of a Self-developed Open and Expandable Underlying Platform to Create a Sustainable Digital Campus Ecosystem

The smart campus management platform has a flexible, open-ended Internet of Things platform and data platform, which support data docking and aggregation with third-party software and hardware such as class signs, access control, monitoring, and IoT control. This design can provide schools with a sustainable underlying platform for school informatization construction.

The platform integrates with class sign system data to determine student attendance, synchronizes classroom and student seat data, and outputs formatted schedule data, student file query data, class publicity data, etc.

It implements a single sign-on with the campus consumption system, obtains student-based consumption flow data, forms related statistical reports, and pushes notifications about abnormal consumption situations to teachers and parents.

It links with the access control management platform to associate basic student and teacher information; reads the original data of the access control process according to the attendance segments; outputs it to class files, grade files, teacher files, student files, and moral education management files; forms flow data; reports data; and notifies relevant personnel.

It synchronizes with the library management system platform's basic data, reads student borrowing data for specific time periods, integrates the data into the student's multidimensional files, and pushes information to students and parents.

Creating a Campus Hardware Infrastructure for "Modernized Teaching, Fun Classrooms, and the Normalized Use of Technology"

The hardware facilities of Hanwen Experimental School are configured with the goal of "modernizing teaching, making classrooms fun, and normalizing the use of technology". The external "electronic class sign" can integrate the management of daily teaching activities such as information release, moral education management, teacher–student mutual evaluation, schedule management, homework release, attendance management, and home–school messages. It can easily achieve home-school connectivity, thereby better promoting home-school cooperative education.

The classroom is equipped with a "smart blackboard" and an "interactive smart tablet" to achieve intelligent multimedia teaching. The smart blackboard supports chalk writing, is waterproof and explosion-proof, and integrates antiglare and highlight filtering technology to protect students' eyesight. The interactive smart tablet uses paper and pen data technology to support the traditional writing experience. The classroom can choose exercises to interact with students, and the system can quickly count the number of questions answered incorrectly during the classroom interaction, and summarize them into a wrong question set. Moreover, every stroke on the interactive smart tablet can reproduce the writing trajectory, which is beneficial in terms of enabling students to check their own writing situation and problem-solving thinking and improving the frequency of teacher–student classroom interaction and the quality of learning evaluations.

Building a Cloud Desktop Office to Improve Campus Office Efficiency and Ensure Data Security

Hanwen Experimental School has built a cloud desktop office system, which centralizes the campus office desktop in the data center; forms a resource pool through virtualization technology; provides business users with thin terminals, soft terminals, smart terminals and other mobile access; and creates a new, safe, convenient, and efficient way of working. The cloud desktop function architecture is shown in Fig. 5.32.

Enhancing security through data and user isolation: The user desktop data originally scattered on various PCs are centralized in the data center to achieve unified security control. Users can access only the changes in the desktop image, and data cannot be removed from the data center.

Enhancing operation and maintenance efficiency through resource centralization: Transitioning from decentralized operation and maintenance to centralized operation and maintenance, administrators can handle most user problems through the backend, reducing the operation and maintenance workload and improving maintenance efficiency.

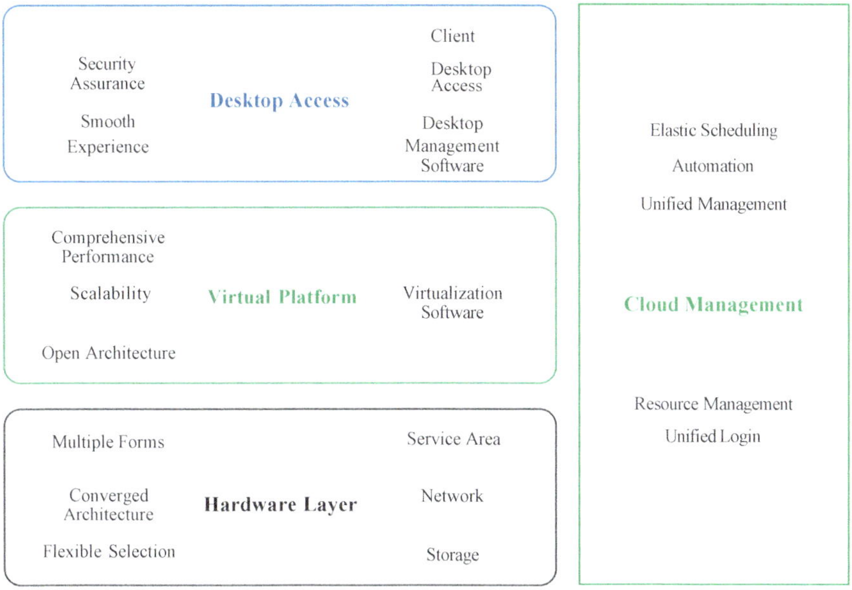

Fig. 5.32 Cloud desktop function architecture diagram

Enhancing office flexibility through network access: Users can access their own desktop office at any time and any place through the network.

Advanced Architecture Supports Future Evolution: Based on the server virtualization architecture, an efficient, optimized, and easy-to-manage desktop cloud architecture can be intensively built.

Introducing the Concept of "Big Science Popularization Innovation", Combined with Metaverse Technology, to Build a Teaching Practice Platform in Eastern Guangdong

Hanwen Experimental School emphasizes the popularization of science and traditional culture. By introducing a series of cultural popularizations, education, and other interactive content, it has built the first "AR + Science Popularization + Culture" digital research base in Eastern Guangdong. Students can check in at a one-stop interactive science popularization classroom covering a variety of knowledge, including the history of the Communist Party; health and wellness; cultural relics; plants; animals; intangible cultural heritage; safety education; waste sorting; space science; modern literature; Chinese history; and the origin of Chinese characters.

In the "AR + Science Popularization Research" digital interactive experience area set up at Hanwen Experimental School, students can experience six AR interactive systems, including smart party building, super student, health and wellness, science popularization education, the cool world, digital intangible cultural experience, and the digital AR zoo, bringing a rich digital interactive experience for students.

As we enter the era of digital intelligence, the promotion of digital transformation in education and the exploration of smart education are attracting worldwide attention. With the development and application of modern technology, the integration and penetration of "Internet+" and education continue to deepen. Informational education is also constantly developing in connotation, depth, and quality. The structure and form of the education system are undergoing changes and transformations, gradually forming a new education concept and model—smart education. Smart education will center on the learning and development of all students, use the logic and technology of "Internet+", create a smart learning environment, provide intelligent and efficient education services for learners, promote structural changes in traditional teaching, and promote personalized growth and the intelligent development of students.

5.4 Smart Tourism: Empowering Cultural Tourism with Technology to Create a New Tourism Model

5.4.1 Changes in Traditional Tourism Methods in the Post-Pandemic Era

"Three years of time, it's like it was stolen. The scenic spots in my memory, once bustling with crowds, are now silent in the deserted city after a night of lockdown. We used to be free, walking whenever we wanted, but now we can only yearn for the distance, and our longing for the world seems to have become a luxurious dream," travel expert Mr. Chen lamented.

Like a storm, the COVID-19 pandemic has resulted in a great shock to the global tourism industry. The lament of travel expert Mr. Chen symbolizes the helplessness of countless people on days when they cannot travel far. Before the pandemic, the total number of international tourists worldwide exceeded 1.5 billion, but in 2020, 2021, and 2022, the total number of international tourists worldwide was 70–75% less than that in 2019. It was not until 2023 that a recovery was observed, with the number of international tourists in the first quarter of 2023 reaching 80% of the prepandemic level.

Although COVID-19 has brought unprecedented shocks to the global tourism industry, it has also given birth to many new trends and changes.

Digitalization and Online Transition: With the implementation of social distancing measures, digitalization and online transition have become particularly important. Online tourism platforms, such as virtual travel and online booking platforms, have experienced significant growth. Many tourist attractions have also launched online interactive and virtual experiences.

Self-driving and Local Travel: Due to the restrictions and risks of international travel, many people choose to drive themselves and explore local attractions. This change has promoted people's understanding and appreciation of their local environment and community.

Emphasis on Health and Safety: Since the pandemic, travelers' emphasis on health and safety has greatly increased. Many tourism organizations and hotels need to provide strict cleaning and disinfection measures, as well as flexible booking and cancellation policies, to attract customers.

Sustainable and Ecotourism: The pandemic has prompted people to rethink their relationships with nature, so sustainability and ecotourism have received more attention. An increasing number of travelers are choosing travel methods that have less impact on the environment and a positive impact on the local community.

Personalized and Customized Travel: Travelers in the postpandemic era are paying more attention to personalized and customized experiences. To meet customer needs, travel companies need to provide more unique and personalized services.

In general, the pandemic has profoundly changed our way of travel, and the future tourism industry will focus more on safety, sustainability, and personalized experiences.

5.4.2 *Cultural Tourism Pain Points Are Obvious, and Smart Tourism Brings New Vitality*

The pain points faced by the cultural tourism industry include low operational performance, homogenized experiences, low consumer stickiness, a lack of attractive services, and low satisfaction. "Smart tourism" can help solve these problems, bringing new vitality to the industry.

Smart tourism integrates the new generation of information technology to provide destinations, governments, scenic spots, tourists, etc., with smart tourism applications based on full-process services and management, thus promoting collaborative development in tourism services, tourism experiences, tourism management, tourism marketing, tourism resource utilization, and industry promotion. In addition, smart tourism has a positive enabling role in decision-making, customer experience, operational efficiency, sustainable development, and crisis management.

Data-driven decision-making: Through big data analysis, tourism companies can better understand customer behavior and needs and formulate more accurate market strategies. For example, data analysis can be used to predict tourism hotspots, optimize resource allocation, and improve service quality.

Enhanced customer experience: Through mobile applications, virtual reality (VR), and augmented reality (AR) technologies, tourism companies can provide customers with richer and more personalized tourism experiences. For example, customers can book hotels, buy tickets, navigate, and obtain local tourism information through mobile applications.

Improved operational efficiency: Digitization can help tourism companies improve their operational efficiency and reduce costs. For example, automation tools can simplify the booking and management process, and artificial intelligence (AI) and machine learning (ML) technologies can improve the efficiency and quality of customer service.

Sustainable development: Digitization can also help the tourism industry achieve sustainable development. For example, data analysis can be used to assess the impact of tourism activities on the environment and communities and formulate more sustainable tourism policies and practices.

Improved crisis management: Digitization can help tourism companies manage crises, such as pandemics, more effectively. For example, real-time data tracking and analysis can be used to understand changes in a situation and respond quickly.

In summary, smart tourism can help the tourism industry better adapt to changes, meet customer needs, improve efficiency, and achieve sustainable development.

5.4.3 Policy and Technology Drive Smart Tourism, Foreshadowing Future Trends

To develop smart tourism, national and local tourism policies must support its rapid development. With the development of communication technology, sensor technology, spatial information technology, and cloud computing technology, the future of informatization presents new construction directions and development models, making the precise collection and resource-based application of tourism information possible. The smart tourism system, which focuses on serving tourists and tourism management departments and promoting tourism marketing, will also usher in a rapid development period.

National Policy Promotes the Intelligent Development of the Tourism Industry

In November 2020, the Ministry of Culture and Tourism of the People's Republic of China, the National Development and Reform Commission, and 10 other departments jointly issued the "Opinions on Deepening 'Internet + Tourism' to Promote High-Quality Development of the Tourism Industry", which is conducive to comprehensively promoting the development of "Internet + Tourism" and promoting the quality, efficiency, and power of the tourism industry.

In April 2021, the "14th Five-Year Plan for Cultural and Tourism Development", issued by the Ministry of Culture and Tourism, proposed deepening the supply-side structural reform of the tourism industry; extensively promoting mass tourism, smart tourism, and "tourism+"; and providing more high-quality tourism products and services. In addition, it proposed strengthening the tourism information infrastructure; deepening "Internet + Tourism"; accelerating the development of smart tourism characterized by digitization, networking, and intelligence; promoting appointment, off-peak, and limited normalization, build scenic spot monitoring facilities and big data platforms; and promoting the digital, networked, and intelligent development of culture and tourism, including the application of 5G, artificial intelligence, the Internet of Things, big data, cloud computing, BeiDou navigation, etc., in the fields of culture and tourism.

New Technology Provides Good Technical Support for the Construction of Smart Tourism

Modern technology, represented by digitization, is profoundly changing people's lifestyles and changing the behavior of tourists in terms of information collection, travel decision-making, organization, and consumption at tourist destinations. The improvements in transportation infrastructure, technological progress, and cultural changes are making people's tourism needs increasingly personalized and fragmented. The current artificial intelligence, big data, cloud computing, Internet of Things, metaverse, and other new technologies and scenarios are changing with each passing day, providing good technical support for the construction of smart tourism.

Innovative Application of Artificial Intelligence in Smart Tourism

The application of artificial intelligence technology in smart tourism can improve the tourism experience and service quality and optimize the allocation and utilization of scenic resources. The main application scenarios are as follows:

Intelligent Navigation: Through artificial intelligence technology, data such as scenic area maps, tour guide information, and tourist preferences are analyzed and processed to provide tourists with personalized route recommendations and detailed tour guide explanations.

Intelligent Personal Services: Through technologies such as artificial intelligence robots, we provide tourists with 24-h uninterrupted personal services, including luggage handling, traffic guidance, and problem solving.

Intelligent Voice Interaction: Through voice recognition and natural language processing technology, intelligent voice interaction between tourists and devices or systems is realized, facilitating tourist inquiries and consultations.

Intelligent Security Monitoring: Through technologies such as facial recognition and video surveillance, intelligent monitoring and management of tourists and items within the scenic area are carried out, improving the safety of the scenic area and tourist satisfaction.

Intelligent Recommendation Marketing: Through big data analysis and machine learning technologies, an in-depth understanding of tourists' interests and consumption behaviors is achieved, suitable products and services are recommended to tourists, and tourism revenue in the scenic area is increased.

Innovative Applications of Big Data Analysis in Smart Tourism

The application of big data analysis technology in smart tourism can help scenic area managers better understand tourist needs and behaviors, optimize resource allocation and utilization in scenic areas, and improve the quality and profitability of tourism services. The main application scenarios are as follows:

Tourist Profiling: Through big data analysis technology, information such as the demographics, interests, and consumption of tourists are analyzed and mined to construct detailed tourist profiles and better understand their needs and behaviors.

Scenic Area Traffic Forecast: Through historical data analysis and trend prediction models, the flow of tourists in scenic areas for a future period is predicted, allowing for advanced arrangements for resource allocation and service preparation.

Intelligent Recommendation: Through big data analysis technology, tourists' personal preferences and historical consumption records are considered with other tourist data to intelligently recommend suitable attractions, restaurants, accommodations, and other tourism products and services.

Resource Optimization: Through big data analysis technology, the use of resources within scenic areas is monitored and analyzed, and the allocation and utilization of resources within these areas are adjusted according to changes in tourist flow and demand.

Intelligent Pricing Strategy: Through big data analysis technology, the consumption levels and habits of tourists of different types and at different times are analyzed to formulate a more reasonable pricing strategy for increasing tourism revenue in scenic areas.

Innovative Applications of the Internet of Things in Smart Tourism

Internet of Things technology refers to connecting various devices through the internet to achieve intelligent interaction and data sharing between devices. The application of this technology in the field of smart tourism can help scenic areas provide more intelligent and convenient services, improve the management level of scenic areas and tourist satisfaction, and promote the development of the tourism economy. The main application scenarios are as follows:

Scenic Area Safety Monitoring: Through Internet of Things technology, real-time monitoring and management of personnel, vehicles, and facilities in scenic areas are carried out to prevent accidents.

Scenic Area Climate Monitoring: Through Internet of Things technology, the meteorological conditions in scenic areas are monitored in real time, providing tourists with accurate weather forecast services so that they can make travel plans in advance.

Scenic Area Resource Management: Through Internet of Things technology, automated management of scenic area resources, including ticket sales, tourist guidance, and catering and accommodation services, is performed.

Intelligent Parking Lot Management: Through Internet of Things technology, intelligent management of parking lots in scenic areas is achieved, providing tourists with a more convenient parking experience.

Scenic Area Environmental Monitoring: Through Internet of Things technology, real-time monitoring of environmental pollution in scenic areas is carried out, and corresponding measures are taken to protect the ecological environment.

Innovative Applications of Metaverse Technology in Smart Tourism

Metaverse technology involves the construction of a virtual reality space based on the digital world through virtual reality technology, blockchain technology, artificial intelligence technology, etc. It can help scenic areas provide more realistic, rich, and colorful virtual experiences; optimize resource allocation and utilization in scenic areas; and improve operational efficiency and tourist satisfaction in scenic areas. The specific application scenarios are as follows:

Scenic Area Virtual Display: Using metaverse technology, a highly realistic virtual display of scenic areas are created for tourists, facilitating online visits to the areas.

Scenic Area Interactive Experience: Using metaverse technology, a rich and colorful interactive experience in scenic areas, including virtual tours and virtual interactive games, is provided for tourists.

Virtual Exhibition of Artworks: Using metaverse technology, a virtual exhibition space is created for artworks in scenic areas, facilitating online appreciation and learning for tourists.

Tourism Product Display: Using metaverse technology, a virtual display space is created for products in scenic areas, facilitating online selection and shopping for tourists.

Digital Management of Scenic Areas: Using metaverse technology, various types of information within scenic areas are digitized and managed, facilitating comprehensive monitoring and management of all tasks by scenic area administrators.

Grasping the Future Direction and Seizing the Trend of Smart Tourism Construction

As government incentives are further released, technology applications need to be more comprehensive and integrated to provide more precise and personalized services. This requirement leads to more innovative and sustainable business models. Smart tourism uses the corresponding technologies of smart cities, allowing the intelligence of tourism to permeate all aspects.

The creation of smart tourism provides three services well: serving supply-side enterprises, serving tourists, and serving government supervisory units (as shown in Fig. 5.33). These services improve the comprehensive management and operational capabilities of tourism businesses, create a high-quality tourism ecological environment, enhance the quality of tourism services, and thereby promote the rapid and healthy development of the regional tourism economy.

Serving Operational Supervisory Units

On the one hand, by providing high-quality experiences and services, corporate brands are established, images are enhanced, and corporate marketing channels are expanded, creating more opportunities for corporate development.

On the other hand, through the creation of smart tourism, safety detection and protection capabilities for tourists and staff are improved, the comprehensive management and monitoring capabilities of scenic areas are enhanced, and the management and service capabilities of tourism businesses are improved, making operations simpler, more convenient, efficient, and orderly.

Serving Tourists

Through the creation of smart tourism, the six elements covered by tourism that drive regional economic development (namely, eating, accommodation, transportation, travel, shopping, and entertainment) are integrated in an orderly manner to provide convenient services for tourists, thereby maximizing the economic effects of tourism.

Can tickets be purchased quickly for park entry? What fun activities are there in the park? What actions should be taken in the case of an emergency? How are order and safety ensured? All these questions from tourists will be well addressed in smart tourism.

Fig. 5.33 Precision empowerment through smartization, leading to the creation of smart tourism

Considering the affluent and leisurely elderly group, the middle-aged group, who gradually prefers short-distance and personalized travel, and children, who are the focus of educational tourism, personalized travel plans are customized with respect to different time periods, dimensions, and subject perspectives to meet the respective needs of these different groups.

Serving the Government

Through the construction of provincial- and municipal-level cultural tourism big data platforms, technical support and information services for cultural tourism industry monitoring, management, and command dispatch can be provided. By aggregating data from scenic areas and tourists, data barriers can be broken down, and local tourism IPs can be created.

Smart Tourism Solutions: Relying on Digital Transformation

Smart tourism involves managing people, vehicles, objects, the environment, and time within scenic areas and matches values with smart scene businesses. It practices the concept of tourism industry transformation; facilitates the high-quality development of the tourism industry; encompasses opening and operation; and considers all elements, such as scenic area resources, human and vehicle safety, scenic area operation, fire protection, and emergencies, to build a complete chain from command center operation to management measure execution and ground data analysis. It promotes service through management, experience through service, and marketing through experience. The architecture of smart tourism is shown in Fig. 5.34.

Smart tourism takes resource protection as the baseline, tourism safety as the foundation, operation management as the core, quality service as the mission, and

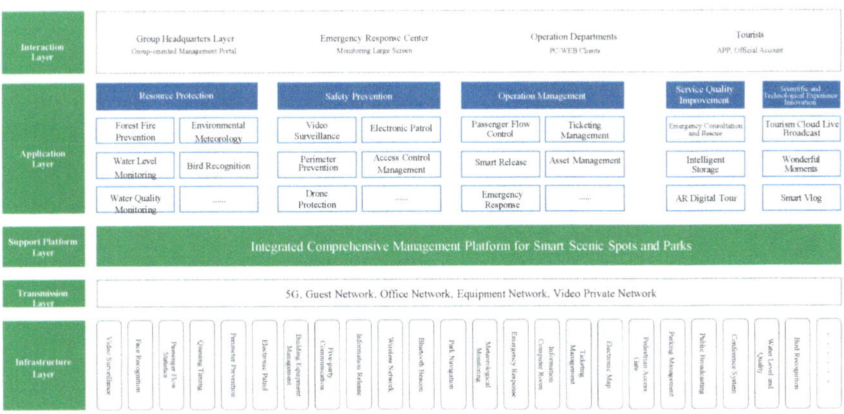

Fig. 5.34 Smart tourism architecture diagram

technology experience as the key to achieve vitality. The main applications include the following:

Resource Protection Management

Resources are the core materials of the core attractions of tourism. Real-time monitoring of the tourism resource status is carried out through video surveillance and IoT sensing, and protection is ensured through monitoring and early warning. This application includes forest fire prevention, water level monitoring, water quality monitoring, environmental meteorology, bird identification, and intelligent perception of forest resources, water bodies and lakes, environmental meteorology, and fauna and flora.

Safety Prevention Management

Safety is the foundation of scenic area operations. Full-dimensional, all-weather, and all-time video intelligence, with automatic alarm and perimeter defense capabilities, ensures the safety of scenic areas and tourists and facilitates worry-free sightseeing.

Personnel Management: Personnel management is carried out through video surveillance systems, passenger flow systems, etc.

Traffic Management: Traffic management is carried out through parking lot management systems, parking space guidance, etc.

Comprehensive security management is achieved through means such as video surveillance, perimeter defense, and drone protection. It includes large-scale monitoring, night surveillance, and anti-drowning audio-visual alarms. By utilizing the precision, persistence, and speed of drones, combined with a visual control platform, we integrate various system resources to create smart, process-oriented, real-time, intelligent scenic area management. We promote the intelligence of security checks to achieve "efficiency improvement, safety assurance, and regulatory promotion" in scenic area safety prevention management.

Operations Management

Through artificial intelligence and smart management, we enhance the management efficiency of staff, data analysis flow control intelligent alarms, improve the entire scenic area response mechanism, and increase business efficiency. This process includes ticket management, asset management, broadcasting and information release, emergency commands, WeChat services, all-in-one cards, etc.

Service Quality Upgrade

Fully utilizing AI intelligence technology and capabilities, we assist tourists in experiencing a worry-free and enjoyable trip and high-quality tourism. This application includes emergency consultation and rescue, the intelligent storage of belongings, AR digital tour guides, digital interactive experiences, and other quality features.

Technology Experience Innovation

Intelligent technology integrates with the entire tourism scene, innovating interaction and engagement during the tourism process and enhancing the experience of technology. It covers the entire journey before, during, and after the visit; keeping up with the times, it is closely linked to the needs and habits of young people. Features such as tourism cloud live broadcasts, capturing exciting moments, intelligent vlogs, etc., are available.

5.4.4 Smart Tourism Practice Case: The Largest Indoor Theme Park in the World

The largest indoor theme park in the world is a smart tourism project undertaken by Gloryview. Located in Hengqin, Zhuhai, it boasts the world's largest fish exhibition pool, the world's largest acrylic glass viewing window, the world's largest live coral tank, the world's largest shark display tank, and the world's largest marine life support system, creating multiple world records. It brings together rare marine animals, novel large-scale performances, and top-notch amusement facilities, providing a more intelligent tourism experience for visitors.

The entire building has a double-layer + mezzanine structure that is approximately 650 m long, with a total area of 400,000 square meters. It looks like a large spaceship from outer space. As the largest indoor theme park in the world, it has a variety of dynamic sculptures accompanied by sound and light effects, and amusement facilities such as the "Deep Sea Submarine" and "Bermuda Adventure" dome dynamic platform cinema. A rich array of entertainment projects provides visitors with an immersive and shocking experience. Fifteen science sharing spaces, including the largest "Ocean University", can accommodate 5000 people at the same time, and use multimedia technology to provide education and entertainment. These spaces are designed and built by integrating global resources under the leadership of Chinese creativity.

Owing to the large flow of tourists who span a wide range of ages, the customer experience faces multiple challenges, such as safety and personalization. Due to the complex equipment and complex park requirements, while both performance and aesthetics are achieved, construction and operation are somewhat difficult.

A large flow of people results in many safety hazards: the park area exceeds 400,000 square meters, including multiple themed areas such as Coral Secret, Song of the Ocean, Ocean University, etc. It is expected that after the official opening, there will be more than 50,000 visitors every day. With such a high visitor volume, ensuring the safety of tourists will inevitably be a challenge.

Tourists cover a wide age range, and they have many different needs: "It's not easy to have time to accompany the family on the weekend. It's inconvenient to travel too far. Bringing children here is a good choice for a short trip." Mr. Zhang, who likes to travel, said, "Here is a combination of eating, living, traveling, shopping, and entertainment. It's relaxing and they can learn knowledge; [it's] not just a cursory visit, it's very good." After the indoor theme park opens, it will attract a large number of parent–child tourist pairs, covering a very wide age range. Different age groups have different needs, so there is a need to provide personalized routes. Ensuring tourists have an immersive experience is a focus of construction.

Complex equipment, difficult operation management: Complex equipment, a large number of information points, and the management of the network and equipment onsite have become considerable challenges for the venue.

The park's requirements are complex, and both performance and aesthetics must be considered. This park will be the world's largest indoor theme park, a playground, and a benchmark for similar venues, so how can both aesthetics and performance be achieved?

Complexity of the Venue Structure: The overall architectural structure of the venue is complex, with a large area on the first floor designed with high and wide spaces, some of which reach a height of 13.85 m. The children's playground and the large fish viewing area are covered with steel structures, reaching a maximum height of 25 m. Owing to the complexity of the building structure and the presence of various mezzanines and platforms, the project can be identified only by elevation levels (−6 h, 0.0 h, 5.5 h, 13.85 h, 20.95 h, 27 h, 32 h, 45 h) instead of floor labels.

Multiple Scene Requirements: The venue includes a rehearsal hall, a musical theater, a dynamic cinema, themed tour passages, themed restaurants, rail stations, indoor beaches, amusement parks, science classrooms, rotating ramps, logistics offices, 66 different-sized marine environment pools and large amusement equipment. The largest stand can accommodate 5000 spectators. The characteristics of scene decorations, lighting conditions, equipment types, and environmental features vary.

Unique Design: The ceilings and walls of the guest areas inside the venue are mostly covered with painted designs, shaped packaging, and decorative ceilings. The selection and installation of equipment in the guest areas must be highly compatible with the decorations.

In response to these issues, Gloryview designed the world's largest indoor theme park using a 5-layer architecture and 3-platform approach (as shown in Fig. 5.35). The park serves both visitors and operators and provides a novel experience in smart tourism.

The following functions are designed based on this top-level planning, which can be divided into two main parts.

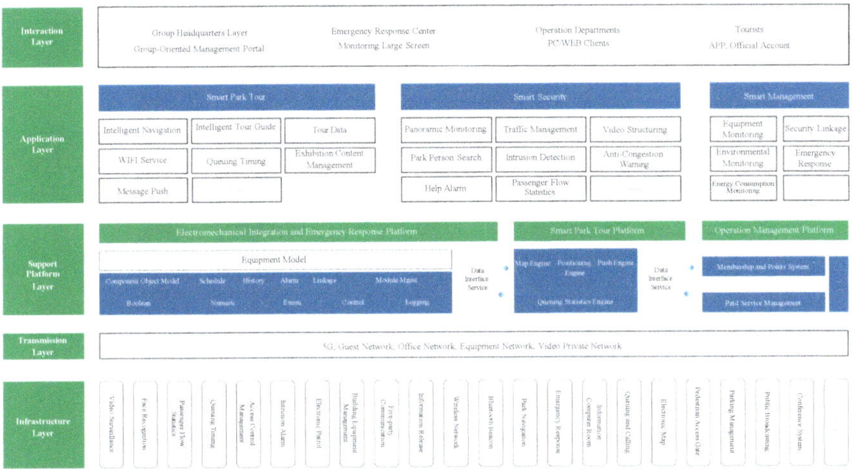

Fig. 5.35 Overall architecture of the world's largest indoor theme park

Visitor Side: Creating a Joyful, Worry-free, and Secure Travel Experience for Visitors

To provide customers with better service, considering safety and experience, we offer a smarter park solution.

Safety without blind spots, imperceptible control

Given the high and wide space construction of the park and the complex cable routing of the staggered layers, to avoid affecting everyone's play experience and to balance the aesthetic and visual coverage needs of the guest areas, more than 3000 monitoring points were set up throughout the venue for hidden deployment. Owing to the different scenes and different focuses of each department, 19 types of surveillance cameras were used to cover the venue, ensuring full coverage of the guest areas and no blind spots.

As most of the walls and ceilings in the guest areas are covered with painted designs and shaped packaging, to balance the aesthetic and visual coverage needs of the guest areas and reduce the impact of equipment installation on the overall aesthetics, we checked the points one by one with the owner and the decoration design unit on site and determined the installation location based on the characteristics of each camera.

Considering the high and wide space of the venue, cameras with large coverage, ultrawide angles, and long viewing distances were selected. The main passages and entrances and exits of attractions were equipped with passenger flow counting cameras to analyze passenger flow and grasp real-time passenger flow information in the venue, providing a basis for emergency handling during peak passenger flow; face capture cameras were used for blacklist prompts, in-park person finding, and quick personnel positioning; and the backend was connected with the public security

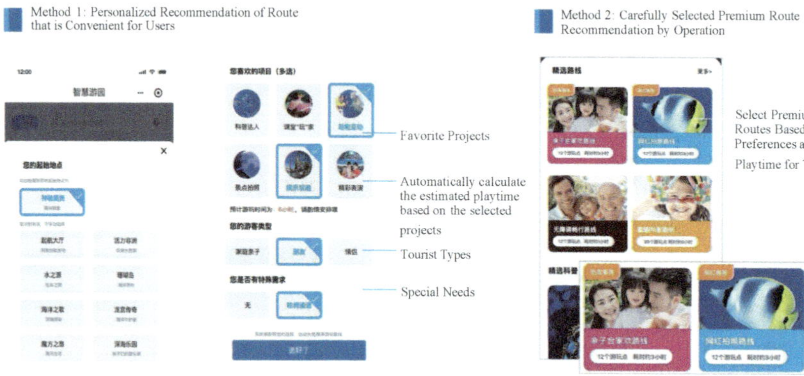

Fig. 5.36 Smart park route recommendation method

system to compare faces with the public security face database, ensuring the safety of visitors in the park.

Smart Park, Enhancing Park Experience

To provide visitors with a better experience, we created a smart park system to enhance the park experience.

The smart park system integrates online and offline methods. Not only can it customize and plan itineraries according to visitors' preferences, but it can also push the best next step based on the visitors' location. The smart park route recommendation method is shown in Fig. 5.36.

The system can not only display the location and navigation of all static facilities in the venue but also provide real-time information on the flow of people and queuing situation in each area based on the AI algorithm of the camera. When visitors arrive at various marine creatures, they experience real-time explanation services, which can also interact with visitors to answer related questions through voice. These applications are all based on the company's indoor high-precision location service technology, video AI, NLP and other artificial intelligence technologies.

A Mobile Phone for the Amusement Park, Starting a New Tourism Revolution

The park manages the entire tour process from start to finish through an app and mini-program product forms. A mobile phone can be used to enjoy the entire park. Visitors can enter the guide mini-program by scanning the code at the entrance. On their way through the park, visitors can scan the QR code or check the surrounding attractions on the map, receive assistance with a voice introduction, quickly locate and share the places they want to visit, and arrange the next itinerary by appointment, all in an orderly manner. During the tour, visitors can earn points to exchange for corresponding coupons, thereby enhancing user stickiness. The five major application scenarios of the smart park are shown in Fig. 5.37.

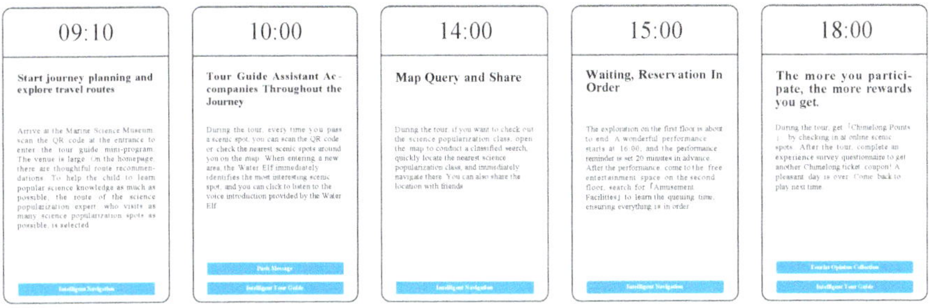

Fig. 5.37 Five major application scenarios of smart parks

Operators: Creating a User-Friendly, Low-Fault, and Easy-to-Maintain Management Experience

Owing to the multitude of departments involved (including security, visitor services, catering, engineering equipment, kitchen, finance, merchandise, park services, total value operations, life support systems, animal management, performance, attraction operations, digital and information departments, and general affairs), construction involves multiple professional construction units. In response to the actual problems and needs on site, point-to-point deepening design is carried out. In addition, intelligent management is integrated into operations according to the needs of each department and each scenario, creating a comprehensive, efficient, and streamlined operational experience. Multidimensional operation management of the venue is implemented to improve the quality of venue services.

Operational management encompasses various functions, including the following:

Tour route management, as shown in Fig. 5.38.

Map management, as shown in Fig. 5.39.

Maintenance of exhibition spaces and animal content, as shown in Fig. 5.40.

Park data display and analysis, as shown in Fig. 5.41.

5.5 Smart Energy: Assisting in Achieving China's Dual Carbon Goals and Sustainable Development

5.5.1 Smart Energy, a Concept Born Out of Crisis

Crisis is the key driving force behind the formation of the concept of smart energy. Crises occur in multiple dimensions: resource scarcity, environmental pressure, energy supply instability, etc. Therefore, we must find a sustainable and efficient energy solution in the face of these difficulties.

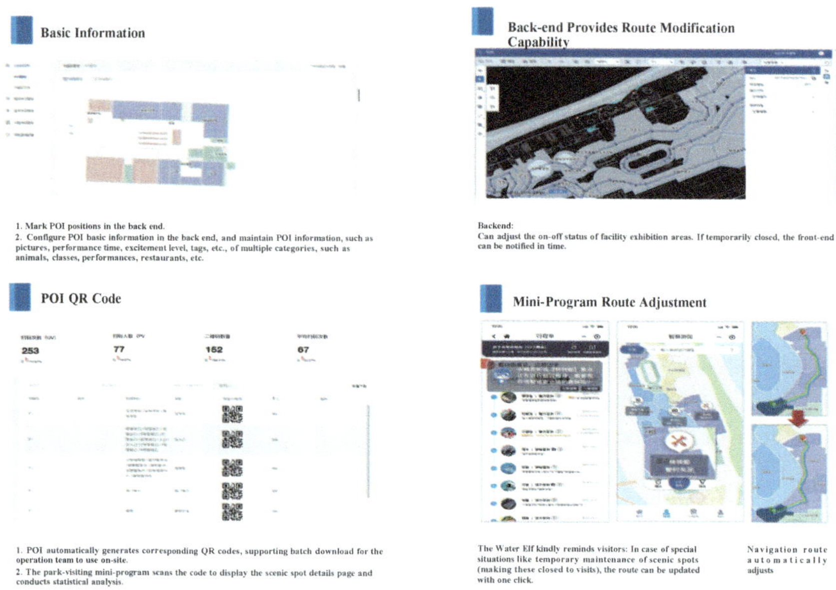

Fig. 5.38 Tour route management

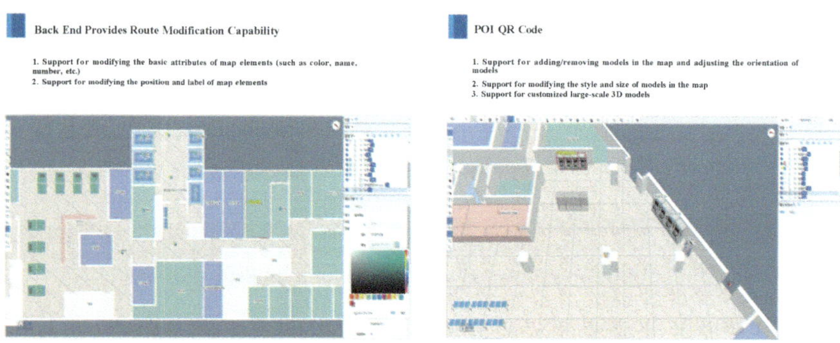

Fig. 5.39 Map management

Resource Scarcity

With humanity's overreliance on traditional fossil energy, the Earth's resources are gradually being depleted. This situation not only leads to instability in the energy supply but also prompts us to rethink our use of traditional energy and our attitude toward energy consumption.

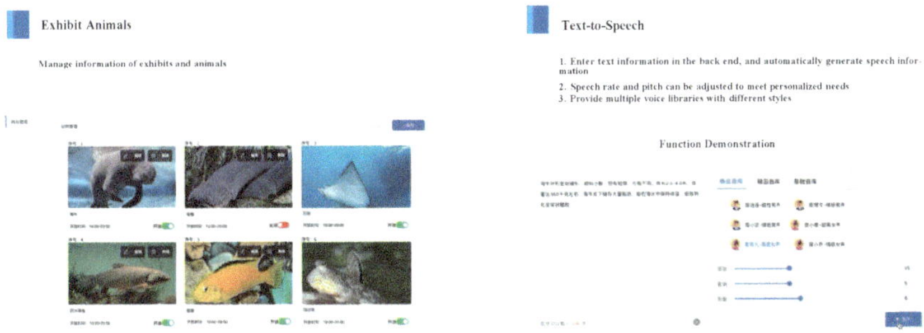

Fig. 5.40 Maintenance of exhibition spaces and animal content

Fig. 5.41 Park data display and analysis

Environmental Pressure

While fossil energy has led to economic development, it has also put tremendous pressure on the environment. The severity of climate change and the rise of environmental awareness have led the international community to gradually reach a consensus on addressing global climate change through emission reduction and carbon reduction. From the United Nations Framework Convention on Climate Change in 1992, the Kyoto Protocol in 1997, and the Paris Agreement in 2016, the international community has collectively established the political and legal basis for the global response to climate change.[4] These agreements encourage countries to reduce the use of fossil

[4] The **United Nations Framework Convention on Climate Change (UNFCCC)** is one of the three major environmental conventions adopted under *Agenda 21* at the 1992 Earth Summit in Rio

energy and increase the use of renewable energy to reduce greenhouse gas emissions and mitigate climate change.

Instability of Energy Supply

Political factors, such as wars, regime changes, trade conflicts, etc., all of which affect the energy supply, can trigger energy supply crises. Moreover, in the context of global economic integration, the uncertainty of the energy supply causes fluctuations in energy prices, posing significant challenges to the economies of all countries.

Smart energy is a concept born out of crisis and offers a solution to energy, environmental, and economic challenges. Smart energy uses the new generation of information and communication technology to monitor, control, manage, and operate all aspects of energy production, transmission, distribution, and use; improve energy efficiency; ensure energy supply; reduce environmental impact; improve service quality; and help form a sustainable energy system. The main significance and role of smart energy include the following:

Enhancing Energy Efficiency: Through the new generation of artificial intelligence and big data analysis technology, smart energy can optimize the mode of energy production and consumption, reduce energy waste, and thus improve energy use efficiency.

Ensuring Energy Supply: Smart energy systems can predict and manage energy demand, ensuring a stable power supply during peak periods and emergencies.

Reducing Environmental Impact: Smart energy can more effectively integrate renewable energy sources, such as wind and solar power, reducing reliance on fossil fuels and thereby lowering greenhouse gas emissions to combat climate change.

Improving Service Quality: Smart energy systems can more accurately bill users and provide customized services, such as demand response and energy efficiency consulting, thereby improving user satisfaction.

Supporting Sustainable Development: Smart energy is an important means to achieve sustainable development. It can help us meet current energy needs without compromising the ability of future generations to meet their own needs.

In summary, the goal of smart energy is to establish a sustainable, efficient, safe, flexible, and environmentally friendly energy system.

---

de Janeiro. It was passed at the United Nations Headquarters in New York on May 9, 1992, and entered into force on March 21, 1994.

The **Kyoto Protocol** is a key legal instrument developed to implement the UNFCCC. It was adopted in Kyoto, Japan, on December 11, 1997, and came into effect on February 16, 2005.

The **Paris Agreement** was adopted on December 12, 2015, at the 21st session of the Conference of the Parties to the UNFCCC (COP21) in Paris. It was signed on April 22, 2016, at the United Nations Headquarters in New York and officially entered into force on November 4, 2016. As the successor to the expired Kyoto Protocol, the Paris Agreement is a globally binding climate accord signed by 178 parties, establishing a unified framework for global climate action beyond 2020.

5.5.2 Future Development Directions and Applications of Smart Energy

At the general debate of the 75th United Nations General Assembly in September 2020, General Secretary Xi Jinping officially announced: "China will enhance its national contributions, adopt more powerful policies and measures, strive to peak carbon dioxide emissions by 2030, and strive to achieve carbon neutrality by 2060." The "dual carbon" goal was subsequently written into the "14th Five-Year" plan, elevating carbon peaking and carbon neutrality to the national strategic level. The policies related to dual carbon and the digital and intelligent transformation of energy released by China in recent years are shown in Table 5.2.

In the industry map of the carbon neutrality mainline, under the guidance of the top-level design of the "1 + N" policy system (as shown in Fig. 5.42) for peaking carbon and achieving carbon neutrality, policy measures have been formed in various fields and industries, including energy, industry, urban and rural construction (building), transportation, agriculture and rural areas, and other key areas and industries. Under policy guidance, we should accelerate transformation and innovation. We need to build a new power system with new energy as the main body, promote industrial electric transportation and improve energy utilization efficiency, promote industrial and industrial optimization and upgrading, promote energy-saving and low-carbon buildings and low-carbon facilities, build a green and low-carbon transportation system, develop a circular economy and improve resource utilization efficiency, promote green and low-carbon technology innovation, contribute solutions and insights to facilitate the achievement of peak carbon and carbon neutrality in China, and accelerate the global low-carbon and zero-carbon transformation.

China is the world's largest energy consumer, carbon emitter and producer of renewable energy, as well as the global leader in the manufacture and use of electric vehicles. According to the China Emission Accounts and Datasets (CEADs), in 2022, China's carbon emissions totaled 11 billion tons, accounting for approximately 28.87% of global carbon emissions. Thus, China is both part of the global climate challenge and an important part of its solution.

Considering the above situation and under the guidance of national policies, we believe that the future development direction and applications of smart energy include the following four main aspects:

Networked smart energy. Starting with the actual needs of energy applications, various energy management systems will be integrated to build a comprehensive system; and various forms of energy, such as conventional energy, new energy, and distributed energy, will be covered. Future energy systems will become smart energy networks, where centralized and distributed power generation and energy storage coexist, transmission, distribution, sales, and use are coordinated, and wind, light, hydrogen, and storage complement each other.

The new generation of information technology promotes the fine management and upgrading of the entire energy chain. Technologies such as the new generation of artificial intelligence, big data, cloud computing, the Internet of Things, and

Table 5.2 Overview of domestic dual carbon and energy digitalization and intelligent transformation policies

Date	Issuing unit	Policy document	Main content
December 2020	Ministry of Ecology and Environment	"Carbon Emission Trading Management Measures (Trial)"	Regulates national carbon emission quota allocation and clearing, carbon emission rights registration, trading, settlement, greenhouse gas emission reporting and accounting activities
February 2021	State Council	"Guiding Opinions on Accelerating the Establishment and Improvement of a Green, Low-Carbon, and Circular Economic System"	Proposes to improve the production, circulation, and consumption systems of green, low-carbon, and circular development, accelerate the green upgrade of infrastructure, build a market-oriented green technology innovation system, and improve the legal and policy system
March 2021	State Council	"14th Five-Year" Plan	Clearly defines the task of "formulating an action plan to peak carbon emissions by 2030", implements a system that controls carbon intensity and supplements total carbon emissions, and supports regions, key industries, and key enterprises that are ready to reach peak carbon emissions first
May 2021	Ministry of Ecology and Environment	"Carbon Emission Rights Registration Management Rules (Trial)", "Carbon Emission Trading Management Rules (Trial)", "Carbon Emission Rights Settlement Management Rules (Trial)"	Further regulates national carbon emission rights registration, trading, and settlement activities and protects the legal rights and interests of all parties participating in the national carbon emission trading market
September 2021	Office of the National Ecological Industrial Demonstration Park Construction Coordination Leadership Group	"Notice on Promoting Carbon Peaking and Carbon Neutrality Related Work in National Ecological Industrial Demonstration Parks"	Clearly defines the four key tasks of optimizing energy and the industrial structure, promoting the application and transformation of low-carbon technology innovation, building a dual carbon goal management platform, and strengthening green and low-carbon concept publicity and education

(continued)

Table 5.2 (continued)

Date	Issuing unit	Policy document	Main content
September 2021	Central Committee of the Communist Party of China, State Council	"Opinions on Fully, Accurately, and Comprehensively Implementing the New Development Concept and Doing a Good Job in Carbon Peaking and Carbon Neutrality"	The main goals proposed include building a green, low-carbon, circular economic system, improving energy utilization efficiency, increasing the proportion of nonfossil energy consumption, reducing carbon dioxide emission levels, and enhancing the carbon sequestration capacity of ecosystems, all to ensure the timely achievement of peak carbon and carbon neutrality
October 2021	State Council	"Action Plan for Peaking Carbon Emissions by 2030"	By 2025, the proportion of nonfossil energy consumption will reach approximately 20%, the energy consumption per unit of GDP will decrease by 13.5% compared with 2020, and the carbon dioxide emissions per unit of GDP will decrease by 18% compared with 2020, laying a solid foundation for achieving peak carbon emissions
January 2022	National Energy Administration	"14th Five-Year Plan for Modern Energy System"	Proposes to accelerate the digital and intelligent development of energy, promotes the deep integration of digital technology with the real economy, and empowers the digital and intelligent transformation and upgrading of traditional industries

(continued)

Table 5.2 (continued)

Date	Issuing unit	Policy document	Main content
May 2022	General Office of the State Council	"Implementation Plan for Promoting High-Quality Development of New Energy in the New Era"	Proposes measures to accelerate the advancement of new energy storage and smart energy. Promotes measurement, testing, and experimental research capabilities to reach world-class levels and active participation in the creation of international standards in the fields of wind power, photovoltaics, marine energy, hydrogen energy, energy storage, smart energy, and electric vehicles
March 2023	National Energy Administration	"Several Opinions of the National Energy Administration on Accelerating the Digital and Intelligent Development of Energy"	In response to the digital and intelligent transformation and development needs of industries such as electricity, coal, and oil and gas, through the integrated application of digital and intelligent technologies, urgent needs are prioritized, easy tasks are tackled before difficult ones, and shortcomings in transformation and development are addressed in a phased manner by industry and by segment, providing effective support for high-quality energy development. By 2030, the innovative application system of digital and intelligent technologies in various sectors of the energy system will be initially established, the potential of data elements will be fully activated, and breakthroughs will be made in a number of common key technologies that restrict the digital and intelligent development of energy

blockchain will promote the dynamic lean optimization and management of all links, including energy production, transmission, and consumption, thereby enabling the energy power system to achieve a "smart" transformation and upgrade. Under this model, the application of information technology in the energy power industry will accelerate, thus achieving the deep integration of information technology and the energy power industry and shifting the energy power industry toward digital and intelligent development.

The use of smart energy will revolutionize the energy supply model. Future energy supply will shift from single-source supply to diversified supply, involving the diversification of supply entities, of energy products, and of business structures. Moreover,

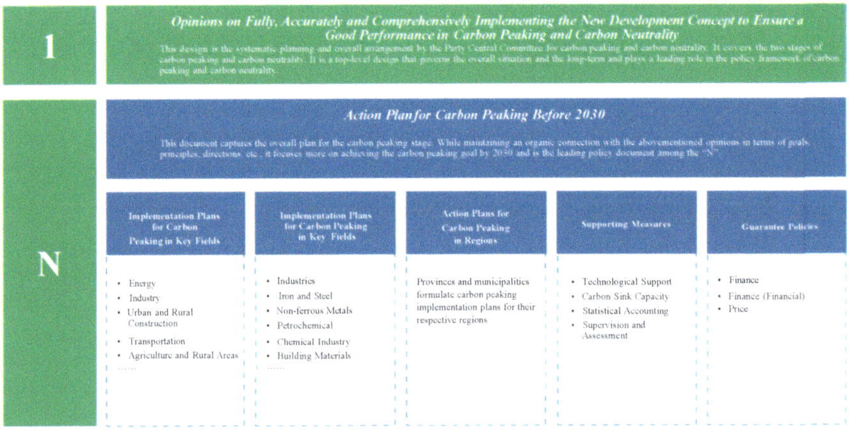

Fig. 5.42 "1 + N" policy system for dual carbon

smart energy will also promote the development of new business forms such as smart power plants and smart grids, providing safer, more intelligent energy production and transmission services, effectively optimizing energy production and supply models, and increasing the proportion of new energy in the total energy supply.

Smart energy will enhance the flexibility and dispatchability of the grid, thereby supporting the integration of large-scale renewable energy. Furthermore, by improving the safety and stability of the grid, smart energy will provide important technical support for marketization reform in the power industry.

This discussion highlights the future development direction and application of smart energy. These developments and applications will make our energy system more efficient, environmentally friendly, safe, and more adaptable to the needs of future societies and economies.

5.5.3 *Smart Energy: Digital Transformation is the Core*

Digital Technology Empowers the Transformation of Energy Management

The deep integration of the digital economy and the real economy will bring new opportunities and challenges to the energy industry. With the development of a new generation of information technology, the integrated application of zero-carbon technology in various links, including production operation optimization, safety and environmental protection control, intelligent equipment maintenance, and digital equipment monitoring, will be achieved. Digital transformation is the core of achieving smart energy. It requires the use of various technical means to establish a digital, intelligent, and efficient energy management system, thereby contributing to the sustainable and green development of energy.

Application of Artificial Intelligence Technology: Artificial intelligence technology can help energy systems quickly diagnose faults, optimize operation management, and improve system efficiency.

Application of Big Data Analysis Technology: Through big data analysis technology, deep mining and analysis of collected data can be carried out to improve the efficiency and accuracy of the energy management system and provide decision-making support.

Application of Cloud Computing Technology: Cloud computing technology can achieve the centralized storage and management of energy data and information, thus providing strong support for energy management.

Application of Blockchain Technology: Blockchain technology can ensure the security and credibility of energy data, prevent data tampering and leakage, and improve the transparency and fairness of energy transactions.

Application of Internet of Things Technology: Through various front-end IoT devices, sensors, etc., various data in the process of energy production, transmission, distribution, and use can be collected, building a comprehensive and accurate data foundation.

By building a smart energy dual-carbon service platform, setting up an evaluation indicator system, combining statistical analysis, dynamic optimization, prediction and early warning, feedback control, carbon trading support, demand response, green electricity trading, etc., centralized control of enterprise energy information, refined energy-saving management of equipment, and systematic management of energy can be achieved.

A full-scene, fully linked support service for the carbon ecosystem should be created. Green and low-carbon lifestyles should be prioritized both upstream and downstream of the industrial and supply chains, and transformations in energy and electricity from high carbon to low carbon and the transition from fossil energy-based to clean energy-based should be jointly promoted.

Smart Energy Digital Business Platform

By building an energy brain, namely, the smart energy digital business platform, the problems of past energy use islands, incomplete energy consumption data, and managers' inability to fully view vertical and horizontal energy consumption data can be solved, including the following: the lack of coordination of the multienergy supply with actual energy use, with each factor acting on its own, and the failure to achieve timely and orderly efficient application; serious energy consumption waste; the inability to achieve regional energy use monitoring and analysis at the platform level; and the inability to provide data support for the next energy-saving transformation. By managing energy consumption, energy supply efficiency can be increased, and energy-saving optimization can be achieved.

The platform is based on edge computing at the edge + cloud data and AI technologies. It connects the energy supply side and the energy demand side, intelligently supervises energy consumption in different time zones and regions, combines energy supply condition (municipal electricity/photovoltaic/energy storage) configurations and local government peak-valley electricity policies, uses smart energy use algorithm strategies, and achieves the best energy use and the greatest benefits. Moreover, through carbon management, digital tools are used to achieve dual-carbon goals.

The architecture of the smart energy digital business platform is shown in Fig. 5.43.

The platform includes ten functions of dual-carbon applications and comprehensive energy control. The following text introduces three of the function modules:

Operation Control

It includes two aspects of energy use operation control and energy supply operation control:

Energy Use Operation Control: This feature includes functions such as real-time energy use monitoring, user energy use overview, energy use unit analysis, energy efficiency indicator analysis, and energy efficiency services (as shown in Fig. 5.44).

Energy Supply Operation Control

Load Forecasting: This feature provides accurate power generation and load forecasting for the big data platform, assisting in the optimization of microgrid scheduling. The main functions include short-term power generation forecasting, ultrashort-term power generation forecasting, short-term load forecasting, and ultrashort-term load forecasting. Based on load characteristics, various types of loads are classified into multiple categories. Combining planning and weather impact

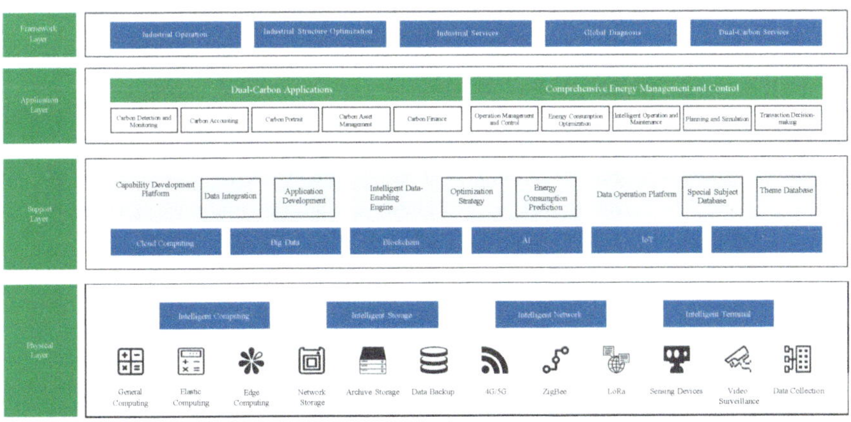

Fig. 5.43 Smart energy digital business platform

Fig. 5.44 360-degree monitoring of energy usage

factors, different algorithms, such as similar days, regression, and neural networks, are used to forecast medium- and short-term loads for classified loads (stable loads, impact loads, intermittent loads, discrete loads, etc.).

Distributed Photovoltaic Management: This feature enables comprehensive analysis of photovoltaic output characteristics, load characteristics, and energy storage system characteristics. The revenue of distributed photovoltaic energy storage is influenced by its scheduling strategy. In each period, the photovoltaic power first supplies the load, and the general strategy for energy storage charging and discharging is as follows:

When the photovoltaic output is sufficient and there is a peak price period, the remaining power is first used to charge the energy storage. If there is a surplus, it is sold to the grid.

When the photovoltaic output is sufficient and there is a valley price period, the remaining power is first used to charge the energy storage. If there is a surplus, it is sold to the grid; if the remaining photovoltaic power cannot meet the energy storage charging demand, power is purchased from the grid.

When the photovoltaic output is insufficient and there is a peak price period, the energy storage is discharged to the load. If it is insufficient, power is purchased from the grid.

When the photovoltaic output is insufficient and there is a flat or valley price period, energy storage purchases power from the grid.

During flat price periods, energy storage neither charges nor discharges. The ultimate goal is to maintain the power balance constraints and battery constraints.

Energy Consumption Optimization

By building an AI expert database, an analysis of energy consumption is conducted, and optimization suggestions are proposed.

Energy Consumption Baseline: Automatically query similar days in history with similar weather, human flow, and other factors to generate an energy consumption baseline, making energy consumption comparisons more scientific.

Fault Diagnosis: Diagnose common problems in electromechanical systems and personnel behavior, analyze causes, assess impacts, recommend detailed operation and maintenance guidance, and explore the potential for energy-saving optimization.

Load Forecasting: Accurately predict future energy consumption loads, providing a basis for the advance deployment of control strategies.

Energy Saving Analysis: Real-time dynamic evaluation of energy-saving performance. The evaluation indicators include three main aspects: energy management, environmental comfort, and equipment operation. Subitem indicators and parameter query functions are established for each aspect. For each indicator, a scientific threshold range is set based on historical data analysis, industry standards, and project regulations. Indicators that exceed the range are flagged. Finally, an overall evaluation of the energy-saving performance level is performed.

Carbon Asset Management

A carbon asset management service is built for the entire lifecycle of the carbon cycle, supporting the review of the carbon footprint, carbon quota, and carbon emissions data at each stage of the carbon cycle's full lifecycle. It provides a powerful means for users to manage quota assets and emission reduction assets and provides information decision analysis for carbon trading.

5.5.4 Smart Energy Practice Case: Nanning Key Energy-Consuming Enterprise Real-Time Energy Monitoring System

Gloryview Technology, which is based on the concept of cloud-edge-end collaboration technology and application innovation in smart cities, has developed a city energy management system for the Nanning Key Energy-Consuming Enterprise Real-Time Energy Monitoring System project (as shown in Fig. 5.45). By dynamically monitoring and diagnosing the energy consumption of key energy-consuming enterprises in Nanning, it achieves full control of enterprise energy, implements targeted measures to reduce energy consumption, and helps Nanning to precisely control energy savings, develop smart energy, build an industrial ecology, provide convenient services for the public and enterprises, and gradually achieve the "carbon neutrality" goal.

Overall Project Architecture

The Nanning City Key Energy-Consuming Enterprises Real-Time Energy Monitoring System is an energy-saving platform based on data collection from enterprises that integrates data display and analysis. The software platform includes a

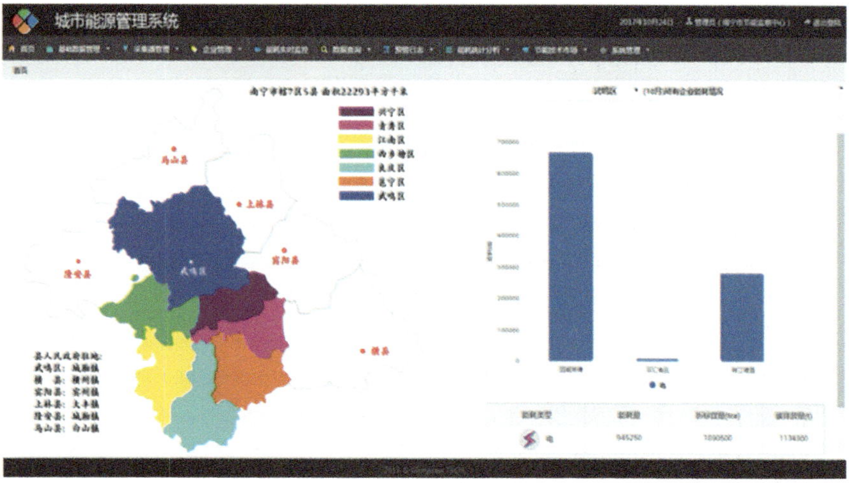

Fig. 5.45 City energy management system for Nanning key energy-consuming enterprise real-time energy monitoring system project

city-level regional platform and an enterprise-level energy consumption measurement platform. The energy consumption data of the enterprise can be stored locally and transmitted in real time to the city-level energy-saving supervision center system server. The enterprise-end energy consumption data call the local server, grant user permissions for the enterprise, and provide an efficient and reliable energy consumption supervision platform for the enterprise. The system enhances the enterprise's energy-saving awareness and further improves the efficiency of the energy-saving management department through information technology. It serves as a basis for city-wide energy-saving assessment, energy-saving supervision, energy efficiency benchmarking, and the promotion of energy-saving products. The architecture of its data collection and transmission system is shown in Fig. 5.46.

City-Level Regional Platform

The system's functions and data are accessed by users, using strict user role hierarchical authorization management. Logged-in users can view only the functions and data within their scope of authorization, thus ensuring the security and legality of system data access.

The city-level regional platform is managed at the city level. The system includes enterprise management, enterprise distribution maps, real-time energy consumption monitoring, energy consumption statistical analysis, district and county energy consumption comparison analysis, energy use behavior analysis, energy consumption time-day-month analysis, energy-saving technology market, basic data management, data collector management, data queries, warning logs, etc. The real-time energy

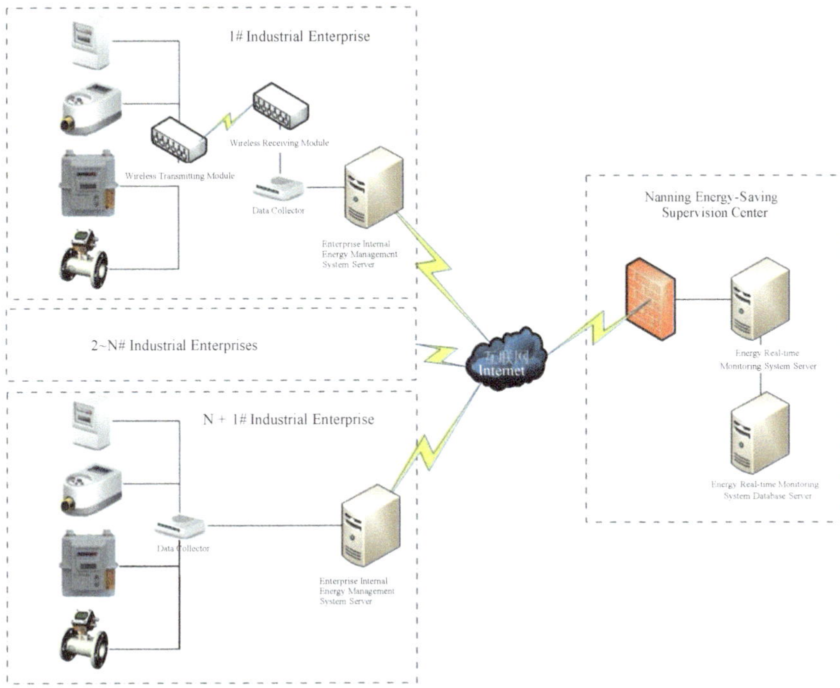

Fig. 5.46 Data collection and transmission system architecture

consumption monitoring diagram of the city-level regional platform is shown in Fig. 5.47.

Enterprise-Level Energy Consumption Measurement Platform

Enterprise users can grasp the current energy consumption status of the enterprise in real time, as well as the comparison analysis results given by the system, which is conducive to quickly and accurately locating the high energy consumption links of the enterprise. Targeted measures are taken to reduce energy consumption and achieve energy-saving goals.

Enterprise-End Energy Consumption Collection Equipment

By implementing real-time perceptions of energy consumption equipment such as mechanical and electrical equipment, electrical equipment, and process equipment through intelligent electric meters, intelligent gas meters, flow meters, and other metering instruments in the industrial enterprise factory area, the unreasonable use of energy can be avoided, the use and utilization efficiency of energy can be improved, and energy-saving services supported by an energy-saving supervision system can be built. The establishment of electricity, gas, coal, and other classification models

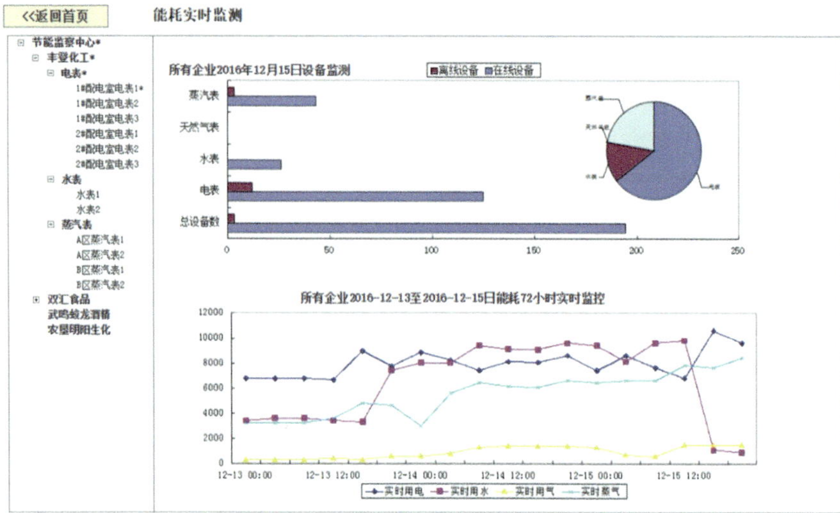

Fig. 5.47 City-level regional platform real-time energy consumption monitoring diagram

through horizontal and vertical two-dimensional comparisons provides scientific support for energy-saving management and energy-saving transformation of the energy consumption supervision system and provides personalized energy-saving guidance suggestions for users.

Enterprise-Level Energy Consumption Measurement Platform

The main functions are as follows:

Real-Time Energy Consumption Data Collection: Real-time energy consumption data collection uses Internet of Things technology to collect data from various energy consumption (electricity, water, gas, coal, etc.) metering devices or existing system database data of industrial enterprises in real time through data collectors. The sampling period can be set to 1 day, 6 h, 1 h, or 15 min and can be as short as seconds. The underlying hardware equipment enables real-time energy consumption data collection and then transmits the data to the data gateway through RS485 communication. After protocol conversion, the data are uploaded to the system database. Managers at all levels can browse all or part of the relevant energy metering information according to their platform permissions. The real-time energy consumption data collection diagram is shown in Fig. 5.48.

Overview of the Energy Consumption Monitoring Scheme: Three-dimensional animation (as shown in Fig. 5.49) shows the overall layout when energy consumption monitoring is conducted for a certain enterprise and explains the setting of the metering points.

Energy Consumption Data Statistics and Analysis: The energy consumption data statistics and analysis module provides hourly, daily, monthly, and annual statistical

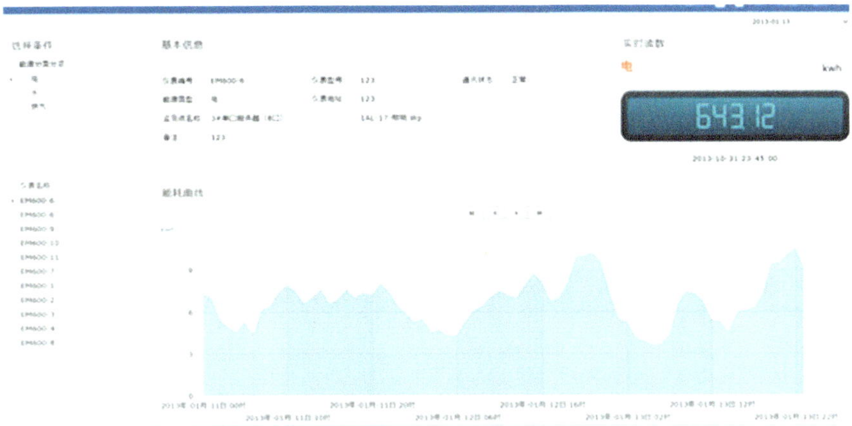

Fig. 5.48 Real-time energy consumption data collection diagram

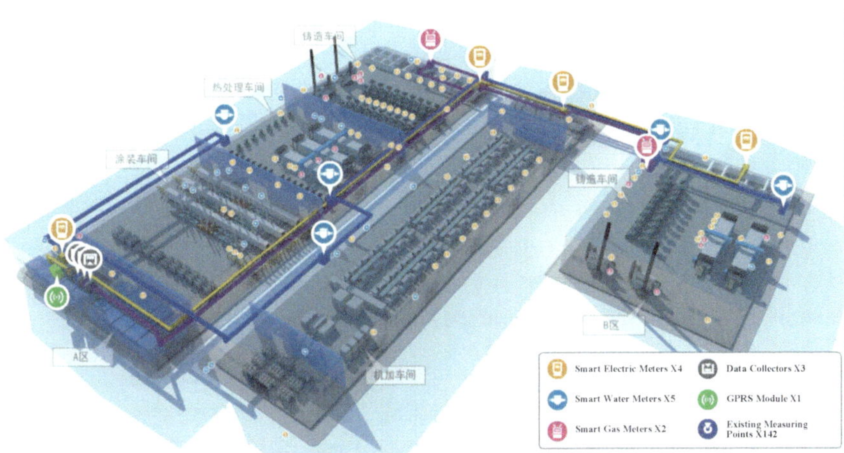

Fig. 5.49 Preview of industrial enterprise data collection

charts and text reports of various classified subitem energy consumption data, as well as charts of various related energy consumption indicators, and conducts year-on-year and month-on-month analyses of historical data. Enterprise managers can conduct year-on-year or month-on-month analyses of the energy consumption of their enterprises, analyze loopholes and unreasonable situations encountered in the process of energy use, and achieve energy efficiency audits and energy savings reduction.

Energy Consumption Data Query: This feature provides real-time and historical energy consumption data query functions for enterprises. Since the energy consumption monitoring object is automatically collected, the energy consumption information collected automatically can be accurate to the hour level; facilitate annual, monthly, and daily queries; and support year-on-year and month-on-month analyses of annual data and monthly data.

Energy Consumption Report Management: The platform supports the export of enterprise energy consumption reports. The three supported report formats are PDF, Word, and Excel files. The results of the energy consumption comparison analysis can be exported in the form of images.

Energy Consumption Warning: A model is used to predict future energy consumption trends for each user's total energy consumption or any classified subitem energy consumption data, considering the impact of seasons and production cycles. This prediction allows users to accurately grasp their energy usage characteristics and trends and promptly issues warnings when abnormal energy usage behavior occurs, enabling users to discover problems early and improve their energy system.

Additionally, the system facilitates energy-saving technology management, supports government departments in pushing various energy-saving policies and technical support information, and has many other functions.

5.6 Smart Urban and Rural: Technology Empowers Rural Revitalization and Promotes Common Prosperity

5.6.1 Reflections on the Imbalances Between Urban and Rural Development

Historical Evolution of the Imbalances between Urban and Rural Development

At the beginning of the founding of New China in 1949, there were weaknesses and poor foundations in all aspects of the country. To accelerate the development of the national economy, the Communist Party of China began to explore the path of industrialization with Chinese characteristics, laying the foundation for the transition from a backward traditional agricultural society to a modern industrial society. However, under the traditional system constituted by the "three pillars", farmers accumulated funds for national industrialization through agricultural taxes and the "scissors gap" in the prices of industrial and agricultural products, resulting in a serious skew in urban–rural relations and the loss of self-development capabilities in agricultural and rural areas,[5] which greatly restricted the development of rural areas. Moreover, cities, as the main locations for industrial development, received the most significant support from national funds, policies, and labor resources. A series of dual urban–rural system policies were subsequently formulated, which strictly distinguished rural

[5] Xiangzhi Kong. The Evolution of Urban–Rural Relations in China over the 70 Years Since Its Founding. Teaching and Research, 2019(08): 5–14.

populations from urban populations and gradually formed the historical environment of the urban–rural division.

After the reform and opening up in 1978, to allow some people and some regions to become rich first and lead the poor, the regional economic scale expanded rapidly, and a large amount of high-quality rural resources flowed into coastal open cities. Although the early stage of the reform and opening up gradually dismantled the urban–rural dual system, promoted the reform of the joint production contract responsibility system, and changed the household registration policy, owing to the stubbornness of the urban–rural dual system and the incomplete reform, the actual work still prioritized industry over agriculture. The quality of elements represented by human capital further differentiated between urban and rural areas, resulting in the inability to increase dividends such as the city opening, technological progress, and ownership reform to benefit rural farmers,[6] and the "three rural" issues (agriculture, rural areas, and farmers) remained severe.

The imbalance between urban and rural development is commonly observed in different economies at specific stages of development. Industrialization and urbanization are inevitable paths, but at the same time, the imbalance between urban and rural development in China continues to worsen. A large amount of resources flow from rural areas to cities, leading to multiple imbalances. The gap between urban and rural areas gradually widens, cities occupy the dominant position in urban and rural areas, and the inherent advantages and values of rural areas are difficult to tap. Thus, the "three rural" issues remain prominent, which can easily trigger deeper and wider social contradictions, making the resolution of the imbalance pattern increasingly complex.

Manifestations of Urban and Rural Development Imbalances: Imbalances in the Population Structure, Economic Development, Ecological Environment, Infrastructure Supply, and Public Service

Although various policies and systems have been introduced since the twentieth century to control the imbalance and uncoordinated situation of urban and rural development and certain results have been achieved, owing to the serious solidification of the historical urban–rural gap, the "three rural" issues are still prominent. The report of the 19th National Congress of the Communist Party of China noted that the main contradiction in Chinese society has transformed from "the contradiction between the people's ever-growing material and cultural needs and backward social production" to "the contradiction between the people's ever-growing needs for a better life and unbalanced and inadequate development". Through the transformation of the main contradiction in Chinese society, we can see the major achievements made since the reform and opening up as well as the resulting development imbalances.

[6] Rui Mao, Xianyi Lin. Promoting Integrated Urban–Rural Development Through Rural Revitalization: Lessons from Major Developed Countries. International Economic Review, 2022(01): 155–173.

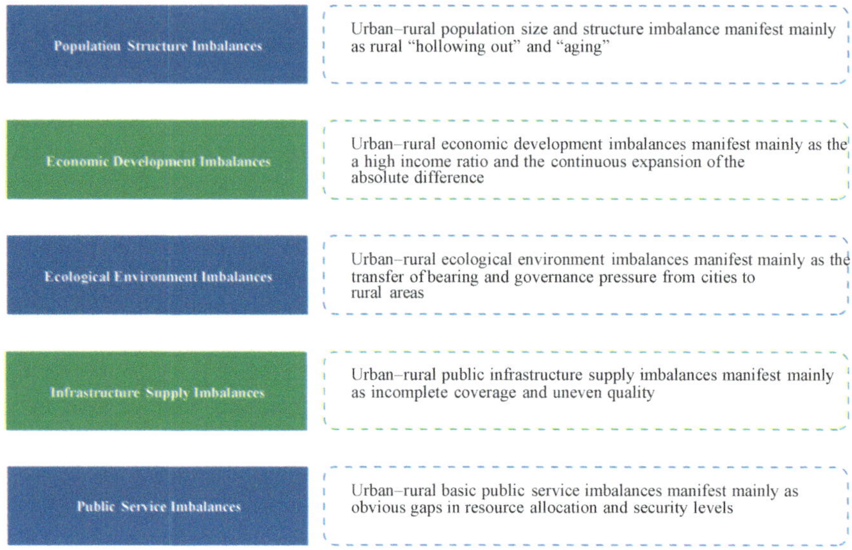

Fig. 5.50 Main manifestations of urban and rural development imbalances (Rui Mao, Xianyi Lin. Promoting Integrated Urban–Rural Development Through Rural Revitalization: Lessons from Major Developed Countries. International Economic Review, 2022(01): 155–173)

The greatest imbalance in China's development is the imbalance between urban and rural development, and the greatest inadequacy in development is the inadequacy of rural development.

The imbalances between urban and rural development are manifested mainly as imbalances in the population structure, economic development, the ecological environment, infrastructure supply, and public services (as shown in Fig. 5.50).

The Digital Divide Brings a New Urban–Rural Development Gap

Given the dual structure of urban and rural areas in China, with the continuous evolution of digital technology, discussions on technological development and digital society often have a strong urban bias and urban standpoint. Rural areas are often marginalized in these discussions, gradually forming a digital divide between urban and rural areas. With the construction of smart cities, new coordinates have been established for the urban–rural digital divide at the levels of technological foundation and application. This process has expanded from the original mobile phones, computers, internet, e-commerce, digital inclusive finance and other ICT access and usage gaps to include industry, reform, governance, and public service gaps based on new generation information technologies such as the Internet of Things, cloud computing, big data, artificial intelligence, and blockchain. These digital divides will lead smart city construction to impact the income gap between urban and rural areas,

Fig. 5.51 The digital divide introduces a new urban–rural development gap

and unlike urban residents, rural residents will be unable to share the new digital dividends brought about by digital industrialization, industrial digitization, digital reform, digital governance, and digital public services[7] (as shown in Fig. 5.51).

If rural areas continue to be excluded from mainstream information exchange and technological innovation, then a series of economic, social, and environmental crises faced by rural development, such as poverty, social marginalization, economic underdevelopment, environmental pollution, lack of infrastructure and services, and low economic capacity, will create a "ripple effect" leading to technological weaknesses, and the inequality of development between urban and rural areas will further increase. **The lack of connection with digitization will become an important inhibiting factor for rural economic and social development and will lead to long-term pressure on economic and social development. Therefore, rural digital construction centered on improving the degree of technological connection has become an urgent issue for rural development.**[8]

5.6.2 Implementation Path of Technological Empowerment for Urban–Rural Integration

To solve the problems of uneven development between cities and rural areas, the unsmooth flow of urban and rural elements, and the unreasonable allocation of public resources, the 2017 report of the 19th National Congress of the Communist Party of China proposed the "implementation of a rural revitalization strategy" and the "establishment of a sound urban–rural integration development system, mechanism,

[7] Yiwu Zeng, Wence Sun, Lili Li, et al. New Coordinates of the Digital Divide: The Impact of Smart City Development on the Urban–Rural Income Gap. China Rural Survey, 2022(03): 165–184.

[8] Yuefei Wu. How Technology Becomes Smarter: Digital Rurality and the Development of Smart Villages in Rural Areas. Theory and Reform, 2022(05): 94–108.

and policy system". In 2018, the Central Committee of the Communist Party of China and the State Council successively issued the "Opinions on Implementing the Rural Revitalization Strategy" and the "Rural Revitalization Strategy Plan (2018–2022)", proposing for the first time the concept of the "digital village". Subsequently, policy documents related to digital villages and urban–rural integrated development have been released every year, providing strong policy support for urban–rural integrated development and rural transformation (as shown in Table 5.3).

Establishing and improving the urban–rural integration development system, accelerating the modernization of agriculture and rural areas, accelerating the construction of digital villages, and achieving the coordinated development of smart city construction and digital village construction are effective ways to bridge the new urban–rural digital divide and promote urban–rural integration development.

Through the use of digital technology, the following can be achieved: promoting the efficient use of rural social resources, coordinating the development of digital villages and smart cities, **using the digital economy to lead the development of rural areas, endowing powerless rural development with a digital engine, rejuvenating new vitality into rural digital development**, continuously expanding the depth and breadth of urban–rural integration development, **resolving the imbalances between urban and rural areas**, and helping rural revitalization and urban–rural integration development. The basic concept of urban–rural integration development is shown in Fig. 5.52.

Application Architecture of Smart Agriculture and Rural Areas under the Cloud-Edge-End Collaboration System

Based on the framework of cloud-edge-end collaboration technology integration and application in smart agriculture, we further integrate smart rural applications to provide a more comprehensive digital engine for rural revitalization and integrated urban–rural development. The architecture of smart agriculture and rural applications based on the cloud-edge-end collaboration system is shown in Fig. 5.53.

By utilizing information technologies such as cloud computing, big data, artificial intelligence, and the Internet of Things, we build platforms for cloud computing, big data analysis, data, the Internet of Things, and edge intelligent computing. These platforms allow for precise data collection, early warning, and analysis, achieving interconnection and resource sharing of agricultural and rural data, collaborative business operations, real-time comprehensive understanding of agricultural and rural development, improving the quality and efficiency of rural management, optimizing agricultural resources, promoting the sustainable development of digital agriculture and rural applications, and driving the modernization of agriculture and rural areas with digitization, all of which provide strong support for comprehensive rural revitalization.

Table 5.3 Policy support for urban–rural integration and the digital village

Time	Unit	Policy document	Related content
May 2019	Office of the Central Committee of the Communist Party of China, Office of the State Council	"Outline of the Digital Village Development Strategy"	Proposes to accelerate the modernization of agriculture and rural areas; focus on the basic supporting role of informatization in promoting the modernization of rural governance systems and capabilities; build a new system of digital rural governance; focus on bridging the urban–rural "digital divide"
December 2019	Ministry of Agriculture and Rural Affairs, Central Cyberspace Administration	"Digital Agriculture and Rural Development Plan (2019–2025)"	With industrial digitization and digital industrialization as the main line of development and the deep integration of digital technology and agricultural and rural economy as the main direction, this plan focuses on building a basic data resource system; strengthening the digital production capacity; accelerating the digital transformation of agricultural and rural production and management services; promoting the interconnection and sharing of government information systems and public data; using digitization to lead and drive the modernization of agriculture and rural areas; and providing strong support for the comprehensive revitalization of rural areas
July 2020	Office of the National Development and Reform Commission	"Notice on Accelerating the Implementation of New Urbanization Construction to Strengthen Weak Links and Orderly Promote the Intelligent Transformation of County Towns"	Given the shortcomings and weak links in county town infrastructure, public services, social governance, industrial development, and digital ecology, digital technologies such as big data, artificial intelligence, and 5G should be used to promote the intelligent transformation and construction of county towns in areas with a certain foundation, focusing on strengthening weak links and achieving practical effects
January 2021	Central Committee of the Communist Party of China, State Council	"Opinions on Fully Promoting Rural Revitalization and Accelerating the Modernization of Agriculture and Rural Areas"	Proposes to develop smart agriculture, establish a big data system for agriculture and rural areas, promote the deep integration of new generation information technology with agricultural production and operations, and strengthen the digital and intelligent construction of rural public services and social governance

(continued)

Table 5.3 (continued)

Time	Unit	Policy document	Related content
March 2021	State Council	"Fourteenth Five-Year" Plan	Proposes to construct smart cities and digital villages, promote innovation in urban and rural development and governance models with digitalization, accelerate the construction of digital villages, build a comprehensive information service system for agriculture and rural areas, establish a universal service mechanism for agricultural information, and promote the digitalization of rural management services
July 2021	Office of the Central Cyberspace Affairs Commission, Office of the Ministry of Agriculture and Rural Affairs, etc	"Digital Village Construction Guide 1.0"	To address the questions of "why to build, how to build, who to build, and what to build", a general reference framework for the construction of digital villages is systematically established, clarifying the construction scope, main construction tasks, precautions and other key elements of various application scenarios
January 2022	Central Cyberspace Affairs Commission, Ministry of Agriculture and Rural Affairs, National Development and Reform Commission, etc	"Digital Village Development Action Plan 2022–2025"	To provide strong support for new progress in rural revitalization, take new steps toward the modernization of agriculture and rural areas, and promote new achievements in the construction of Digital China, this plan focuses on developing the digital economy in rural areas; improving the digital literacy and skills of farmers; enriching the network culture in rural areas; and enhancing the efficiency of digital governance in rural areas

(continued)

Table 5.3 (continued)

Time	Unit	Policy document	Related content
March 2022	National Development and Reform Commission	"Key Tasks for New Urbanization and Urban–Rural Integration Development in 2022"	Promotes urban–rural integration development, with the county as the basic unit, extends urban infrastructure to rural areas, and covers public services and social undertakings in rural areas
January 2023	Central Committee of the Communist Party of China, State Council	"Opinions on Doing a Good Job in Fully Promoting Rural Revitalization in 2023"	Proposes to develop modern facility agriculture, implement actions to upgrade the modernization of facility agriculture, deepen the implementation of digital rural development actions, promote research and development on and the promotion of digital application scenarios, accelerate the application of big data in agriculture and rural areas, and promote the development of smart agriculture

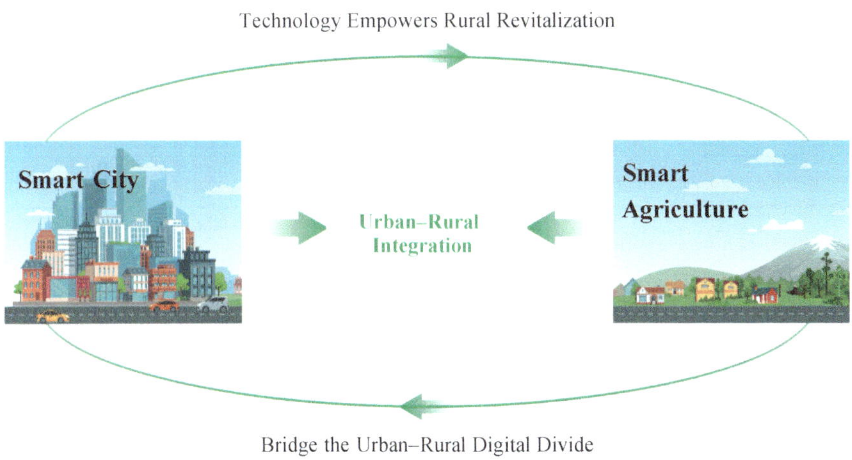

Fig. 5.52 Urban–rural integration development concept

The following sections share how to empower urban–rural integration through technology from the perspectives of smart villages and smart agriculture, achieving improved quality and efficiency in agriculture, increased income for farmers, and effective rural governance and narrowing the digital divide between urban and rural areas.

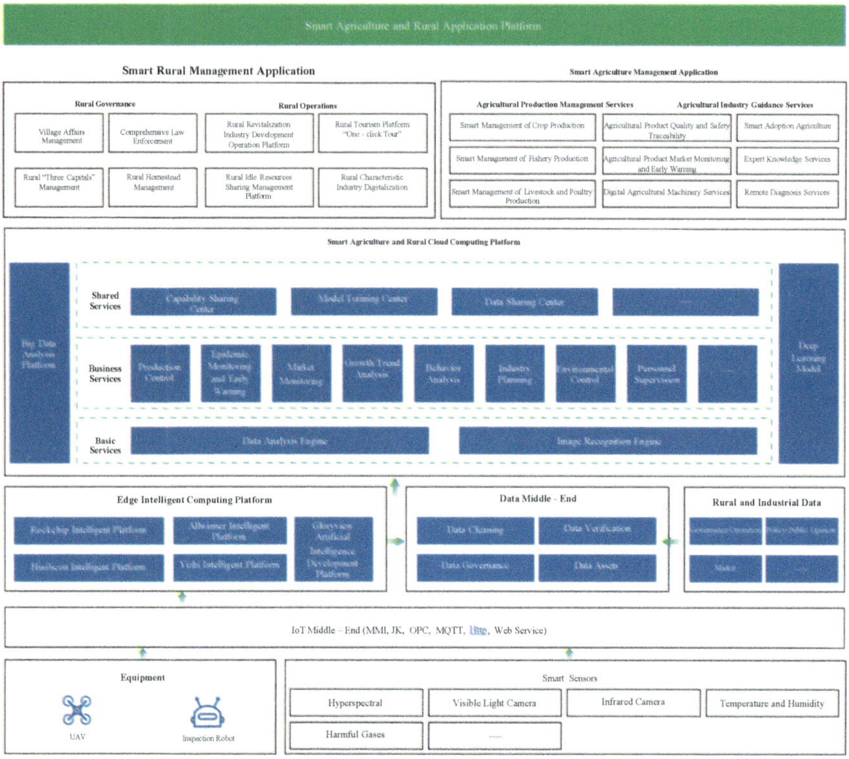

Fig. 5.53 Architecture of smart agriculture and rural applications based on the cloud-edge-end collaboration system

Smart Village: Digitization of Rural Governance and Operations, Narrowing the Urban–Rural Digital Divide

Smart villages improve rural management efficiency and revitalize rural resources with the support of digital technology, thereby increasing the vitality of rural development. The content of smart village construction is shown in Fig. 5.54.

Smart Village Governance: Improving the Level of Rural Governance through Digitization

Digital Village Affairs: Building applications for information publication, village affairs handling, rural governance, idle resources, neighborly assistance, industrial development, and public services are used to strengthen rural governance, serve the public, and facilitate public participation. It is the official channel for the village committee to share information dynamics and handle village affairs by appointment. Villagers can view public affairs, the latest news, and guidance on handling affairs through mini-programs or apps and can use online appointments, video surveillance,

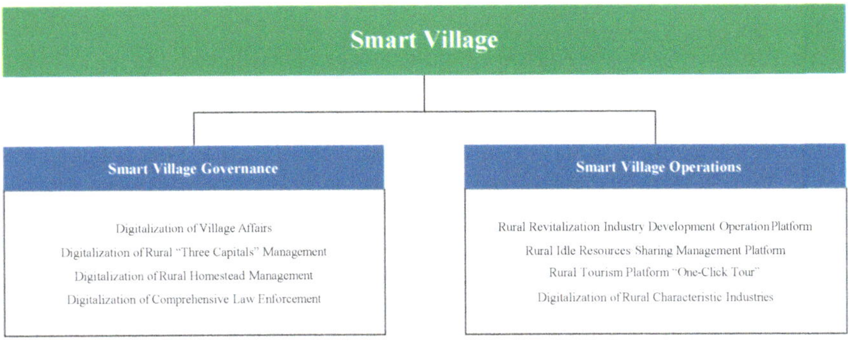

Fig. 5.54 Smart village construction scope

charity activities, legal consultation, idle resource docking, housing rental, and agricultural science popularization applications, making it faster, more service-oriented, and better for rural residents to handle affairs.

Comprehensive Digital Rural Law Enforcement: Drone technology is used to provide intelligent, efficient, and convenient management methods for addressing comprehensive rural law enforcement matters such as illegal construction investigations, drought and flood disaster prevention, farmland operation inspections, and road occupation rectification (as shown in Fig. 5.55).

Digital Management of Rural Collective "Three Capitals": To promote the digitization of rural collective asset inventory and capital verification information; establish electronic ledgers for the registration, custody, use, and disposal of collective assets; build a large database of rural collective assets; strengthen the supervision of agricultural asset possession, use, income, and disposal; guide the standardized circulation and transaction of rural property rights; standardize data collection, transmission, use, and sharing standards; promote the docking and sharing of rural collective asset data with basic land resource data, basic population resource database, etc.; provide comprehensive supervision and early warning analysis of rural collective assets, resources, and funds; facilitate the data submission, statistics, quality inspection, and update management of rural collective asset inventory and capital verification, member definition, equity quantification, organization establishment, etc.; and provide rich collective property rights supervision and operation business

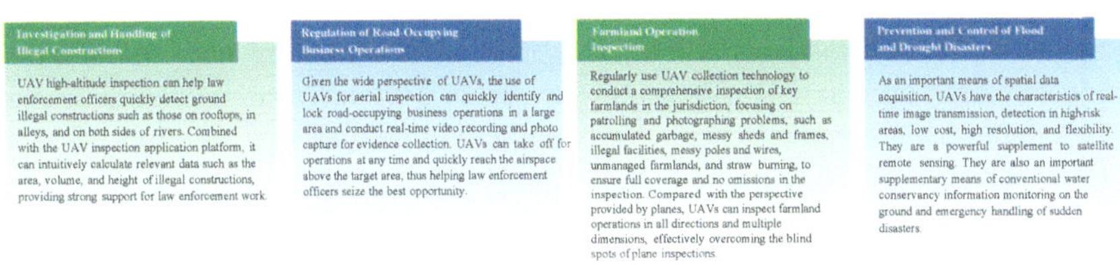

Fig. 5.55 Comprehensive digital law enforcement

processes, such as equity setting, dividend management, asset operation supervision, and statistical analysis, to meet the information management needs of rural collective property rights at all levels of rural management departments. Multiple dimensions can be used to comprehensively summarize, analyze, and display multisource heterogeneous data such as rural land resource data, cooperative data, rural financial data, family farm data, three-capital public information, inventory and capital verification information, equity reform information, etc., in agricultural management work, provide multiangle and comprehensive data for rural economic management work, and provide a data basis for rural economic management decision-making.

Digital Management of Rural Homestead: By establishing a rural homestead management system, we can achieve a seamless flow of information for rural homestead applications, construction reporting, approval, certification, management, and investigation. Moreover, the system organically integrates data on current land use, land use planning, urban and rural planning, village planning, real estate registration survey data, shareholder information, household registration information, etc., to improve the efficiency of homestead approval and integrate the homestead approval database and management. Using data from the third national land survey, satellite remote sensing, and other information, combined with the registration and certification of homestead use rights and surveys on the current status of rural homesteads and rural houses, a rural homestead database is constructed, covering information such as homestead units, spatial distribution, area, ownership, restrictions, and usage status. The informatization of homestead allocation, approval, transfer, use, supervision, statistics, etc., should be promoted, and basic data should be improved and updated in a timely manner.

Smart Village Operation: Creating a New Pattern of Rural Industry Operation for Rural Revitalization

Rural Revitalization Industry Development Operation Platform: The Rural Revitalization Industry Development Operation Platform is a new generation digital platform designed to cultivate and manage new rural industries, solidify rural industry interest linkage mechanisms, innovate new models of rural industry income sharing, enhance village collective income and promote income growth for villagers. The platform revolves around the operation of new rural industries (courtyard economy, market economy, new rural retail, etc.). Guided by rural industry revitalization, with the cultivation and development of new rural industries at its core, and the digital operation of rural industries as its goals, it offers a comprehensive solution for the integration of online and offline rural industry development. It facilitates functions such as unified collection, logistics distribution, automatic accounting, and data analysis; connects village committees, village enterprises, and villagers through a digital platform; creates an integrated online + offline operation model; and achieves effective industry integration. The rural revitalization operation model is shown in Fig. 5.56.

Rural Idle Resource Sharing Management Platform: The Rural Idle Resource Sharing Management Platform is a digital platform that helps rural areas manage idle resources such as farmhouses owned by village collectives and individuals. The

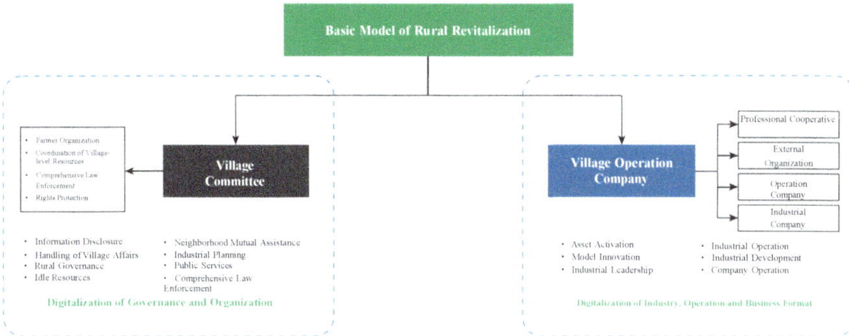

Fig. 5.56 Rural revitalization operations model

village committee and third-party operating companies actively explore diversified utilization models of idle rural resources, increase the development of rural residential homestays, and accelerate the integrated development of agriculture, commerce, culture, and tourism. It encourages and guides capable farmers to lease idle farmhouses to skilled operators to develop courtyard economies, homestay economies, rural leisure and elderly care, and other projects.

Rural Tourism Platform "One Machine Tour": Given the opportunity for comprehensive digital development, various regions in the country are actively exploring and practicing smart agriculture, culture, and tourism. Among them, "One Machine Tour", which is a tourism platform under the new ecological model, has demonstrated outstanding performance in providing effective supervision for local managers of village cultural and tourism services and promoting the transformation and upgrading of local agricultural industries.

Digitalization of Rural Specialty Industries: Through digital marketing and e-commerce platforms, new modes of assisting agriculture are created for rural specialty industries: digital marketing includes establishing regional cooperative short video libraries, short video production training centers, digital marketing leaderboards, etc.; digital e-commerce includes building brand live broadcast platforms, data sharing platforms, product docking systems, etc., to achieve full-chain docking of agricultural products.

Smart Agriculture: Digital Technology Promotes Quality Improvement and Increased Income in Agricultural Production

Traditional agriculture has objective problems that restrict agricultural development, such as high agricultural labor costs, a lack of scientific management of agricultural products, constrictions in agricultural labor, low labor productivity, and weak agricultural innovation. Therefore, to solve these problems, there is an urgent need to shift toward digital and smart agriculture via technologies such as the Internet of

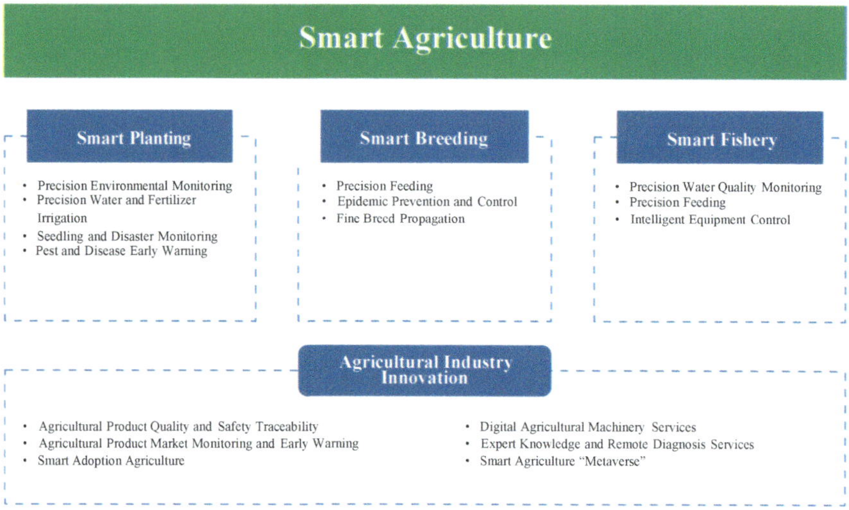

Fig. 5.57 Smart agriculture construction scope

Things, big data, blockchain, cloud computing, artificial intelligence, etc.; organically combine high-tech with geography, agriculture, ecology, plant physiology, soil science and other basic disciplines; construct a smart agriculture system and model; and achieve simulation, monitoring, judgment, prediction, and recommendation throughout the entire process of agricultural production, from crop planning, input, production, to agricultural product harvesting, processing, marketing, etc., to improve resource utilization, reduce costs, and improve agricultural production efficiency and product quality.

The scope of smart agriculture construction is shown in Fig. 5.57. The following describes how to achieve agricultural intelligence from the aspects of smart breeding, smart planting, smart fisheries, and agricultural industry innovation:

Smart Breeding: Intelligent Precision Breeding to Improve Livestock and Poultry Breeding Yield and Quality

By constructing systems for environmental control, precision feeding, and disease prevention, we enhance the integrated application of intelligent perception, intelligent analysis, and intelligent control technologies and equipment in the livestock industry. This enhancement allows for effective management of the entire livestock production process, thus improving production efficiency.

Precision Environmental Monitoring: Through the construction of an automated precision environmental control system, we can achieve automatic control of the livestock and poultry breeding environment, replace manual adjustments, and improve the accuracy of environmental control. By coupling environmental data with production efficiency data for analysis, we can build an agricultural production environment regulation model, identify the optimal coupling value, and provide feedback control-related equipment. These capabilities ensure that the indoor temperature, humidity, ammonia concentration, and other environmental controls are within the optimal range for livestock and poultry growth, thereby improving production performance. The hardware of the automated precision environmental control system consists of water curtains, fan equipment, sensors, environmental controllers, lighting systems, etc. The system uses sensors to detect temperature, ammonia, light intensity, and carbon dioxide concentration and controls the fans, water curtains, and lighting equipment in the breeding house, providing the best growth environment for livestock and poultry. The logic of environmental monitoring is shown in Fig. 5.58, and the function of smart lighting in breeding is shown in Fig. 5.59.

Precision Feeding: By constructing a digital precision feeding system, we replace traditional manual feeding methods with automated feeding methods. Feed storage and transportation use a bulk form, with equipment configured for feeding, thus reducing packaging and bag handling processes. Automatic feeding allows for simultaneous feeding and precise positioning, improving feeding efficiency and effectively reducing livestock and poultry stress responses. At the same time, the system collects feeding and weighing data from the feeding equipment, uploads these data to the data

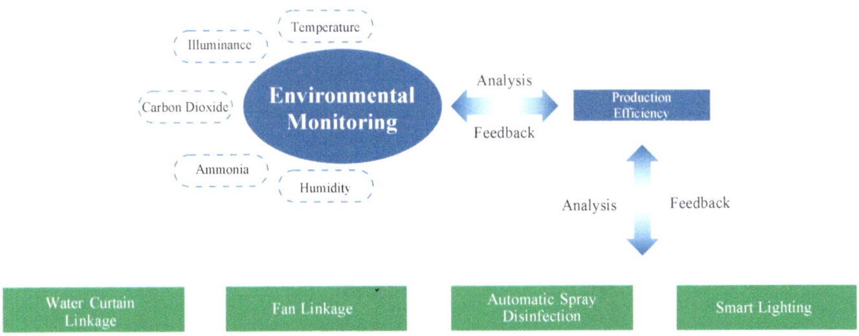

Fig. 5.58 Precision environmental monitoring

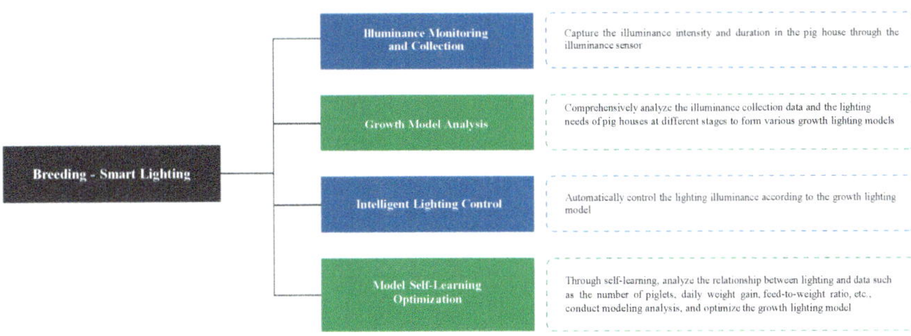

Fig. 5.59 Precision environmental monitoring—smart lighting (taking pig breeding as an example)

Fig. 5.60 Precision feeding

center for analysis, guides precise feeding ratios, and improves the feed-to-weight ratio. As livestock and poultry have different requirements for feed ratios and usage at different production stages and levels, the system can analyze the feed-to-meat ratio and livestock and poultry growth records, automatically generate feeding plans, automatically output reminders concerning feed types, and control feed delivery through intelligent feed weighing. It also generates feed demand plans and automatically links to the feed processing management system. By using front-end intelligent weighing equipment and water consumption monitoring equipment, it collects feed intake data, thereby comprehensively analyzing the feed-to-meat ratio, disease status, etc., of the pig herd. The precision feeding function is shown in Fig. 5.60.

Disease Prevention and Control: This system includes modules for disease monitoring and early warning, disease record management, disease knowledge management, and prevention reminders (as shown in Fig. 5.61). The disease monitoring and early warning module deploys intelligent AI surveillance cameras and uses intelligent video analysis algorithms to monitor the behavior, body temperature, excrement, etc., of livestock and poultry in real time. Through big data analysis, it automatically identifies the disease infection status of livestock and poultry and provides timely warnings and reminders of abnormal situations, effectively preventing the spread of diseases. The system integrates environmental data, food intake data, water intake data, etc., collected from the environmental control system and feeding management system for analysis, judges whether there are abnormalities in livestock and poultry, and dynamically calculates fluctuations in various indicators. When abnormalities occur, it issues warnings and reminders through the system platform to control livestock and poultry diseases in a timely manner. The disease record management module provides functions such as the construction of livestock and poultry disease archives, records of disease onset, and vaccination records and provides more source analysis data for disease monitoring and early warning modules. A knowledge

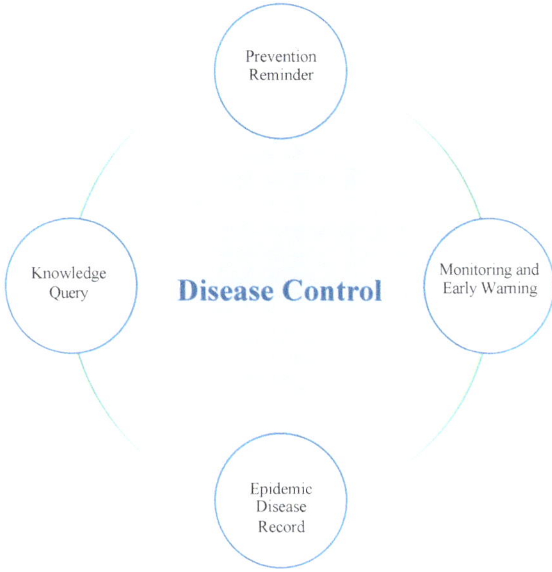

Fig. 5.61 Disease prevention and control

query accumulates data from the disease expert knowledge base, providing disease knowledge queries for managers.

Quality Breeding: This system achieves the pedigree information management for livestock and poultry breeding, including the record management of all livestock and poultry parents' pedigree information and growth conditions. Through the paperless collection of various environmental data, growth data, feeding data, and production performance data of breeding livestock and poultry, the system analyzes pedigree information, formulates balanced breeding plans, uses data analysis to maximize production performance, physical fitness, feed consumption, and other indicators, improves the breeding/hatching rate and chick quality of breeding livestock and poultry, improves the production performance and genetic stability of breeding livestock and poultry, and achieves precision breeding. Taking pig breeding as an example, the function of quality breeding is shown in Fig. 5.62.

Intelligent Agriculture: Technology Transforms Dependence on Weather into Informed Decisions

Intelligent agriculture, which is based on technologies such as blockchain, the Internet of Things, and artificial intelligence, involves the construction of systems for environmental monitoring, the precise application of water and fertilizers, and warning of pest disease. These systems transform traditional agriculture's "dependence on weather" into intelligent agriculture's "informed decisions". The intelligent agriculture platform dynamically monitors soil moisture, crop growth, and pest conditions, promptly issues warning information, enhances the informatization level

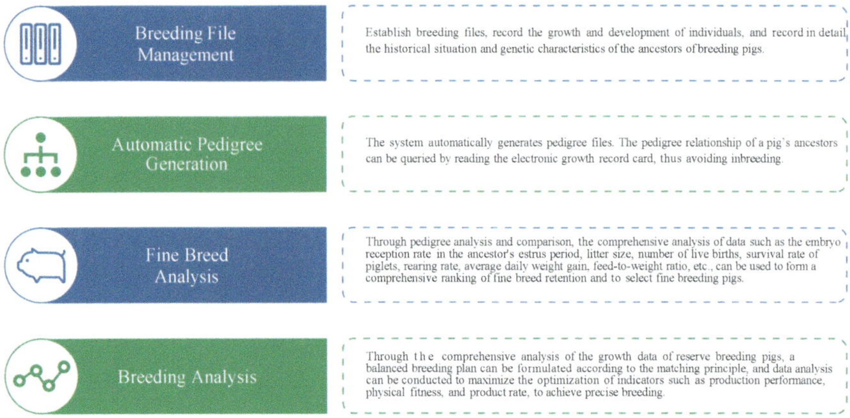

Fig. 5.62 Quality breeding (taking pig breeding as an example)

of agricultural production management, and improves crop yield and quality. The intelligent seed platform is shown in Fig. 5.63.

Soil Moisture Monitoring: Automatic soil moisture monitoring involves monitoring the soil water content and measuring the volumetric water content of the soil through moisture sensors. Additionally, it can be extended to configure soil temperature sensors, soil conductivity sensors, soil compaction meters, in situ soil salinity analyzers, soil (fertilizer) nutrient rapid testers, and other equipment to monitor all soil parameters. Soil moisture monitoring can provide effective guidance for plant operations. For example, in the cultivation of mandarins, the acidic nature of the

Fig. 5.63 Schematic diagram of the intelligent agriculture platform

soil can lead to soil hardening and yellowing of the mandarins. Through intelligent technology, real-time precise monitoring of soil pH can be carried out. The system can issue timely reminders to farmers to "reduce acidity" and "reduce fertilization" or link with the water and fertilizer integration system to automatically control the soil pH, thereby ensuring the quality of agricultural products.

Precision Microclimate Monitoring and Warning: In traditional agricultural planting, the largest problem troubling farmers is that "agriculture depends on the weather". Many natural disasters, such as drought, high temperature, and frost, occur every year and constantly affect the growth of crops. In addition to the impact of extreme weather, factors such as light, heat, water, and carbon dioxide constantly affect the development of crops. Therefore, it is particularly important to predict and monitor the changing climate environment effectively and to be prepared for rain to prevent economic losses to the greatest extent and to greatly improve crop yield and quality. The precision microclimate monitoring system uses front-end sensors to monitor local rainfall, air temperature, air humidity, atmospheric pressure, wind speed, wind direction, etc., combines with global meteorological satellite organization data fitting, achieves precise monitoring and warning of the local microclimate environment, and can combine with the occurrence rules of pests and diseases and growth environment requirements to establish a warning mechanism to guide agricultural production (as shown in Fig. 5.64).

Precision Environmental Monitoring and Control: For greenhouse cultivation, a precision environmental monitoring and control system is constructed. Greenhouse environment monitoring mainly monitors the indoor temperature, humidity, light intensity, carbon dioxide, etc., providing data support for the control of indoor environment control equipment. The environmental control system provides various controls, including manual control, threshold control, and automatic control. The controlling equipment includes fans, internal shading nets, external shading nets,

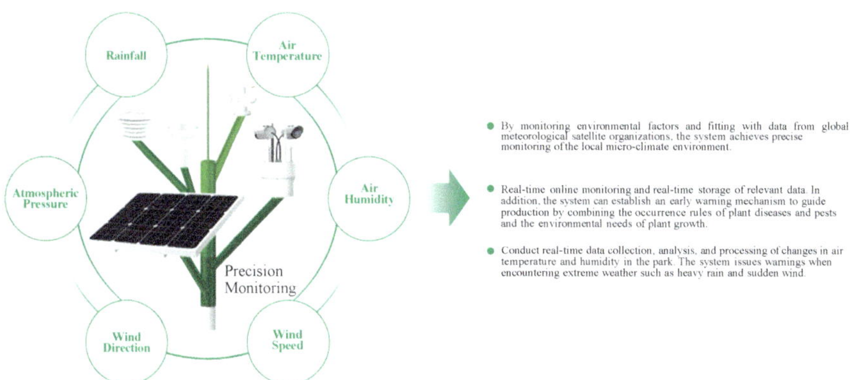

Fig. 5.64 Precision microclimate monitoring and warning

wet curtains, sprinklers, supplementary lights, etc. After completing the fusion analysis of environmental data, soil moisture data, growth information, etc., through information fusion technology, an environmental control strategy is formulated, and related control instructions are automatically issued to water and fertilizer equipment, fans, wet curtains, shading, etc., thereby achieving precise control of facility agriculture.

Pest Disease Warning and Control: By deploying pest monitoring equipment, the automation and intelligence of pest trapping, classification statistics, real-time reporting, remote detection, pest warning, and control guidance are achieved. The system uses image recognition algorithms, cross-references the information with a large number of pest disease image libraries, quickly and accurately identifies pests, and outputs related control methods, thereby reducing losses by control and identification.

Precision Irrigation with Integrated Water and Fertilizer: The integration of water and fertilizer is a new agricultural technology that combines the processes of irrigation and fertilization. With the Internet of Things, big data, cloud computing, and sensor technology, according to soil moisture and environmental climate monitoring data, the water and fertilizer use of crops can be analyzed and predicted, preset thresholds can be used as control conditions for operating irrigation equipment, timely irrigation and fertilization strategies can be formulated, intelligent irrigation can be achieved, water management can be strengthened, and water leakage can be eliminated, thereby reducing channel water loss, achieving scientific scheduling and reasonable water and fertilizer distributions, realizing fine and timely irrigation, and liberating the labor force to develop efficient agriculture systems. The architecture of the intelligent control system of the integrated water and fertilizer system is shown in Fig. 5.65.

Integration of Water and Fertilizer: By effectively connecting the smart water and fertilizer machine with the field irrigation network, intelligent integrated water and fertilizer irrigation is carried out based on the analysis of the soil moisture in the farmland, meteorological rainfall forecasts, and AI algorithm analysis of crop nutrient requirements. This process achieves precise irrigation and fertilization. The system supplies water and fertilizer through a controllable pipeline system, which, after blending, is sprayed through pipes, spray guns, or nozzles in a uniform, timed, and quantified manner in the crop growth area. This step ensures that the main growth area of the soil always maintains a loose and appropriate water content. Additionally, according to the fertilization characteristics of different crops, soil environments, and nutrient content conditions, water and nutrients are timed, quantified and directly provided to crops in proportion.

Precision Irrigation: By collecting information returned by sensors, the soil water demand is analyzed, and automatic irrigation is carried out. When the set threshold is reached, irrigation is stopped, achieving the goals of water conservation and precision irrigation.

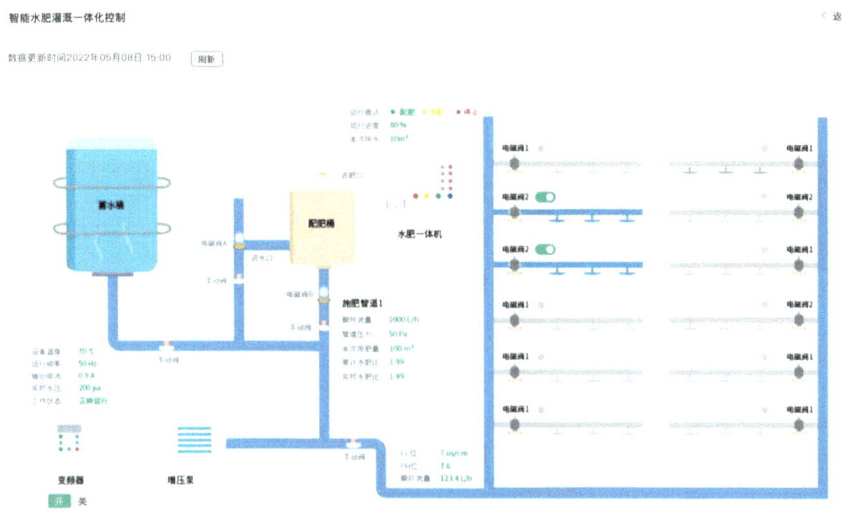

Fig. 5.65 Architecture of the intelligent control system for the integrated water and fertilizer system

Remote Central Control: This function supports remote control, manual control, automatic control, timed control, and other work modes and can control all irrigation equipment, thus reducing labor costs.

Seedling and Disaster Monitoring: By deploying a video surveillance system and combining it with artificial intelligence (AI) algorithms, the growth and disease conditions of crops can be analyzed and identified, and the growth status of crops can be determined in real time. In combination with human and object recognition algorithms, abnormal personnel behaviors, rodent damage, and other abnormal situations can be detected, thus ensuring the safety of biological assets. Video surveillance includes high-definition cameras and AI cameras. High-definition cameras play a role in security and video inspection, whereas AI cameras are used for seedling and disaster identification and monitoring.

Smart Aquaculture: Promoting the High-Quality Development of Aquatic Products

The development of aquaculture in China is rapid, but there is still a large gap in precision breeding, scientific management, disease control, product safety, and efficient aquaculture base management. By constructing an Internet of Things-based aquaculture production and management system, the traditional extensive and experience-based breeding mode has been transformed into a refined, networked, and intelligent management modern aquaculture mode. This new mode achieves scientific management in water quality monitoring, precise feeding, disease prevention and control, and quality traceability, effectively increases yield, shortens the breeding cycle, reduces breeding risks, lowers production costs, and enhances the ability to control the quality of aquatic products. Moreover, it can reduce labor intensity, expand the production

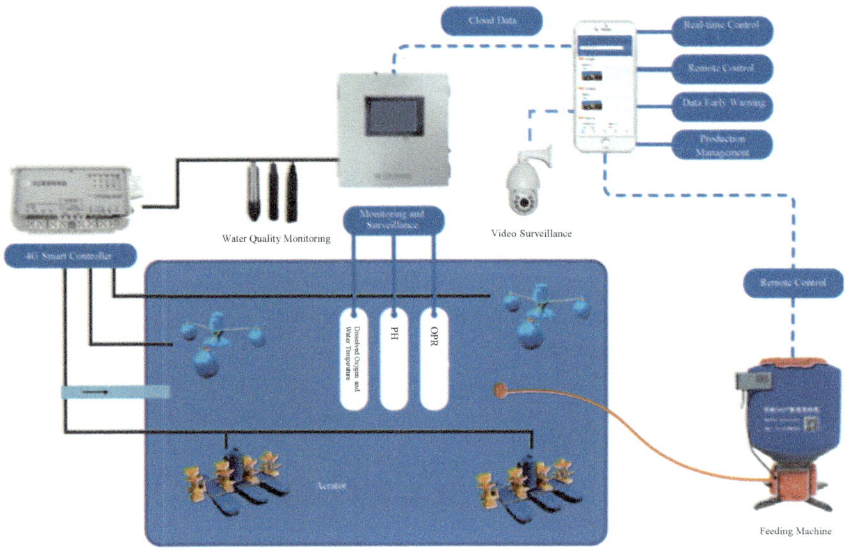

Fig. 5.66 Smart aquaculture system diagram

scale, and increase the automation level of production management, which plays an important role in promoting the development of modern aquaculture.

The system consists of an online aquaculture monitoring platform and an intelligent water quality monitoring system (water temperature, water level, pH, dissolved oxygen, ammonia nitrogen, salinity, turbidity sensors, etc.), an oxygenation/new water control system, an automatic feeding system, an intelligent monitoring system, etc. (as shown in Fig. 5.66).

Water Quality Monitoring: In various water quality monitoring methods (oxygen, temperature, and pH values), the traditional method requires regular manual onsite water quality measurement, and the measurement value varies due to human errors in operation. This system uses 24-h unmanned wireless monitoring to measure water quality and automatically records and saves the measurement results at regular intervals. The water quality monitoring system achieves 24-h real-time monitoring of dissolved oxygen, water temperature, ammonia nitrogen, pH value, salinity, ORP, etc. The data are accurate and can be viewed in real time on the mobile app. When the data are abnormal, the system issues an alarm in a timely manner and can connect to the intelligent controller to automatically control the switch of the breeding equipment.

Precision Feeding: In terms of feeding methods, the traditional method uses manual broadcasting, which requires considerable labor and uneven broadcasting, and the total amount of feed is not easy to control, resulting in the waste of feed. The precision feeding system uses automatic feeding equipment to feed the fish pond at a fixed time and quantity according to the system settings. The total amount of feed, feeding time, and uniformity are controlled. The system uses a pneumatic feeder to transport materials by air. The feeding equipment floats on the surface of the aquaculture pond. Owing to the breeding characteristics of aquaculture, clustering in one location can destroy the soil and damage the breeding base. The feeding system uses four-direction uniform feeding to guide the spread of aquaculture evenly, avoid clustering, destroy the breeding base, and achieve precise feeding, thereby reducing unnecessary loss of feed.

Intelligent Control: The control layer of the intelligent control system includes mainly oxygenation pump control, circulation pump control, camera control, and feeder control. Through the intelligent control system, the automation mode can be freely configured, and centralized, remote, and linked control of the environment within the aquaculture pond can be achieved through circulating water. In terms of the start/stop method of oxygenation equipment, the traditional method cannot determine the oxygen content of the water in real time and can initiate oxygenation recognition based on experience only, which wastes a large amount of electricity due to long-term operation. This system intelligently controls the oxygenation equipment based on the set and measured values, delays equipment wear, reduces production losses, and saves considerable electricity. When the data monitored by the sensor exceed the set safety threshold, multiple devices can be automatically linked to start and stop without human intervention, ensuring safe water quality. This system enables automatic control of water changes, oxygenation, heating, feeding, and other equipment operations according to the preset conditions of aquaculture, meets the stringent requirements of aquaculture environmental conditions, reduces unnecessary losses, and can save electricity and reduce production costs.

Scientific innovation provides new momentum for the high-quality development of the entire agricultural industry chain.

Traceability of agricultural product quality and safety: Using information technology, the entire process of agricultural production is included in the traceability system, including tracking and recording the quality and safety information of agricultural products such as the production and operation entities, the production process, and the flow of agricultural products, to meet the needs of supervision and public inquiries. This approach makes every production link of agricultural products transparent and visible. The information on the production process and the flow of agricultural products (taking vegetables as an example) includes the following: input materials (such as seeds, fertilizers, pesticides, etc.)—plant protection link—harvesting link—processing link—storage link—transportation logistics link—wholesale and retail link, etc. All the data can be traced back to the origin (as shown in Fig. 5.67). Thus, the system achieves the traceability of the source of agricultural products, trackability of the flow, early warning of risks, product recalls, and accountability of responsibilities, effectively promoting green agricultural production and ensuring safe public consumption.

Agricultural product market monitoring and early warning: Industry data, agricultural statistical data, public data from authoritative websites, etc., are combined to build a market index system for production areas and various varieties, including the

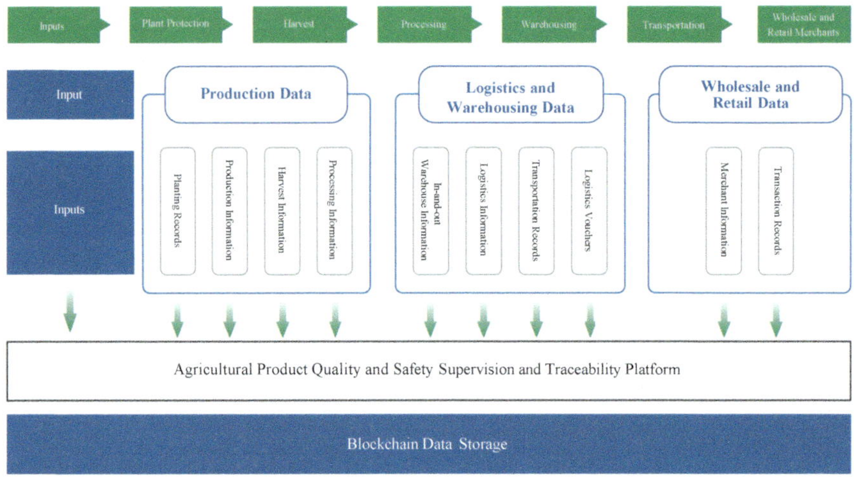

Fig. 5.67 Traceability of agricultural product quality and safety (taking crop farming as an example)

price index, supply and demand index, and industry health index, for big data analysis to fully perceive market trends. Moreover, based on industry policies, industry research reports, emergencies, expert comments, etc., an agricultural product public opinion monitoring system is built to perceive changes in market sentiment; collect and analyze reports (daily, weekly, annual, professional reports); perform monitoring and early warning of agricultural product prices and trends; and provide a decision-making basis for market regulatory bodies, agricultural production and operation entities, and consumers. The market monitoring and early warning of multisource data fusion are shown in Fig. 5.68.

Smart Adoption Agriculture: Smart adoption agriculture is a new rural industry where consumers pay production costs, and producers provide green, organic food. By establishing a risk-sharing and benefit-sharing cooperation model between producers and consumers, the direct connection between rural areas and cities and between land and tables can be established. The operating entities that adopt agriculture carry out digital transformation according to their actual conditions; achieve the intelligent, digital, and remote control of agricultural farming and breeding; and visualize, digitally display, and informatize the management of agricultural objects, the environment, and the entire production process. Consumers can use apps, mini-programs, etc., to perform online adoption and real-time monitoring and view the results of pastoral planting, livestock, fruit trees, and fish ponds.[9] The functions of smart adoption agriculture are shown in Fig. 5.69.

[9] Secretariat of the Cyberspace Administration of China; General Office of the Ministry of Agriculture and Rural Affairs; General Office of the National Development and Reform Commission; et al. *Digital Rural Development Guide 1.0*. 23 July 2021.

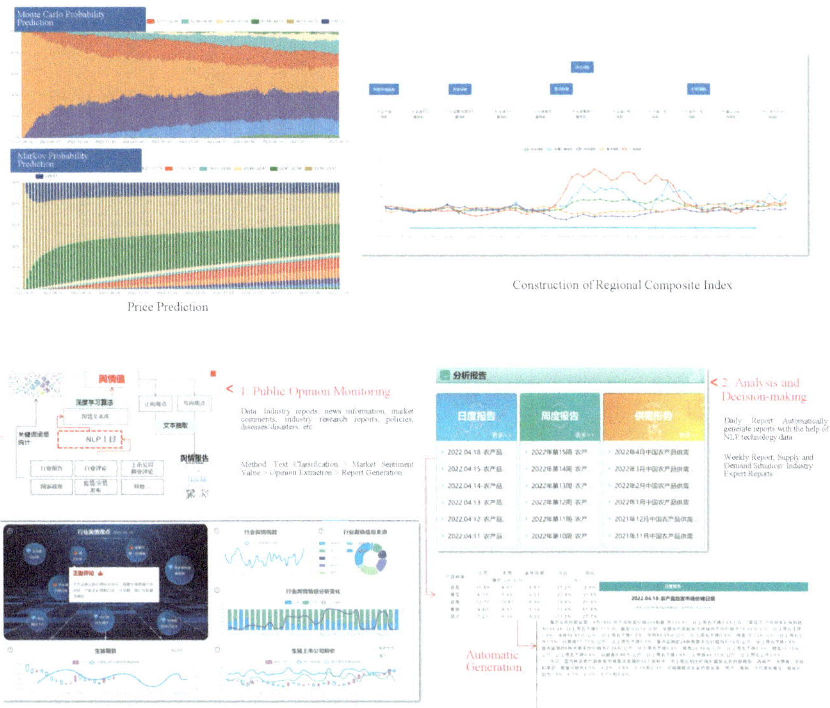

Fig. 5.68 Market monitoring and early warning of multisource data fusion

Fig. 5.69 Smart adoption agriculture

Digital services for agricultural machinery: Operating entities such as farmer cooperatives, family farms, ranches, fish farms, and agricultural enterprises invest in intelligent agricultural machinery equipment such as unmanned aerial vehicles for plant protection, remote-controlled crawler rotary tillers, self-propelled reel sprinklers, unmanned tractors, unmanned rice transplanters, and unmanned harvesters in the planting or harvesting links. They carry out automatic instrument and intelligent instrument plowing and harvesting operations; perform satellite leveling, precise spraying, variable fertilization, automatic grain yield measurement, etc.; reduce labor

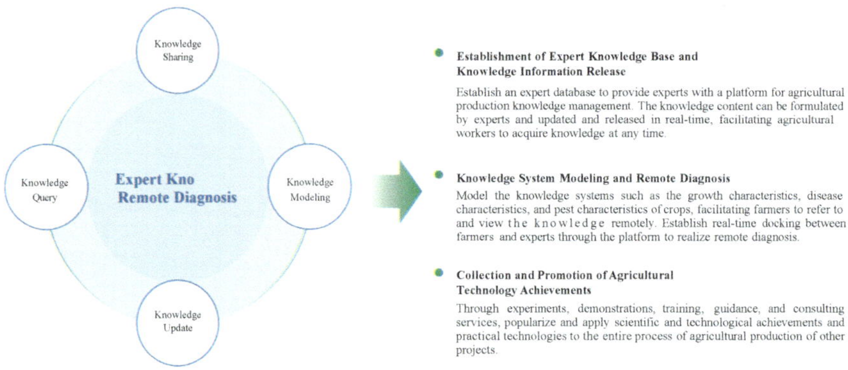

Fig. 5.70 Expert knowledge and remote diagnostic services

input; and ensure product quality while improving input–output efficiency. Operating entities use mobile terminals to publish demand information, and social service organizations or agricultural machinery operators accept orders and visit the designated field to work, accomplishing online ordering, remote monitoring, precise control, and online settlement and helping farmers complete work without leaving their homes.[10]

Expert Knowledge and Remote Diagnostic Services: Through digital technology and the integration of expert resources, we have established a business model of "platform + expert + service" to promote agricultural science knowledge and enhance the popularity of online cultivation. We have set up databases for production technology, pest control, expert resources, and variety knowledge, providing farmers with professional guidance services such as agricultural technology knowledge inquiry, natural disaster prevention knowledge inquiry, and disease prevention knowledge inquiry. Moreover, farmers and experts can interact online on the platform to discuss issues and needs arising from agricultural planting and breeding processes, such as pests and epidemics. The content of the exchange includes messages, audios, pictures, videos, etc. Users can find experts they are interested in through the platform, ask questions, seek help from these experts, and view other users' evaluations of the experts. They can also view frequently asked questions. In addition, experts can diagnose and guide farmers' problems remotely through the platform. The functions of expert knowledge and remote diagnostic services are shown in Fig. 5.70.

The Smart Agriculture "Metaverse"

In agricultural planting, the metaverse can simulate the entire life cycle of plants, simulate pests, simulate extermination, and collect the required plant growth data in a short time. Through quantitative indicators, the traditional agricultural production method has changed and has shifted toward intelligence.

[10] Secretariat of the Cyberspace Administration of China; General Office of the Ministry of Agriculture and Rural Affairs; General Office of the National Development and Reform Commission; et al. *Digital Village Development Guide 1.0*. July 23, 2021.

In the breeding aspect of animal husbandry, the metaverse can also simulate genetic-related information. From organs, tissues, systems to the whole animal, all the related data can be obtained by simulating various growth rates and reactions under different breeding methods, which can provide excellent references for breeding and even the breeding process.

In the agricultural metaverse, a virtual digital person can also be created. People in the real world can interact with digital people in the virtual agricultural world and create agricultural product promotion IPs, which can serve as spokespersons and image ambassadors for agricultural products, increasing sales and contributing to the brand influence of agricultural products. Peripheral products can also be developed through this IP, and the development of the agricultural, cultural, and tourism industries can be linked.

5.6.3 Smart Urban–Rural Integration Practice Case: Digital Technology Empowers Pig Breeding

The digital agriculture construction pilot project of Guangdong Dongrui Food Group Co., Ltd., approved by the Ministry of Agriculture in 2017, is located in Zhifulin Ranch, Heyuan city, covering an area of 1745.04 acres. It adopts a high-bed fermentation-type ecological pig raising model with a breeding scale of 100,000 commercial pigs per year. The project transforms traditional pig houses into high-bed fermentation-type pig houses and fecal pollution treatment-integrated pig houses and builds automated precision environmental control systems, digital precision feeding management systems, safe production video surveillance systems, harmless fecal pollution treatment systems, food safety traceability systems, digital agriculture breeding information management systems and other systems. The Dongrui digital agriculture integrated information platform is shown in Fig. 5.71, and the onsite construction effect is shown in Fig. 5.72.

Scientific Management of the Entire Pig Breeding Process

This project realizes the intelligent and smart control of the entire pig breeding process, from the environmental control of the pig house to digital feeding and the treatment of feces.

Automated Precision Environmental Control: Through the deployment of front-end sensors, the temperature, humidity, ammonia, carbon dioxide, and illumination parameters of the pig house are automatically collected and monitored. Owing to the different environmental needs of pigs at different stages, the system automatically controls onsite fans, water curtains and other equipment to optimize the onsite environment, thereby effectively improving the growth performance of pigs.

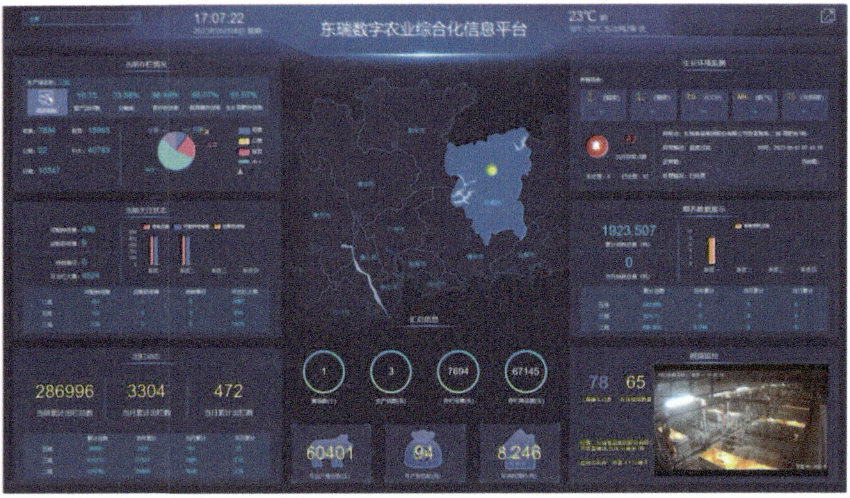

Fig. 5.71 Dongrui digital agriculture integrated information platform

Fig. 5.72 On-site photo of the farrowing house

Digital Precision Feeding Management: Through precision feeding equipment, simultaneous feeding and timed and quantified feeding are achieved, improving the efficiency of feeding. Moreover, the feeding data are integrated and analyzed to simulate and predict the pig growth process, adjust the feeding process during pig growth, and evaluate the feed-to-meat ratio benefit, thus achieving scientific and precise feeding.

Harmless Fecal Pollution Treatment System: In this new high-bed fermentation-type ecological pig raising model, pig raising production and waste treatment are combined. Using the principle of the aerobic fermentation of microorganisms, the automatic turning and treatment of pig manure and automatic scraping of pig manure are carried out through automatic turning and scraping manure equipment. Pig waste is converted into solid organic fertilizer, achieving zero wastewater discharge and solving the pollution problem of pig waste.

Food Safety Traceability System: This system records information about the entire process of pig farming, starting from the source of pig raising. It records detailed growth-related information of each batch of pigs during the breeding process in the system, up to the point of the sale of the batch of pigs. It retains detailed information about each pig batch, allowing the breeding information of the batch to be directly interfaced with the traceability interface.

Highly Integrated Digital Agriculture Information Platform: Through seamless data interface integration, various subsystems, such as automated precision environmental control systems, digital precision feeding management systems, harmless manure treatment systems, and food safety traceability systems, are integrated. This platform effectively manages the entire production process of pig farming, processes and analyzes data, realizes centralized data management and visualization, achieves smart breeding, and improves production efficiency.

Construction Effect: Precision Breeding Improves Pig Production Benefits

The project empowers pig farming through digital technology, achieving intelligent pig farming, information management, data management, and online services, improved pig production benefits, reduced epidemic risk, and improved breeding management and work efficiency.

In terms of breeding management efficiency, according to the data presented in the book "Entering the Internet of Things", which was published by the Machinery Industry Press in 2017, ordinary farmers raise 1000–2000 pigs a year using traditional methods. However, through smart breeding, an average employee can easily raise 10,000 pigs. As in this project, Dongrui Pig Farm has effectively improved its breeding management and work efficiency through intelligent digital breeding methods.

In terms of pig production benefits, major internet companies have also seen the potential for improvement in smart pig farming and have laid out plans for the smart pig farming industry. For example, JD's AI pig farming is expected to shorten the time to market for pigs by 5–8 days and reduce the feeding cost of each pig by 80 yuan. Although the smart pig farming industry has only begun to develop in recent years and the data on its production benefits are not yet sufficient. We believe that the collection of data in the future will better prove the enormous benefits that digital technology brings to the breeding industry.

Chapter 6
Smart City Scale Applications

6.1 Smart Buildings: The Evolution from "Intelligent" to "Smart"

In the 1980s, with the development of computer technology and communication technology, intelligent buildings emerged in the United States, quickly influencing the world and leading the development trend of the construction industry. The American Intelligent Building Institute (AIBI) defines an intelligent building as "a highly efficient, high-functioning, and comfortable building obtained by comprehensively integrating structures, systems, services, operations, and interconnections to achieve the best combination". Comfort, function, and efficiency were the main goals for intelligent buildings at that time. Since the 1990s, China's economy has taken off, cities across the country have been built on a large scale, and the real estate industry is booming. Traditional buildings that were used only for living and working can no longer meet the development needs of society. In addition, people have higher requirements for buildings, such as safety, comfort, automation, and efficiency. To ensure market competitiveness, real estate developers are increasingly applying intelligent technology to buildings. Video surveillance and intrusion alarm systems have become standard configurations for buildings; air conditioning and fresh air systems have become essential for people's comfortable living and working; and building equipment management systems have enabled timed or condition-triggered automatic start–stopping of equipment, as well as remote monitoring and fault alarm functions by equipment, greatly improving management and operation efficiency.

However, with the in-depth development of the construction industry, people have gradually found that intelligent buildings also have significant limitations and that they cannot meet people's additional functional requirements for buildings.

The building intelligence industry has evolved from the automation stage to the intelligent stage and is now moving toward the development of smart buildings (as shown in Fig. 6.1). The "Smart Building Design Standard" (T/ASC19-2021) and

F. Liu, *Smart and Digital Cities*, https://doi.org/10.1007/978-981-95-2243-9_6

Fig. 6.1 The ongoing transition from intelligent buildings to smart buildings

the "Smart Building Evaluation Standard" (T/CECS 1082-2022) provide a standard basis for the development of smart buildings.

6.1.1 Lack of Wisdom in Intelligent Buildings

Monitoring that watches but does not act: In terms of security prevention, video surveillance is an indispensable system in intelligent buildings. However, traditional video surveillance systems play a passive monitoring role. That is, the course of events can be recovered only by viewing the surveillance video afterward, which is inefficient and lacks early warning functions. They cannot identify and warn personnel before a security event occurs to reduce losses or prevent the event from happening.

Inflexible Equipment Monitoring: Currently, intelligent buildings monitor and control systems such as lighting, fresh air, heating and cooling sources, and water supply and drainage through building equipment management systems. These systems operate according to preprogrammed fixed procedures. As long as the operating conditions and parameters of the program are not modified, the operation effect of the system will remain unchanged. Moreover, this effect is closely related to the debugging experience of the programmer, and the setting of system parameters has not been calculated and demonstrated scientifically, which can lead to problems with the energy use system of intelligent buildings such as being inefficient or difficult to use. For example, intelligent lighting systems usually adjust the switch and brightness according to the time period. The program defaults to the maximum daylight illumination at noon, and the lights are all turned off at a fixed time every day according to the program. If there is no illuminance detector set or if the curtains are fully drawn and natural light cannot be collected, manual operation is required to turn on the lights in dark weather or when natural light cannot be used. If the system defaults to 9 pm as the time for employees to leave work, the automatic shutdown of lights will inevitably affect overtime workers, because the system is only accurately executing the program and cannot make conditional judgments and execute the correct actions based on actual conditions.

Energy-Unaware Buildings: Comfort is a major feature of intelligent buildings. Bright lights throughout a building can create a bright mood; air conditioning that is warm in winter and cool in summer and a fresh air system that continuously provides fresh air can keep people in a high-efficiency working state. However, maintaining this comfortable environment requires a large amount of electrical energy. According to relevant data statistics, the three main types of electricity-consuming equipment in commercial buildings are air conditioning, which accounts for approximately 47% of energy consumption; lighting, which accounts for approximately 45%; and elevators, which account for approximately 8%. Considering the distribution of functional areas within the building and application scenarios, how much of the consumed electrical energy is fully and effectively utilized? After a meeting, system equipment, air conditioning, and lighting are still running at full load; in the cafeteria during nonmeal times, air conditioning, TV, and background music continue to run; and underground parking lots, corridors, and public areas are brightly lit at midnight. This energy consumption is difficult to control in a timely and effective manner through manual means.

6.1.2 Intelligent Buildings with Multielement Coupling, Multisource Computing, and Multidimensional Thinking

In the shift from traditional buildings to intelligent buildings, people have gained safety, comfort, and efficiency. Therefore, what will the next technological innovation in intelligent buildings bring? In 2017, Alibaba organized a smart building summit, proposing that smart buildings lead the arrival of the new era of Building 3.0, and presented the concept of smart buildings to people. People do not know the difference between smart buildings and intelligent buildings and may even think that these are just two different terms for the same thing. Therefore, Alibaba's "Smart Building White Paper" provides a new interpretation of smart buildings, distinguishing them from intelligent buildings. The white paper suggests that with the advent of the big data era, smart buildings will become a "living entity" with perception and always online. They will be self-evolving intelligent platforms with a brain and an open ecosystem deeply integrated with humans, machines, and things, which can integrate all innovative technologies and products that serve humans. Compared with traditional intelligent buildings, smart buildings are more like "living entities" that can think, analyze and learn a large amount of data and grow on their own, whereas intelligent buildings are more likely to have the ability to store information but cannot perform independent data analysis and learning. In addition, intelligent building systems emphasize technical content more, focusing on the development and application of information and communication technology, automation technology, etc. In contrast, smart buildings emphasize green energy savings and environmental protection, with a focus on the development of the user experience, paying more attention to improving users' quality of life, environmental friendliness, and energy savings.

In 2021, the "Smart Building Evaluation Standards" (T/CREA 002-2020 T\CBIMU14-2020) released by the China Real Estate Industry Association and others defined "smart buildings" as follows: the use of Internet of Things, cloud computing, big data, artificial intelligence and other technologies to achieve automatic perception, ubiquitous connection, timely transmission and information integration, with autonomous learning, autonomous diagnosis, assisted decision-making and execution capabilities to create a safe, reliable, green, ecological, efficient, convenient, and economical building environment.

In 2021, the Architectural Society of China issued a notice, approving the "Smart Building Design Standards" (T/ASC 19-2021) as the standard of the Architectural Society of China; defining "smart buildings" as aiming to build convenient, comfortable, safe, green, healthy, and efficient buildings; fully reflecting data integration, analysis and judgment; controlling decision-making in the aspects of concept planning, technology application, management operation, and sustainable development; and having overall adaptive and self-evolving capabilities of new building forms.

In 2022, the China Engineering Construction Standardization Association announced the approval and release of the "Smart Building Evaluation Standards" (T/CECS 1082-2022), in which the term "smart building" is defined as the comprehensive application of new generation information technology to build a smart building integrated management platform; achieve automatic perception, ubiquitous connection, autonomous learning, autonomous inference, proactive decision-making and other functions; achieve synergy between people, buildings, and the environment; integrate with the functions of smart cities; and provide people with a safe, healthy, low-carbon, convenient environment within high-quality buildings.

Although the above definitions of terms vary, they all mention a common point: smart buildings possess autonomous learning, analysis, and decision-making capabilities, creating safer, greener, and more convenient spatial environments.

Through technologies such as the Internet of Things (IoT), big data, and artificial intelligence, buildings are empowered to evolve from intelligent to smart, endowing them with capabilities for multielement coupling, multisource computation, and multidimensional thinking (as shown in Fig. 6.2).

Internet of Things: Realizing the Coupling and Connection of "People", "Objects", and "Spaces" in Smart Buildings

Currently, the Internet of Things has permeated various industries, and its integration with architecture is becoming tighter, providing a significant impetus for the development of the construction industry and becoming the sensory nerve endings of smart buildings.

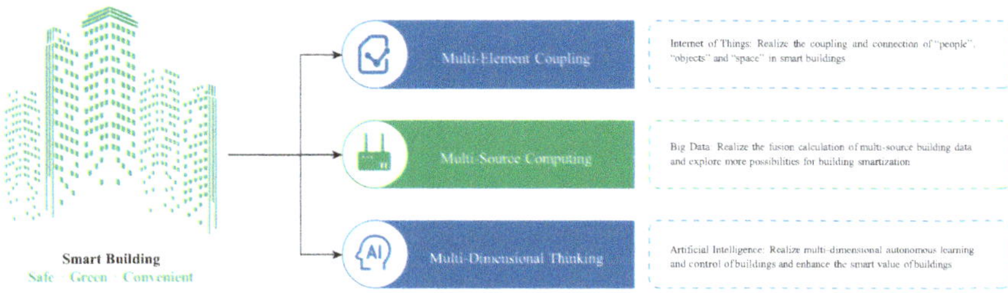

Fig. 6.2 Smart capabilities of smart buildings

Unlike intelligent buildings that program systems to meet the best intentions of users, smart buildings have appropriate sensing and processing capabilities. They can autonomously detect their surrounding environment, analyze and compute the data obtained and then execute what they consider the best results. To achieve smart management, buildings must first acquire sufficient basic building data and external environmental data. For this purpose, buildings must have relevant perception capabilities. The emergence of the Internet of Things allows buildings to fully perceive internal information and recognize the current status index of the building, such as temperature, humidity, and illumination, and transmit these data back to the building's "brain". The machine learning algorithms in the brain process the information it receives, thereby jointly adjusting air conditioning and lighting equipment.

In smart buildings, all intelligent systems and devices need to achieve the interconnection and remote sharing of information through the Internet of Things, including sensors, cameras, trigger devices, smart water and gas meters, fire detectors, etc. Together, they form a networked structure to constitute the sensing network of the building's "smart" control system. Then, the building's invisible state is clearly presented to users through data visualization, providing users with a more intuitive understanding of the status of the building and equipment.

Smart buildings and the Internet of Things develop while promoting each other. The various needs of smart buildings provide a basis for the research and development of Internet of Things products, and networked products make the functions of smart buildings more complete and systematic, thereby greatly improving the overall functioning of smart buildings. The Internet of Things platform can fully perceive, reliably transmit, and intelligently process the internal information of a building, construct a coupling model of "people", "objects", and "spaces", and realize the multielement collaborative ability of smart buildings (as shown in Fig. 6.3).

Big Data: Enabling the Fusion Calculation of Multisource Data in Buildings and Exploring More Possibilities for Building Intelligence

There are a large number of data sources in buildings, including relatively stable data, such as basic building data, property, and fixed assets, as well as dynamic data collected by the Internet of Things, such as equipment operation status parameters,

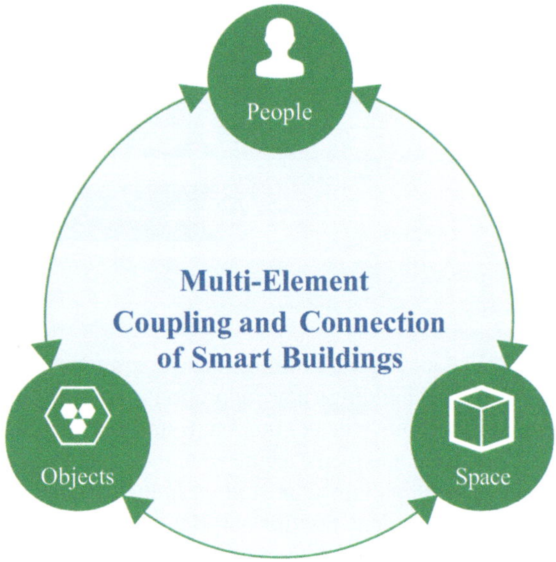

Fig. 6.3 Multielement coupling and connection of the internet of things in smart buildings

personnel and vehicle information, face databases, surveillance videos, etc. If these isolated data cannot be utilized, they are meaningless numbers. The smart building management system processes and analyzes these collected data, shares the originally independent system data, mines the relationships and rules behind these data through data processing capabilities, helps managers obtain valuable and actionable guidance, facilitates scientific decision-making, maximizes the operation efficiency of the building, safely and reliably uses data assets, and improves the experience of people in the building.

In smart buildings, no piece of data exists in isolation, and the systems are interconnected. The significance of a piece of a data may be as a reference object affecting the operation strategy of the system, or it may be an important indicator related to the chain reaction of multiple systems. Taking digital facial recognition as an example, when the facial camera recognizes the identity of a person, it immediately opens the channel gate and simultaneously triggers the attendance system to clock in the person. The elevator control system automatically stops at the employee's office floor. Upon reaching the reserved shared workstation, it automatically registers occupancy and displays employee information on the workstation LCD screen. The intelligent lighting defaults to the employee's preferred color temperature and brightness. The smart system pays attention to every habit and preference and silently arranges everything for the person.

Artificial Intelligence: Enabling Multidimensional Autonomous Learning and Control of Buildings, Thus Enhancing the Smart Value of Buildings

At the "AI Empowerment - Smart Building" forum held in Shanghai in August 2019 during the "2019 World Artificial Intelligence Conference", the "Smart Building Evaluation Index System" of the Shanghai Building Technology Research Association for the first time integrated intelligent buildings, green buildings, cloud computing and other technologies with comprehensive building management, forming a "smart building" (SMART). The concept of an "intelligent building" was developed, leading to evaluation methods and content for four major systems: green building, automated integration, modern property management, and integration into "smart cities". This development has attracted widespread attention and support both domestically and internationally.

Artificial intelligence empowers comprehensive building management, serving as the brain that allows the building to learn and manage itself. This function allows the building's systems to operate in the best and most energy-efficient manner without human intervention and increasingly replaces the need for humans to perform various tasks.

The management of traditional buildings relies mainly on the operations of managers, who achieve energy-saving effects by adjusting system parameters. Building equipment management systems rigidly execute set programs and cannot effectively solve the problem of high energy consumption. In response to this issue, several building equipment management manufacturers have proposed energy-saving solutions through AI. First, they use simulation tools to predict various energy consumptions within a building; then, they use past building data and data from other buildings at different times to train the model. AI combines user habits and internal and external building data learned from training to perform real-time simulations, continuously performs comparisons and optimizations to select the best parameters, and adjusts the system operation status to maximize energy savings. For example, the lights, air conditioning, and system equipment in a conference room will be turned on in advance according to the reservation time. However, if the system does not detect any personnel in the conference room at the meeting time, the reservation will be cancelled, the room will become idle, and the system and environmental equipment will be automatically shut down.

In the era of smart buildings, artificial intelligence is increasingly being used in meetings and gradually changing the mode of meetings. For example, in daily meetings of government departments, two special seats are reserved: one for the conference room administrator, who is responsible for switching and operating system equipment, and the other for a clerk or secretary, who is responsible for recording meeting minutes. Both of these roles have been replaced by AI voice recognition. Meeting equipment control no longer requires special training. A participant merely needs to express their intentions through language, such as "turn on the projector", "switch on the computer screen", "pull up the curtains", or "turn off the machine after the meeting", etc., and the system will execute the corresponding operations. Meeting minutes no longer need to be recorded and noted, as the intelligent voice transcription system can automatically convert meeting speeches into text, with an accuracy rate of over 99%. It can accurately recognize the voiceprints of different speakers and assign different role codes, generating comprehensive and accurate meeting minutes and greatly improving the efficiency of meetings. AI can even recognize English and dialects, translate English and display Chinese subtitles on the big screen. Thus, future meetings involving foreign guests will no longer need simultaneous interpretation systems and translators.

6.1.3 Case Study of Smart Building Practice: Tencent's Exploration of Smart IoT Collaborative Buildings

Tencent's Guangzhou headquarters is located in a building in the Pazhou Internet Innovation Cluster, with a height of 206.75 m and a total area of 173,975.1 m^2. The building is divided into four underground floors, two podium floors, eight floors in the north tower, and 38 floors in the south tower. There are four single-sided cantilever platforms on the 14th, 23rd, 32nd, and 38th floors, with the 38th floor having the largest cantilever area of 28 m and a total area of 1657.6 m^2, approximately the size of four standard basketball courts, and weighing 750 t, making it the largest single-sided cantilever structure in the country.

The building is a self-use scientific research office building. It is designed according to the concept of "life, communication, and openness" and includes scientific research office areas, employee restaurants, sports facilities, education colleges, as well as coffee shops, bars, convenience stores, charging stations and other supporting commercial facilities. As a key project in Guangzhou, it has been included in the Guangzhou City Quality Engineering Cultivation Project and will strive to win excellent engineering awards at the national level, China Construction Engineering Luban Awards, two-star green building recognition, the American LEED building gold award certification and other honorary awards.

Features of Tencent's Smart IoT Collaborative Building Construction

Special Requirements for IoT Collaboration Docking

Tencent's Guangzhou headquarters building is constructed according to the requirements of a smart IoT collaborative building. The local intelligent system is connected to Tencent's headquarters through Tencent's standard rules (as shown in Fig. 6.4). The number of docking points of the national security system can exceed 20,000, and the platform clearly stipulates the interfaces and protocols of each subsystem/device.

For example, the thermal imaging cameras and intelligent cameras deployed in this project are connected to Tencent's smart management platform through the network.

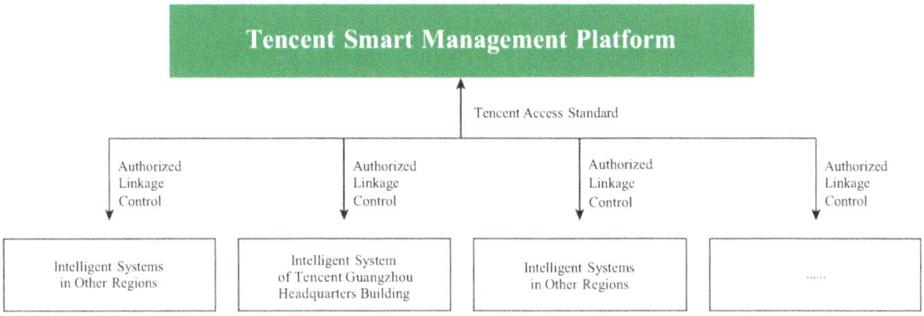

Fig. 6.4 Schematic diagram of Tencent IoT collaboration

The monitoring display end does not use a video decoder but uses software decoding to decode the monitoring image from the platform. The cameras used in the project have some built-in intelligent analysis functions, but the video analysis function of the smart management platform is the main function. Through the platform's capabilities, AI monitoring is implemented, such as traffic flow statistics, blacklist settings, crowd monitoring, illegal parking of vehicles, intrusion of defense zones, personnel loitering, illegal climbing, and fire warnings. When an abnormal event is recognized, it automatically triggers the local related system to respond.

Furthermore, the Tencent Guangzhou headquarters building project places great emphasis on security management, especially with respect to the entry of external personnel into the core office area, which requires passing through multiple security lines, including barriers, elevator controls, and access controls. The access control card and visitor management of this project are directly connected to the Tencent Smart Management Platform, thereby eliminating the need for separate access control and visitor management software, all of which are authorized and controlled by the platform.

The network cabling requirements for ensuring the security and stability of data transmission are complex.

The integrated pipe network is complex: the indoors includes mobile signal coverage trunking, communication network trunking, and intelligent trunking with three types of complex routing; the outdoors has a small municipal pipe network, along with ceiling and floor cushion trunking; finally, the problem of excessive distance needs to be solved, such as floor slab penetration and undercushion trunking.

The building space and floor plan layout are complex: security zones are divided, and defense levels are set according to the flow of people, which is unique.

The network requirements are complex and include an office/test network, a wireless network, an IP telephone network, and a control network.

High-stability requirements exist mainly for data centers, trunk optical cables, and office routing.

Meeting requirements are high and specific, creating a convenient and humanized office environment.

There are numerous meeting rooms, including multifunctional halls (300 square meters), meeting rooms (115), and large training rooms (4).

Meeting rooms need acoustic simulation based on the fine decoration design scheme: the acoustic index is designed according to the “multipurpose class one” standard of the “Hall Sound System Design Specification” (GB 50371).

There are multiple types of professional coordination in meeting rooms, such as intelligent, power distribution, air conditioning, and LED screens, which coordinate with each other.

The meeting system connects to the Tencent Meeting Rooms platform, which manages it uniformly.

There are multiple BAS interfaces, providing a basis for green energy-saving buildings.

With a two-layer distributed architecture, I/O modules access DDC via the BACnet/Lon communication protocol, DDC transmits via TCP/IP, and access is gained by the control network.

The integration of third-party systems/devices through standard interfaces includes intelligent lighting, building energy efficiency supervision, cold source group control, FCU, multiunit air conditioning, intelligent window opening, elevators, etc.

BAS provides an open interface upward, facilitating data upload and command issuance.

Construction Effect

The Tencent Guangzhou headquarters building project is a smart IoT collaborative building that uses IoT and artificial intelligence technologies, integrates digitalization and intelligence, and combines the powerful analysis and processing capabilities of the Tencent Smart Management Platform and the well-constructed intelligent system. This project will be a representative smart building that is safe, efficient, comfortable, convenient, and green. After the building is completed, the tower lobby, second-floor platform, basement art corridor, and first-floor overhead layer will be fully open to the public, becoming a large garden for citizens to view and relax in, thus allowing more citizens to experience the charm of a smart city.

6.2 Smart Park: An Important Foothold for the “Digital China” Strategy

The “Digital China Construction Overall Layout Plan”, which was issued by the Central Committee of the Communist Party of China and the State Council in February 2023, noted that building a digital China is an important engine for promoting modernization in the digital age and strongly supports the establishment of new national competitive advantages. Accelerating the construction of a digital China therefore has importance for and far-reaching impacts on the comprehensive construction of a socialist modern country and the comprehensive promotion of the great rejuvenation of the Chinese nation.

As the basic unit of the city connecting individuals, enterprises, and the government, the park is an important component of the "Digital China" strategy. Parks are the most important population and industrial agglomeration areas. However, the scope of park management is increasing, the number of management objects is increasing, and the businesses they support are becoming increasingly complex. The contradiction between the limited infrastructure and service capabilities of parks and the growing production and living needs of the people is becoming increasingly prominent. Parks face challenges in terms of safety, experience, cost, and efficiency. The construction of a smart park is an effective way to solve these problems.

In accordance with the guidance of the Ministry of Housing and Urban–Rural Development, the National Smart Standards Committee, in collaboration with nearly 20 domestic "industry, academia, research, and application" units, compiled the "Research Report on the Standard System of Smart Industrial Parks (2022)". The report defines a smart park as a deep integration of new generation information technologies such as cloud computing, big data, the Internet of Things, artificial intelligence, 5G, and digital twins with industries. It integrates park manufacturing resources and third-party service capabilities to achieve layered resource sharing, industry-linked development, real-time environmental perception, full-process event visibility, automatic production adaptation, full-time equipment utilization, and community value association. It extends the industrial value chain and enhances the intelligent management and social integration capabilities of the park. In December 2021, the "14th Five-Year Plan for Digital Economy Development" issued by the State Council clearly proposed promoting the digital transformation of industrial parks and industrial clusters, marking a new role of parks as an important part of new infrastructure construction and digital economy construction.

Under the guidance of national strategies in recent years, various national ministries and industry institutions have successively issued a series of policies and guidance related to smart parks, providing guidance for the intelligent construction and digital transformation of parks and creating a favorable policy environment. The rapid development of new generation information technology has become a core element of smart parks, empowering personnel, enterprises, and equipment in the park.

6.2.1 *Problems Encountered in the Intelligent Construction of Parks*

There are many types of parks. Considering that the intelligent construction, upgrading, and renovation of parks require unified coordination and comprehensive control by the operating entity, parks are classified according to the operating entity, namely, government-led development-type smart parks, industry real estate-led smart parks, large enterprise smart parks, and small and micro smart parks.

The essence of a smart park is to create an "intelligent" park. It uses information technology as a means and intelligent applications to fully integrate the internal and external resources of the park and build a carrier and platform for networked infrastructure construction, refined construction management, professional service functions, and intelligent industries. However, the pain points faced by smart parks in the development process are becoming increasingly prominent.

Lack of Top-Level Design

The intelligent construction of a park is a complex systemic project that involves simplifying, concretizing, and formalizing complex objects from uncertainty to certainty. More specifically, it involves multidimensional technology integration, service unions, management linkages, and collaboration in implementation. The core purpose of top-level design is to help park owners identify and clarify project construction goals based on the actual situation of the park and business development needs. However, in actual project proposals, different types of parks have different emphases in their construction ideas. Some parks lack a top-level design in the construction planning process, do not aim to meet user needs or provide commercial value, do not comprehensively consider the various components, various roles, and various influencing factors of the park, and do not start from a strategic perspective to plan and design blueprints and scenarios for the intelligent transformation of the park, leading to a series of problems in the later stage.

Lack of Data Sharing, Low Management Efficiency

Traditional parks have independent information systems for different departments/businesses and lack unified system control. The "chimney-style" system deployment, lack of data sharing, business nonlinkage, and inefficient services make it difficult to achieve information management. Park services have an "island effect", and different services are led by different departments, which can easily lead to a fragmented user experience.

Over time, the number of intelligent application services will increase, especially with the popularization and application of the Internet of Things. First, a large amount of data will be generated. How to manage, store, and apply these data. Second, the lack of standards for data formats and data interfaces makes it difficult to share data between the same business systems, which in turn makes information collaboration processing of various system applications difficult and prevents convenient data exchange services between them.

Lack of Industrial Operations Management, Leading to Low Industrial Operation Efficiency

Currently, many development parks focus only on attracting investment and the construction of basic infrastructure in the process of construction and development. They do not consider the operations management of the park industry. They neglect the service of industrial operations, have few service resources, and exhibit slow economic growth. Most development parks experience low efficiency, and the problem of strong construction and weak operation is common in current development parks. How to improve the overall service level of a park, promote the intelligent operations management system of the entire park, and achieve efficient development of the industrial park is a major challenge facing the construction of many parks.

Unable to Achieve Centralized Management, Leading to High Operating Costs

In traditional development parks, there is no interaction between the park management platform and the enterprise platforms within the park. The park lacks a unified information platform system, and data collection from park enterprises is isolated, making it difficult to obtain a comprehensive grasp of information. For example, various reports designed for economic development and various data involved in social development all reflect isolated data collection, making it difficult to link information data systems for the entire park enterprise. On the one hand, this issue not only adds considerable repetitive labor to daily work but also makes it extremely easy to misjudge the operations monitoring of enterprises. Therefore, a complete smart park operations system is needed to establish feasible operation modes and profit spaces for the park.

Lack of Efficient Investment Attraction Management Leads to Longer Operation Cycles

Owing to the lack of overall operations planning and smart monitoring for the park industry, most development parks still use traditional methods to attract investment. Currently, the effect of traditional investment attraction methods has greatly decreased, but the cost continues to rise. Owing to outdated investment attraction methods and lagging means, the investment attraction cycle for introducing enterprises is long and the process is complex, resulting in a longer operations cycle for the entire investment attraction link. Therefore, how to innovate investment attraction methods has become a pressing issue.

The problems encountered in the development of smart parks are shown in Fig. 6.5.

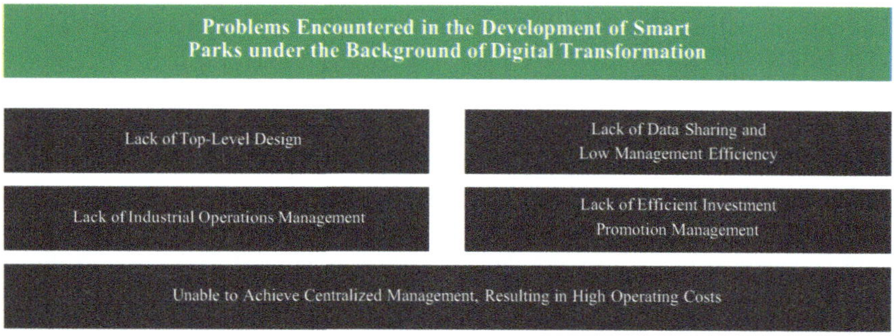

Fig. 6.5 Problems encountered in the development of smart parks

6.2.2 Analysis of Key Points in the Construction of Smart Parks

Although various parks have many differences in terms of digital transformation and smart construction, such as industry foundations, regional locations, and functional matching, they also have many common needs. We believe that in the new era, the digital transformation of parks must fundamentally change "formalism", carry out tasks in stages, demonstrate effective top-level planning and design, and pursue the digital transformation of park characteristics under limited resources.

Based on the problems encountered in the construction and development of smart parks, combined with the experience of Gloryview Technology in the construction of industrial parks, venues, commercial complexes, and other parks in recent years, we believe that the key points of smart park construction are as follows.

Top-Level Design as the Blueprint, Aiming for Efficient Operation

The top-level design of a smart park is the conception and action goal of the park's informatization and smart construction for the next 5 to 10 years. Smart parks need to use a top-level design as a blueprint and aim for efficient operation. They should form a vibrant organization by the construction subject, bring together a wide range of ecological resources, based on an open architecture and a user-centered service concept, and implement this design methodically. The basic process of the top-level design for smart parks is shown in Fig. 6.6.

Centralized Data Management, Data Source Integration and Data Sharing

Through the Internet of Things platform and data platform, data docking with various independent subsystems and related business systems is achieved, breaking through

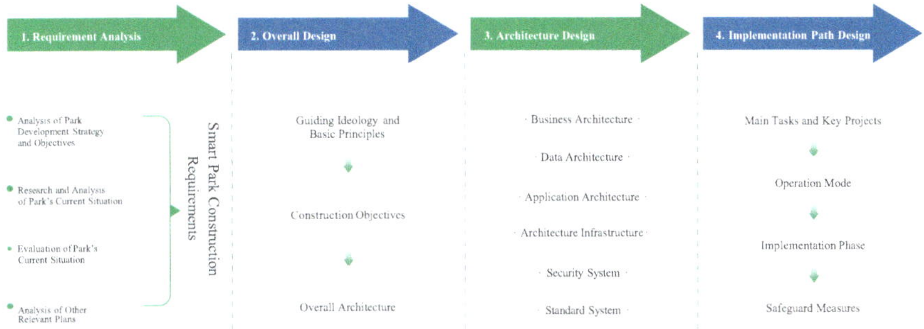

Fig. 6.6 Basic process of top-level design for smart parks

the pattern of independent and noncommunicating data of the original systems. Data integration is carried out by docking with the heterogeneous data of various independent subsystems, and a data sharing mechanism and norms are established. Through a unified data service interface, data exchange and sharing between systems are accomplished. Through the data platform, data from various subsystems are collected, managed, and stored, achieving centralized data management and laying the foundation for data query, statistics, analysis, and display functions.

The construction of a smart park should enable dispersed and isolated data to flow and generate value in the process. It should ensure the security and consistency of data, provide a unified data service, and be widely and conveniently called by various applications. The main obstacle to the construction of a smart park is no only technology but also thinking and methods. In the process of constructing a smart park, we need to shift from the perspective of engineering project construction to that of digital transformation for thinking, preparation, and construction. Engineering construction pays more attention to the content, cost, and cycle of construction; and digital transformation pays more attention to the goals and business value of construction.

Integrating the Concept of "Dual Carbon", Smart Parks Are the Main Focus

The Central Committee of the Communist Party of China and the State Council issued the "Opinions on Fully and Accurately Implementing the New Development Concept and Doing a Good Job in Carbon Peaking and Carbon Neutrality" and the "Action Plan for Carbon Peaking Before 2030" in September and October 2021, respectively. As the core content of the "1 + N" policy system for carbon peaking and carbon neutrality, these documents further clarify China's overall goal of achieving peak carbon emissions. They propose a comprehensive green transformation of economic and social development, deep adjustments to industrial and energy structures, low-carbon construction in transportation and urban–rural areas, and the enhancement of carbon sinks in ecosystems as key measures. They also emphasize the role of technological innovation as the core driving force, promoting green, low-carbon urban construction and deep energy conservation throughout the entire socioeconomic system. As the main carriers of people's production and life, industrial parks are the main footholds for implementing national strategies. In January 2021, the Ministry of Science and Technology issued the "Implementation Plan for the Special Action for Green Development of National High-tech Zones", which proposed enhancing the green development governance capacity of industrial parks, reducing energy consumption and carbon emissions, and achieving carbon neutrality in some high-tech zones. The orderly construction of zero-carbon industrial parks has been officially put on the agenda.

By leveraging information technology and big data in industrial parks, the precise prediction and efficient management of park personnel activities can be achieved. Through the effective control of information flow, energy supply and consumption can be automatically matched, enabling intelligent scheduling of various energy infrastructures, such as distributed power sources, energy storage, air conditioning, charging piles, etc., at different time and space scales within the park. This process reduces energy consumption and carbon emissions, which increases cash flow savings for park enterprises and contributes to carbon neutrality. Moreover, through integrated energy services, the operating income of the park itself can also be increased.

Operations Management is the Implementation Phase of the Entire Lifecycle of a Park

The operation of a smart park is the implementation phase of the entire lifecycle of the park. It is the stage where park services are provided, value is created, and intelligent empowerment is realized. It is the most important stage. It can solve problems such as reliance on human management in traditional parks, poor employee experience, and high daily equipment and facility maintenance costs. Park operators need to carry out continuous operations based on business scenarios and integrate shared park data to improve the value of smart parks.

Application of New Technologies

Smart parks can apply a variety of new generation information technologies, such as artificial intelligence, big data, digital twins, and the Internet of Things, to improve park governance efficiency, optimize the industrial structure layout, strengthen infrastructure construction, and improve the environment.

Application of artificial intelligence in smart parks.

Park enterprises are gradually developing to include high technology, new fields, and specialization. In the future, parks will become the basis for high-tech enterprises to concentrate research and development, product incubation, and production. Parks

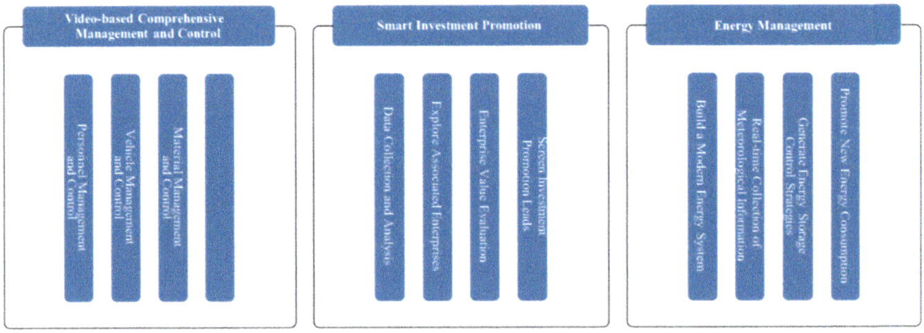

Fig. 6.7 Application of artificial intelligence in smart parks

have a large scale and diverse business formats. Smart parks will gradually transition from a management model that provides users with water, electricity, gas, and other supporting facilities to a layout of standardized buildings or building groups that can meet the needs of specific industries and scientific experiments. Under this requirement, building an AI platform, integrating various resources, assisting the digital transformation of the park, and achieving intelligent operation, intelligent work, and quality and efficiency improvement are necessary. The application of artificial intelligence in smart parks is shown in Fig. 6.7.

Comprehensive control based on video: Through the integration of the video system and other business subsystems and with the help of machine vision algorithms, the visualization of personnel, vehicles, and material data can be achieved to ensure coordinated management. This capability facilitates user analysis and decision-making, effectively improving problems such as low resource utilization, management challenges, and complex reuse in the user's informationization process. More importantly, it organically combines with the user's actual business to optimize business processes, promote improvements in the overall management level of the park, and help users optimize traditional security management methods.

Smart investment promotion: Precise investment promotion has always been the core demand of park investment promotion. Accurately obtaining investment promotion clues has become the key to investment promotion. Smart investment promotion can collect and analyze data, such as industry intelligence, industry exhibitions, investment and financing intelligence, and patent data, and excavate high-value associated enterprises from this information. After the target enterprises are obtained for investment promotion, the enterprise value evaluation model can be used to conduct a 360-degree evaluation of the enterprises, comprehensively and objectively evaluate the comprehensive strength and potential risks of the enterprises, and finally screen out high-value investment promotion clues.

Energy management: The "Fourteenth Five-Year Plan" proposes that accelerating the digital and intelligent upgrade of the energy industry is a support for to achieving the "dual carbon" goal. As the most concentrated space for urban carbon emissions,

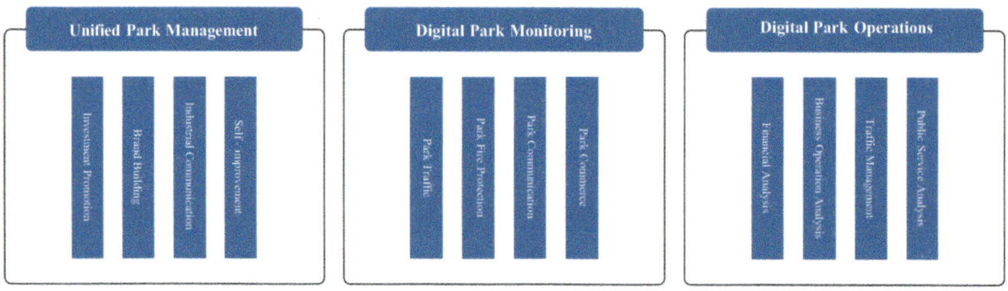

Fig. 6.8 Application of big data in smart parks

parks have rich business types and diverse business models and can serve as important footholds for implementing energy digitization and low-carbon transformation.

Using artificial intelligence algorithms centered on clean, low-carbon, safe, efficient, and intelligent open energy digital features, a modern energy system centered on electricity is constructed for the park. Real-time meteorological information is collected and input into the power prediction model to improve the accuracy of photovoltaic power generation prediction. Combined with photovoltaic power generation and loading, energy storage control strategies are automatically generated to promote the consumption of new energy and maximize benefits.

Application of Big Data in Smart Parks

Smart parks fully utilize big data to perform intelligent management of the park, promote the economic development of the park, improve the quality of park services and management capabilities, provide intelligent data applications and services for enterprises, enhance the overall competitiveness of park enterprises, and achieve industrial agglomeration and upgrading. The application of big data in smart parks is shown in Fig. 6.8.

Unified Park Management: By establishing a smart park display experience center, investment promotion, brand building, industrial communication, and self-improvement can be integrated. At the same time, the center can highlight the park's international positioning and the pursuit of innovative concepts. Short-term batch reception can improve the efficiency of investment promotion, and new types of experience methods can help potential users better understand the advantages of smart parks, thereby improving the success rate of investment promotion.

Digital Monitoring of the Park: Through cross-platform, cross-department, and multitype data fusion, the park integrates data from various systems for comprehensive management and monitoring of the overall operation of the park, including data fusion presentations of park traffic, fire protection, communication, business, etc. Analyzing key indicators in specific fields provides visual monitoring of the park's operational situation, improves the efficiency of problem handling, and reasonably dispatches park resources.

Digital Operation of the Park: Multiple indicator systems are constructed to support multidimensional analysis of park operation management data, including financial analysis, business operation analysis, traffic management, and public service analysis, to achieve dynamic monitoring of park operation management data, providing data support for intelligent park management.

Application of Digital Twins in Parks

Digital twin parks can accurately reproduce the external environment, buildings, the industrial distribution, the internal structure of buildings, and specific equipment operation conditions of the park based on digital twin scenarios while effectively integrating various information resources of park operations. Comprehensive monitoring and analysis of key indicators of park integrated operations, security, traffic, facilities, energy efficiency, environmental space, and other management businesses and unified management of park people, events, and objects can assist managers in comprehensively perceiving and comprehensively judging the operational situation of the park, achieving refined management, scientific decision-making, and efficient services.

Application of the Internet of Things in Smart Parks

Smart parks integrate park buildings, security, energy consumption, equipment, offices, and other infrastructure with information systems through IoT technology. building a safe, efficient, and convenient information resource exchange network, achieving high sharing and utilization of various park resources, as well as intelligent management through the coordination of multiple systems, multiple users, and multiple businesses.

6.2.3 Smart Park Practice Case: Digital Empowerment to Gui Valley, Creating a High-End Carrier for Enterprises

Case Features: Digital Base Connecting Basic Systems and Smart Park Business Innovation

Gui Valley Technology Park in Guangzhou was jointly initiated by the Silicon Valley Technology Association, Guangdong Zhonghuan Investment Group Company, and the Silicon Valley Technology Advisory Committee. The technology park is located at the intersection of Kaitai Avenue and Kefeng Road in the Science City of the Guangzhou Development Zone and has convenient transportation. The total land area is 73,522 square meters, and the total construction area is 304,000 square meters. As one of the key industrial projects in the Guangzhou Development Zone, the park includes apartments, offices, and commercial supporting formats.

Gloryview Technology is responsible for undertaking the in-depth design and implementation of the intelligent and smart platform of Gui Valley Technology Park.

By collecting and organizing the intelligent needs of various roles in the project, the vision, business service scenarios, operation and service platform, technical solutions, and implementation strategies and plans for the smartization of Gui Valley Technology Park were prepared. Gui Valley Technology Park is positioned as a demonstrative park that is intelligent, interconnected, shared, and integrated.

In addition to the basic data center and intelligent building facilities, Gui Valley Technology Park adopts the "3 + 1" concept in the construction of intelligence, as shown in Fig. 6.9, where "3" represents three platforms, namely, the park operation management service platform, the Internet of Things platform, and the big data analysis platform, and "1" represents a park visualization system. Through this concept, the project has achieved open connections and comprehensive integration between devices, people and devices, and people; complete integration of digital infrastructure and basic systems; and innovative combinations of smart park business and operations.

Innovative digital infrastructure to solidify the foundation.

Based on the three connection technologies of "Wi-Fi6, passive optical network, Ethernet", we achieve architecture integration, reduce network levels, and lower deployment costs. We deploy a unified lightweight network management system to achieve operation and maintenance management integration and improve the user connection experience through unified authentication and seamless roaming.

The application of artificial intelligence technology ensures the safe operation of the park. Security management is an important safeguard for park operations. The intelligent security system built by Guigu Technology Park can perform park security monitoring through image recognition technology, facial recognition technology, etc., perceive security risks in real time, issue early warnings and responses, and achieve seamless access to ensure safety.

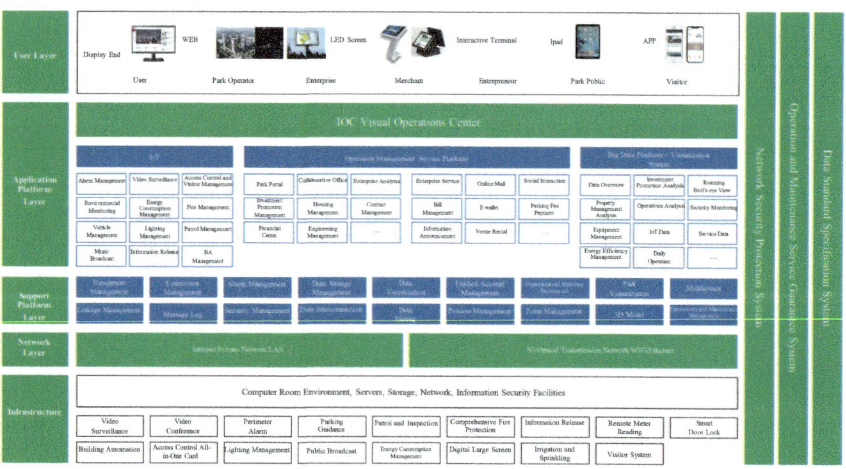

Fig. 6.9 The "3 + 1" concept of Guigu for building a smart park

Visualization management of data centers. The data center is the operating environment of the main servers of the smart park. It is necessary to ensure the normal operation of the data center through visual management.

Internet of Things platform allows data to move from isolation to integration.

Guigu Technology Park uses the Internet of Things platform as the connection core, supports various business applications through service-oriented support, and connects various terminals downward. It changes the traditional park intelligence integration method and system architecture and "decouples and reconstructs" the traditional integration method through "connection + platform", thereby changing the spider web-like application integration relationship between traditional intelligent subsystems. The Internet of Things platform supports multinetwork access and multiprotocol access, solves complex and diverse device access and fragmentation problems, and achieves fast, unified access and control of devices. Through condition triggering, which is based on preset rules, it can trigger the coordinated response of multiple devices, link devices, and achieve intelligent control. It establishes Guigu Park's own IoT standard specifications, simplifies device integration, and saves considerable time and costs in park construction.

The Internet of Things platform we provide in the Guigu project supports common MQTT (Message Queuing Telemetry Transport) protocols, HTTPS (Hypertext Transfer Protocol Secure) protocols, COAP (Constrained Application) protocols, NB_IOT (Narrowband Internet of Things) device connections, etc., as well as various protocol clusters. The architecture of the Internet of Things platform is shown in Fig. 6.10.

The operations management service platform provides the park with sustainable development capabilities.

The operations management service platform of Guigu Technology Park integrates park data and communication, supports multidimensional intelligent analysis

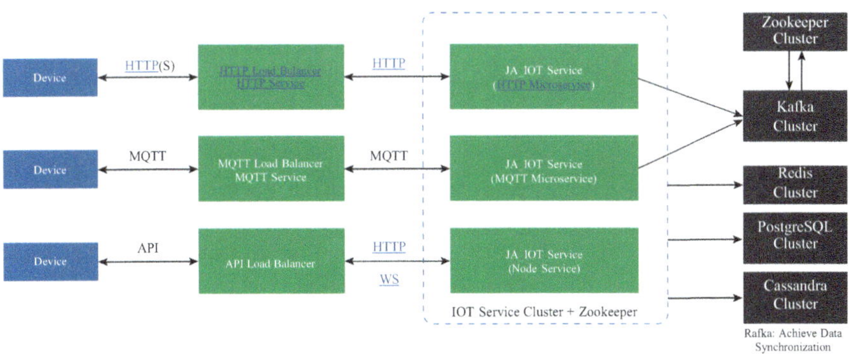

Fig. 6.10 Architecture design of the internet of things platform

and data processing under the project, and enables full-media communication collaboration. It creates a safe, efficient, flexible, green and energy-saving innovative environment for management staff, enterprises, and the public, making project management and operations visible, manageable, controllable, more intelligent, environmentally friendly, efficient and refined. The architecture of the operations management service platform is shown in Fig. 6.11.

The operations management platform has the following main functions:

Park Information Management: Divided into internal park information and external park information. Internal park information reflects the basic status of the park and economic activity information within the park. Park status includes

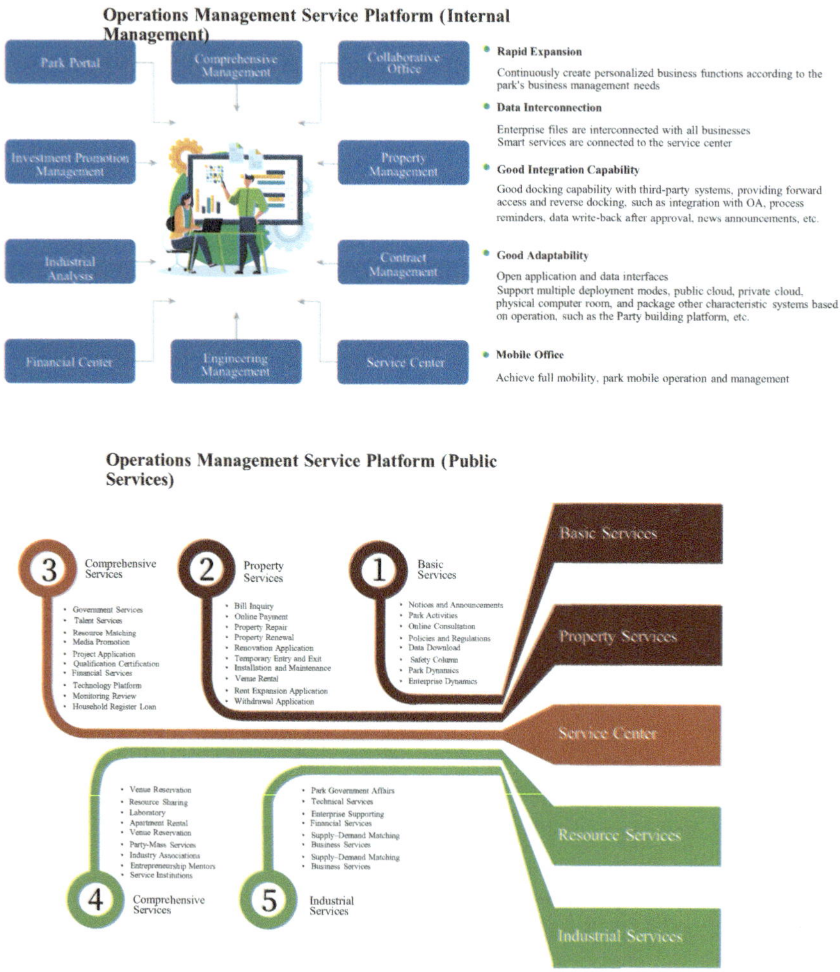

Fig. 6.11 Architecture of the operations management platform

information about personnel, finance, materials, and the scale of the park, whereas economic activity includes information about supply, production, sales, and financial accounting. External park information includes information about the macro social environment, such as politics, the economy, culture, social conditions, and the legal environment. This external information has a significant effect on the development and operation of the park.

Industry Services: Concentrated in the fields of technology intermediaries and modern services, covering the following aspects: park government affairs, technical services, enterprise support, financial services, enterprise management, marketing services, supply and demand docking, business travel services, and human resources. The service platform is based on the service portal, providing notices, news (including industry news and enterprise news), policies and regulations, park activities, online consultations, safety columns, data downloads, etc. According to the needs of the park, the content can be changed and added, and it supports changing the park, full-text search, account login, display of enterprise information, and safe exit.

Park Services: Includes park product displays, park activities, park circles, cleaning services, plant services, office supplies procurement services, etc. Park services also support conference room reservation systems, venue reservations, and advertising space rentals, providing a multifaceted service experience for park enterprises and the public.

Investment Management: The investment management module displays information about commercial investment projects, investment industries, and preferential investment policies. It also includes all the basic information about the location, area, building number, code, and whether it is occupied, along with the tenant name, rent, assets, etc., of the office or shop.

Property Management: Provides users of the property management department with a comprehensive set of services, including a property overview, rental situation, upcoming contract expiration, overdue nonpayment, property renewal situation, property fee payment situation, etc., presented in an intuitive chart form. Users can view the corresponding information by clicking on the drop-down menu. In addition, the system provides property management, energy management, contract management, lease management, etc., to meet the needs of the property management department.

Smart Office: Realizes the online management and approval process of daily business management within Guigu Park, making internal work more convenient, more efficient, and traceable. This function enhances the office efficiency of Guigu Park management.

Platform-based Management and Public Service: Platform-based and mobile park management is designed based on the work management of various functional departments in the park and the service requirements of enterprises. The platform provides common business functions required for daily office work, message reminders, and internal communication to cross-departmental business collaboration, as well as investment management, property management, service centers, industry analysis, equipment management, etc. By achieving comprehensive informatization and mobilization, park operations are more efficient. The public service platform (including the app, public account, and mini-program) is people oriented, meets the public's needs for work and life in the park, provides high-quality services, and improves the public's sense of happiness and security in the park. By adopting the platform + operation model, the park's function is transformed from management to service, enhancing the park's soft service capabilities. By serving the public, this model promotes enterprise development, driving industrial upgrading, which in turn promotes local economic prosperity.

Digital Twins, Increasing the Park's Vitality

The Guigu Science and Technology Park project is based on the construction of a digital base and data connection of the smart park, enabling the park space and equipment to be digitally managed through digital twins, which efficiently supports the operational goals based on a single map. The project fully integrates all facilities and equipment in the park, four major types of projects, more than 20 subsystems, and more than 10,000 device terminals.

Big data analysis and prediction technologies are used to achieve the intelligent management of Guigu Industrial Park, enabling park management to assist in decision-making analysis of basic park data, industry situations, investment management, property situations, operational situations, park traffic situations, and financial situations.

The park uses a three-dimensional architectural model built with UE4, importing onsite facility and equipment data, videos, etc., into this three-dimensional model, constructing a digital twin of the real-world Guigu Science and Technology Park in the digital world. This twin is not static but rather vibrant, and on the platform, real-time data of people, vehicles, venues, industries, etc., can be intuitively obtained. The three-dimensional visualization interface of Guigu Smart Park is shown in Fig. 6.12.

Fig. 6.12 Guigu smart park three-dimensional visualization interface

Construction Results: Safe, Green, Smart Guigu Science and Technology Park

We have always placed great importance on the research of park solutions and have made innovative achievements in the application of equipment, methods, and systems in specific scenarios.

With an emphasis on international first-class standards, Guangzhou Guigu Science and Technology Park sets a benchmark for science and technology parks based on the operational, management, and service needs of new smart parks. With the smart park as the entry point of planning, it builds a "1" unified information entrance (data center, shareable), "3" major service targets (park management, park enterprises, and park public), "1" map (three-dimensional simulation map), and "1" unified control system (park operation management service platform, Internet of Things platform, big data analysis platform, and visualization system). It focuses on creating a diversified smart park that includes smart management, smart offices, and smart life, leading to the innovative development of smart parks with an open and forward-looking mindset.

With respect to the goal of building a safe park, Guigu Science and Technology Park has constructed a three-dimensional security defense system for the physical space and information space of the park, with physical defense as the foundation, technical defense as the support, and human defense as the mainstay, forming a visible, manageable, and controllable comprehensive security situation in the park to ensure park safety. The platform + AI equipment is used to issue early warnings, facilitate control during the event, and enable second-speed precise positioning of abnormal events, which serves park safety management and enhances the park's safety image. The emergency response time of the conventional park is 7 min. Guigu Science and Technology Park's emergency response time can be reduced to 1.5 min, greatly improving efficiency.

Guigu Technology Park is committed to building a green park. It explores new models of comprehensive energy operation in the park in terms of carbon reduction and carbon savings, enabling the digitization of the carbon reduction process and helping the park achieve its "dual carbon" goals. By creating an Internet of Things platform and integrating all energy equipment in the park, it can refine management and reduce operating costs. For intelligent control of ventilation and air conditioning, it adjusts HVAC system operation based on indoor and outdoor environments, ensuring high comfort and health while improving energy efficiency. For lighting equipment, it provides adjustment strategies based on human experience, controls the state of the lighting equipment, and achieves optimal operation. For power supply system protection, it uses telemetry, remote signaling, remote control, remote adjustment, and video surveillance functions to achieve unmanned or minimally manned operation of the transformer and distribution room. For energy consumption monitoring, it can automatically record, summarize, and analyze building energy consumption. The use of expert systems can provide energy optimization suggestions. Equipment abnormalities provide intelligent alarms through the comprehensive use of various detectors, detecting system abnormalities in a timely manner and addressing them promptly.

We digitally map the space and equipment of Guigu Technology Park, accomplishing the twinning of spatial data based on the unified digital base of the smart park and supporting the unified integration, management, and application of business system data. This map provides a multidimensional perspective for park management and operations, allowing managers to understand the status of the park in real time, optimize operational mechanisms and strategies, improve asset management efficiency, and reduce overall operating costs.

6.3 Smart Cities: Improving City Management Effectiveness and Enhancing People's Quality of Life

6.3.1 Formation of Urban Disease

Since the reform and opening up, with the development of the economy and society, China's urbanization process has been continuously advancing. The urbanization rate increased from 17.92% in 1978 to 30.48% in 1996 and then to 60.6% in 2019. The stage of increasing the urbanization rate from 30 to 60% took approximately 180 years in the UK, approximately 90 years in the US, approximately 60 years in Japan, and only approximately 30 years in China. China's urbanization development is rapid. Currently, in 2022, China's urbanization rate is 65.22%, while the urbanization rate of developed countries is approximately 80%. There is still a gap between China's urbanization level and that of developed countries, and China's urbanization process will continue to advance. China's annual urbanization rate is shown in Fig. 6.13.

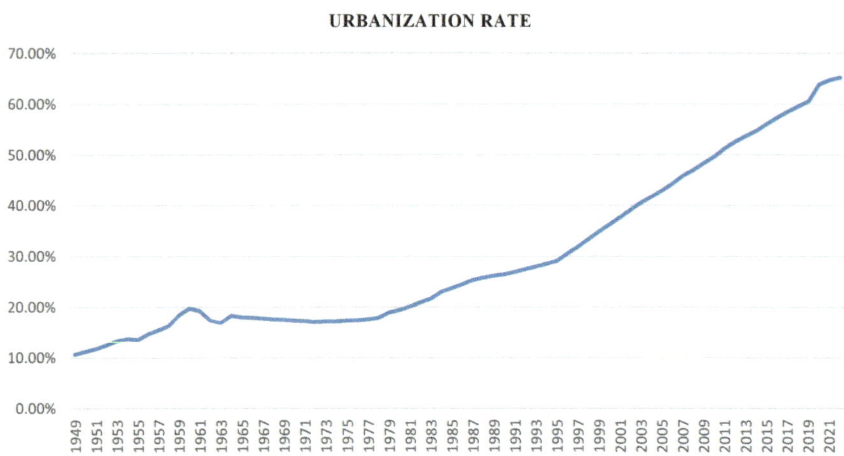

Fig. 6.13 China's annual urbanization rate

The previous text analyzed the problems of imbalance in urban and rural development and the large gap in urban and rural development. Therefore, why should we continue to develop urbanization when the gap between urban and rural areas is already so large? Will urbanization affect the development of agriculture and rural areas?

Urbanization (or townification) is a social development trend that inevitably accompanies industrialization and modernization. It is an important symbol of national modernization and a powerful engine for sustained economic development. It facilitates the transition from a small-scale peasant economy to a market economy. The migration of the rural population to towns can expand the consumer group, upgrade the consumption structure, release consumption potential, accelerate the transformation and upgrading of the industrial structure, and solve the contradictions of a low agricultural land utilization rate and the inability to mechanize production on a large scale, which is conducive to modernizing agriculture on a large scale.

However, in the rapid development process of urbanization, the current situation of political and economic resource concentration attracts and drives public resources, including finance, education, and medical care, to concentrate in large cities, causing an imbalance in the allocation of public resources between cities at different levels. This imbalance leads to a tilt of population and socioeconomic activities toward large cities. When a city develops to a certain extent, the population size exceeds the maximum capacity of the city, and the city's development planning, infrastructure, public service support, and city management capabilities cannot keep pace with the actual development speed of the city. Problems such as traffic congestion, housing tension, difficulty seeing a doctor, difficulty in employment, difficulty in schooling, and environmental pollution occur, and "urban disease" gradually emerges.

"Urban disease" is a negative issue that hinders the healthy development of cities and accompanies urbanization. Compared with that of developed countries, China's urbanization process is fast, and its late development advantage is obvious. The rapid advancement of urbanization has played an important role in promoting socioeconomic development, but the characteristics of "concentrated development" have also led to a series of negative "urban disease" problems.[1]

6.3.2 A Smart City is the Best Solution to Urban Disease

In recent years, with the rapid development and innovative application of new generation information technologies such as the Internet of Things, cloud computing, big data, and the mobile internet in China, the country's urban modernization process has ushered in new development opportunities. The key issues that urgently need to be addressed by urban builders and managers everywhere include how to increase the innovative application of information technology in urban management, services,

[1] Huirong Zhang, Shijie Zhang. A Review of the Causes and Formation Mechanisms of "Urban Diseases" [J]. Journal of Tongling University, 2019, 18(02): 14–18.

and operations, enhance the comprehensive competitiveness and sustainable development of cities, effectively solve and prevent problems such as the imbalance between urban population growth and carrying capacity, which leads to increasing pressure on resources and environmental ecology, and the contradiction between government public management and public demand, which results in poor coordination of urban management and low efficiency of work; and how to improve urban operation management and social governance, as well as social management and public services.

With the construction and development of modern cities, traditional ways of managing cities manually have gradually been phased out. Urban management innovation has become a new topic in modern urban management, and **smart cities represent a new concept and model for global urban development and the best way to solve "urban disease" problems.** According to the national standard "Smart City Terminology" (GB/T 37043-2018), a smart city is an innovative city that uses information and communication technology to effectively integrate various urban management systems, facilitate information resource sharing and business collaboration between various city systems, promote the intelligent management and services of cities, enhance urban operation management and public services, improve the happiness and satisfaction of urban residents, and achieve sustainable development.

Smart cities are the inevitable direction of modern urban development, and the modern urban "big city management" model characterized by informatization, networking, digitization, automation, and intelligent technology applications has emerged. Informatization vigorously promotes the innovation and transformation of urban management systems and management processes. By integrating urban element information resources, logically centralized physical distribution data can meet the development needs of modern cities at the data level, quickly obtain information from various aspects of the city, and enable comprehensive analysis and decision-making.

Smart cities aim to achieve efficient urban governance, convenient services for the people, prosperous industrial development, green and livable ecology, and coordinated regional development (as shown in Fig. 6.14). They apply new means, new technologies, new models, and new systems to achieve unified management of urban information infrastructure and resources, improve urban governance capabilities, and ensure sustainable and efficient urban development.

Smart cities are massive system engineering projects, with applications covering all aspects, from the clothing, food, shelter, and transportation of each individual to the planning and management of the city. Moreover, they will continue to expand and become enriched during the development process. According to the development goals of smart cities, the applications of smart cities are classified into **smart governance, smart livelihood, smart industry, smart ecology, and smart regional collaboration**. Since smart cities cover a wide range and each region has distinct characteristics and needs, the development direction of smart cities in different places will have different emphases. Hence, planning and construction should be carried out according to local conditions. The scope of smart city construction is shown in Fig. 6.15.

Fig. 6.14 Smart city goals

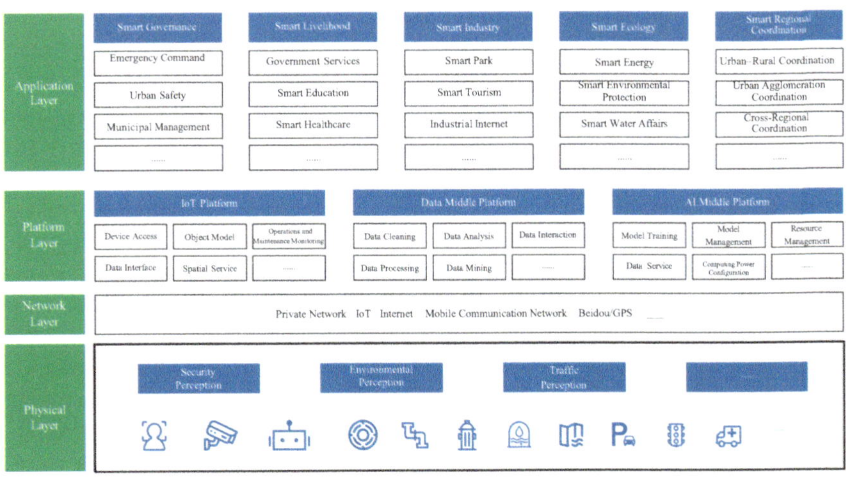

Fig. 6.15 Smart city construction scope

Smart Governance

The efficient governance of smart cities involves identifying the information resources of various government departments in areas such as urban safety, emergency response, and municipal management; clarifying their respective businesses and resource data; breaking down data barriers between departments; achieving dynamic monitoring of comprehensive signs of urban operation; providing comprehensive, visual management tools to ensure the good operation of the city and the healthy operation of government agencies; creating favorable conditions for the transformation of government management functions and improvements in the urban management level; and improving the efficiency of urban management.

For example, Haidian District uses AI algorithms to identify and analyze muck trucks, achieving a nearly 30% increase in the number of muck trucks captured, an accuracy rate of over 95% for the system's identification of illegal features such as muck truck license plate obstructions, and an accuracy rate of over 60% for the prediction and analysis of muck truck travel trajectories. Through AI-assisted muck truck management, the violation rate of muck trucks has decreased from 26 to 4%, effectively reducing management costs and improving governance efficiency.

Smart Livelihood

Technology that benefits and facilitates people's lives is the basic requirement to meet the social public service needs of smart cities. On the one hand, it should organize government information resources, focus on information that the public generally cares about, aggregate social management and social service content resources, build unified and diversified information service channels, provide rich information content and optimal functions, ensure the public's enjoyment of the convenience brought by digitalization, transform businesses from "running multiple times" to "running once, running zero times", and improve the public's experience of handling affairs. On the other hand, in the fields of education, health care, elderly care, employment, parking, and other public services and convenience services, information technology, the Internet of Things, the internet, and other technologies can be used to eliminate information barriers and provide convenient and efficient services for the public.

For example, the "Urban Brain" of Hangzhou city has launched convenient parking, comfortable medical treatment, and joyful tourism scenarios, reflecting a people-centered development philosophy. The comfortable medical treatment service offers mobile appointment registration, "see a doctor first, pay later", "pay only once", "family account", "Internet hospital" and other services, thus greatly improving the medical experience.

Smart Industry

A thriving industry is an important force for the economic and social development of a city. On the one hand, by integrating and analyzing the data generated by various industrial modules in a city, an industry centralized control system can be built to portray regional industrial clusters and industrial vitality, assist government decision-making, continuously exploit the effects of industrial clusters, generate industrial vitality, promote comprehensive industrial development, and facilitate the coordinated, stable, sustainable, and harmonious development of the socioeconomic operations system. For example, Hengzhou city in Guangxi is the world's jasmine capital. It prioritizes the informatization of the entire jasmine industry chain. Through the digital construction of the "Digital Jasmine" greenhouse, "Digital Jasmine" trading platform, "Digital Jasmine" supply and demand system, "Digital Jasmine" big data platform, etc., high-quality development of the entire jasmine planting, production, and sales industry chain in Hengzhou city has been realized. In 2020, the comprehensive annual output value of the jasmine (tea) industry in Hengzhou city surpassed the 12.5 billion yuan mark, and the comprehensive brand value reached 20.685 billion yuan.

On the other hand, information technologies such as the internet, Internet of Things, cloud computing, and artificial intelligence promote the digital upgrade of industries, drive the development of emerging industries, enhance the level of industrial intelligence, and strengthen the endogenous driving force of urban development. A technology innovation mechanism dominated by enterprises and led by the market should be established, the transformation and structural optimization of traditional enterprises should be accelerated, and the core competitiveness of related industries and enterprises should be enhanced through technological innovation, organizational innovation, and service innovation.[2] For example, Moutai, as a traditional physical manufacturing enterprise, has actively opened new tracks in digital innovation transformation. The iMoutai and Xunfeng apps became the focus as soon as they were launched. iMoutai explores the S2B2C online and offline integrated liquor sales model, while "Xunfeng Digital World" is based on virtual reality research and development technology. The apps apply interactive experience engines, digital twins and other core technologies, represent the Moutai brewing environment in detail in the virtual world, explore Moutai's historical culture and craftsmanship, create a "parallel world" encompassing Moutai and users, form a digital ecosystem linking Moutai's upstream and downstream industry chains, enhance the user social experience through digital transformation, and create digital value.

---

[2] National Industrial Information Security Development Research Center; China Industrial Internet Development Alliance; Key Laboratory of Industrial Big Data Analysis and Integrated Application, Ministry of Industry and Information Technology; et al. *Co-Creating New-Type Smart Cities Through Intelligent Services: 2022 White Paper on Smart Cities*[Z]. May 24, 2022.

Smart Ecology

The sustainable development of a city is the basic task of smart city construction. It involves integrating an ecological city with a smart city, helping sustainable thinking to permeate urban development and construction, building a green, low-carbon, harmonious, liveable, and eco-friendly sustainable development system, and generating new opportunities for a city's continuous, healthy, and high-quality development. The development of low-carbon cities, smart energy, smart water conservancy, smart water affairs, smart environmental protection, etc., has become the main direction of the current development of ecologically friendly cities. Through technology empowerment, ecological environment management problems can be solved, and a healthier and more liveable urban environment can be created.

For example, the Sino-Singapore Tianjin Eco-City has formed a smart, full-process control system of "source reduction, classified collection, sealed transportation, and comprehensive treatment" through the "Waste-Free City" information management platform. It can accurately capture the household waste classification data of each household, achieve full coverage of intelligent recycling terminals, and fully implement household waste classification. Moreover, it has built a public service platform for recycling resources, a public participation platform, and a social resource cooperation development platform; created professional recycling and disposal channels for online and offline resource waste; promoted the implementation of "waste classification + resource recycling"; launched a waste classification app; and enabled residents to experience waste classification, large waste appointments, query recycling terminals, participation in public welfare activities, etc., through mobile phones, making residents' lives more intelligent and environmentally friendly. It has also introduced a smart credit evaluation system, increased the sense of honor of residents as "civilized citizens", greatly enhanced the happiness and sense of gain of residents, and truly achieved harmonious coexistence and intelligent sharing between people, nature, and economic activities in the eco-city.[3]

Smart Regional Collaboration

The connotation of a smart city not only concerns the internal development of a single city but also from a macro cross-regional perspective, forms a regional economic layout and territorial spatial system of complementary advantages and high-quality development, solving the problem of unbalanced regional development and achieving smart collaborative development between regions. In terms of urban–rural collaboration, urban group collaboration, and cross-regional collaboration, information technology is used to facilitate the rational flow of elements across regions, strengthen the connection of infrastructure and industrial supporting cooperation between regions,

[3] Cyberspace Administration of China-Singapore Tianjin Eco-City. China-Singapore Tianjin Eco-City: Building a National Model Zone for Smart City Development [J]. China Construction Informatization, 2022(01): 42–45.

enhance the overall coordination of regional development, and form a global and overall regional development strategy.

For example, in terms of urban–rural coordination, the urban–rural integration model of Xiaoshi Village in Chengdu encompasses four dimensions: industry, the market, culture, and ecology. For example, it prioritizes agricultural development and promotes agricultural planning, industry, urbanization, and rural development as a whole while developing large-scale agriculture. It explores the connection between local advantages and characteristic industries, cultivates agricultural sightseeing tourism, camping industries, etc., and expands the ecological and economic functions of agriculture. It continuously stimulates the commercial vitality of suburban villages through market-oriented construction, develops homestays with the concept of the "shared economy", and transforms idle resources into characteristic assets. It constantly improves grassroots organizations, strengthens and enhances the subject status of villagers, develops new rural community management and service functions, and continuously narrows the gap between villagers and urban residents in terms of lifestyle, employment methods, etc. It introduces the concept of urban digital governance, develops the "three intelligent systems", builds the first smart community in Pengzhou city, provides precise intelligent management of all villagers' information, and comprehensively improves the level of rural governance services and work efficiency.[4]

6.3.3 Empowering Smart City Construction with New Technologies

Against the backdrop of the integration of digital technologies such as artificial intelligence, big data, cloud computing, blockchain, the Internet of Things, and the metaverse, smart cities have undergone rapid development. Through the integration of new technologies, smart cities have achieved a series of practical innovations, with typical applications in government affairs, transportation, environmental protection, energy, health care, logistics, etc., which have generated significant social, economic, and ecological benefits.

Application of Artificial Intelligence in Smart Cities

With the continuous advancement of urbanization, cities are facing increasing challenges, such as population growth, traffic congestion, and resource scarcity. To better solve these problems, artificial intelligence technology is gradually being widely used in the construction of smart cities.

[4] Wu Xiaoting. Urban–Rural Integration in the Outer Suburbs of Megacities: Development Models and Insights from Field Research in Xiaoshi Village, Chengdu [J]. Journal of Beijing Vocational College of Agriculture, 2023, 37(02): 38–44.

Smart City Transportation: Through artificial intelligence technology, city traffic flow is monitored and managed, facilitating services such as intelligent traffic lights and intelligent public transportation.

Smart City Environment: Through artificial intelligence technology, real-time monitoring of environmental pollution and meteorological conditions within a city can be conducted, providing support for urban environmental decision-making.

Smart City Safety: Through artificial intelligence technology, public safety monitoring and early warning in cities are enabled, helping cities detect and resolve safety hazards in a timely manner.

Smart City Management: Through artificial intelligence technology, automated management and monitoring of facilities within a city, such as street light brightness and noise detection, are realized.

Smart City Livelihood: Through artificial intelligence technology, convenient services and intelligent living experiences, such as smart homes, smart communities, and smart parks, are provided for city residents.

Application of Cloud Computing and Big Data in Smart City Traffic Governance

The importance of transportation to city construction and development is self-evident. It is essential infrastructure for ensuring the normal operation of a city. Therefore, it is necessary to manage city traffic in a scientific and reasonable manner. Congestion and traffic accidents are common problems in urban traffic operations and affect city development to an extent. Smart traffic can greatly reduce the problems associated with city traffic. Cloud computing and big data technology can be applied in city traffic management. By comprehensively analyzing various types of city traffic information, including road information, traffic flow, and GPS data, the travel patterns of a city can be predicted, and the results can be uploaded to the relevant management departments. The relevant management departments can then formulate corresponding measures based on this information, further improving and optimizing the city traffic situation, with the aim of achieving smart city traffic management.

In practical applications, cloud computing technology monitors various types of information in the process of managing city traffic operations in real time and captures the flow of city traffic in each time period. On the basis of this information, it reasonably adjusts the response time of traffic lights. Even in the absence of manual control, intelligent timing of traffic lights can still be achieved, thereby fully using the role of traffic lights in city traffic and alleviating city traffic pressure while improving the efficiency of smart traffic management.

Furthermore, by using cloud computing and big data technology, real-time monitoring and intelligent analysis of the entire city's traffic situation can be achieved. This capability allows for the precise prediction of traffic conditions at specific intersections within a city. Based on real-time traffic information reflected by the system, the most convenient route for citizens' travel can be determined. During their journeys, citizens can automatically change their driving route according to real-time traffic information, bypass congested sections, and thus ensure smooth travel. This approach not only saves a significant amount of travel time but also minimizes travel costs and provides effective guidance for vehicle parking. In terms of traffic accidents, early warnings can be made through big data, providing reliable data support for traffic police to handle sudden incidents. Hence, the emergency handling capacity of the traffic police department can be improved, and road congestion caused by traffic accidents can be promptly addressed.

Simultaneously, through cloud computing and big data technology, comprehensive statistics and analyses of historical traffic operation data can be conducted. Emergency information such as accident-prone roads can be promptly published on the traffic platform for rational diversion of vehicles at the earliest possible time. The traffic network in smart cities can be improved and optimized, providing timely and accurate traffic information for city residents' daily travel, improving the level and quality of urban traffic management, and ensuring the orderly operation of traffic in smart cities.

Application of Digital Twins in Smart Cities

Digital twin cities can accurately reproduce large-scale urban management elements based on the full integration of information resources in various urban fields. They can perform multidimensional visual analysis of data indicators in segmented business areas, achieving a holistic dynamic perception of the city's operational situation from a global perspective to a microscopic perspective. For example, city managers can more efficiently and conveniently understand the economic development of various districts, identify differences in traffic flow in the morning and evening, and make comparisons of population heat in urban areas.

Digital Twin Police Application: By comprehensively applying technologies such as 5G, big data, cloud computing, AI, IoT, and digital twins, with the latter as a foundation, various police business systems can be connected to achieve multidimensional data integration and intelligent management of all police elements. Focusing on practical application requirements, multiple business topics are established to provide intelligent and efficient "superbrains" for public security command centers to perform intelligence monitoring and early warning, police action handling, police commands and dispatch, intelligence analysis and judgment, and other practical operations.

Digital Twin Emergency Application: The smart emergency application based on digital twin technology can effectively combine IoT, cloud computing, big data, digital twins, artificial intelligence, and other technologies to fully leverage the supporting role of modern technology in disaster prevention, mitigation, and relief.

Multiple business areas, such as urban safety monitoring, production safety monitoring, natural disaster monitoring, and emergency handling, are covered, environmental data are collected in real time, and big data analysis and artificial intelligence algorithms are used to monitor and issue warnings of natural disasters (such as earthquakes, floods, typhoons, etc.) in real time, thereby providing timely disaster prevention and mitigation information for government departments. In emergency situations, digital twin technology can enable real-time monitoring and dispatching of emergency resources (such as personnel, materials, and vehicles), ensuring the rational use of various resources and improving emergency response efficiency. On the basis of digital twin technology, government departments and enterprises can thus formulate targeted emergency plans, test the effectiveness of the plans through simulation drills, and continuously optimize the emergency response process.

Digital Twin Traffic Management Application: The smart traffic digital twin platform can fully integrate various types of data, such as geographic information, traffic conditions, police handling information, and vehicle driving information, providing support for business scenarios such as comprehensive traffic situation monitoring, 122 police handling monitoring, road condition monitoring, and major event traffic security. Through the "one map" command operation, the platform provides strong support for users to perform traffic supervision, situation judgments, service operations, police supervision, and other business applications, thus offering comprehensive, all-weather, and fully processed dynamic perceptions of traffic situations. For example, intelligent signal control involves real-time monitoring of traffic flow, congestion status, and other data, and the digital twin platform can perform intelligent signal control, automatically adjust the duration of traffic lights, optimize traffic flow lines, effectively alleviate traffic congestion, and improve road traffic efficiency. By optimizing the analysis of current and historical traffic data, the digital twin platform can identify the causes and patterns of traffic congestion, provide strong support for urban traffic planning, and improve traffic conditions in a targeted manner.

Digital Twin Pipeline Network Application: Based on digital twin technology, a three-dimensional geographic information system can be used as the foundation to provide a three-dimensional display of the overall distribution of large-scale pipeline networks to local pipe galleries and integrate multisource pipeline network operation data to monitor and analyze the above-ground and underground distribution situation, coordinates, direction, operation of various types of pipeline networks, health status of pipe galleries and pipelines, working conditions of components, etc., in real time. The operation and management of pipeline network "points, lines, and areas" are visible throughout the whole process, assisting users in efficiently mastering the operation of large-scale pipeline networks, transitioning from passive management to proactive, intelligent management, and effectively improving the level of pipeline network safety management.

Digital Twin Watershed Application: Constructing a realistic three-dimensional watershed, the system meticulously reproduces the full panorama and all-element situation of the watershed. By integrating the data resources of existing information systems from various water resource departments and combining the capabilities of the CIM base and dynamic simulation deduction, it organically combines information, technology, equipment, and water management needs. This application covers multiple business areas, such as watershed overviews, flood monitoring, reservoir monitoring, and water dispatching, fully empowering user business applications and effectively improving the efficiency of cross-departmental decision-making and resource coordination in water resources.

The Application of the Metaverse in Smart Cities

Metaverse technology can be used for urban emergency drills, improving the efficiency and safety of emergency rescue. For example, in fire emergency drills, virtual reality technology can be used to construct realistic fire scenes, allowing participants to practice in a virtual environment and thereby improving the effectiveness of the drill and reducing its cost.

In addition, metaverse technology can be used to simulate various emergency situations, including natural disasters such as earthquakes and floods, as well as man-made disasters such as terrorist attacks and fires. This functionality can help emergency rescue personnel practice and train, improving their ability to deal with various disasters. Moreover, metaverse technology can be used to develop a virtual training system for emergency rescue, allowing more people to participate in emergency rescue and thereby improving the overall emergency rescue capability of society.

Metaverse technology can play an important role in earthquake emergency drills. Earthquakes are unpredictable natural disasters, and their location, time, and scale are difficult to predict. Therefore, to address earthquake disasters, we need to conduct earthquake emergency drills to improve our ability to respond to earthquakes. Metaverse technology is an important tool that can help us achieve this goal.

Specifically, using metaverse technology, we can construct realistic earthquake scenes and simulate various situations when an earthquake occurs, such as ground shaking and building collapse, allowing participants to conduct earthquake emergency drills in a virtual environment, thereby improving their ability to respond to earthquakes. Moreover, metaverse technology can be used to simulate earthquake situations in different regions, helping participants to understand the earthquake characteristics of different regions and improving their overall ability to respond to earthquakes.

In addition, metaverse technology can be used to develop a virtual training system for earthquake emergency rescue, allowing more people to participate in earthquake emergency rescue and thereby improving the overall emergency rescue capability of society. Moreover, metaverse technology can be used to analyze data and generate statistics of earthquake emergency drills, analyze and summarize the problems and deficiencies in the drill process, and further improve the effectiveness and quality of earthquake emergency drills.

6.3.4 Smart City Practice Case: Smart City Construction Boosts the Rapid Development of the Kunming Airport Economic Zone

The Kunming Airport Economic Zone smart city informatization construction project has increased the innovative application of information technology in city management, services, and operations. By doing so, it has enhanced the city's overall competitiveness and sustainable development, effectively solving and preventing problems in the construction process, such as the imbalance between city population growth and carrying capacity, the contradiction between government public management and public demand, poor city management coordination, and low work efficiency, thus improving the city's operation management and social governance capabilities and increasing the level of social management and public services.

Gloryview Technology, which is guided by Xi Jinping's vision of socialism with Chinese characteristics for a new era and oriented toward building a digital China and a smart society, is based on the needs of economic and social development, with reform-oriented thinking and innovative measures. It has established a new model of city management and governance services driven by big data, the Internet of Things, and artificial intelligence for this project. It promotes the integration and in-depth development of information resources, promotes the integration and shared use of government information and IoT data, and the collaborative reconstruction of business processes; builds a "digital government" to meet high standards; constructs a big data center based on big data and artificial intelligence; integrates government, IoT, and internet data across the region; and promotes data governance and data openness on this basis, advancing the modernization of the government governance system and governance capabilities. The overall plan is "1 + N + 9 + N" (as shown in Fig. 6.16), that is, one city operation center, N smart nodes for city management, nine topics for refined city management, and N informatization supporting constructions for commissions and bureaus.

The overall architecture of the project is shown in Fig. 6.17.

Converge sensing resources and establish a coordinated sensing system: This stage includes achieving the integration of sensing resources under various network systems, such as industry-specific networks, intranets, the internet, the Internet of Things, etc.; establishing city sensing planning, construction, and evaluation systems; and continuously standardizing and guiding the construction of city sensing.

Govern sensing resources and consolidate the foundation for reliable applications: Enhancing the innovation stage. The focus of this stage is to promote the application of advanced technologies such as artificial intelligence, big data, and the Internet of Things and to explore and implement innovative applications in various fields. Based on the achievements of the previous two stages, we will further improve the intelligent operation and management system of the city, promote the deep integration of various types of data, and explore the innovative application of intelligent perception in various fields, such as smart prevention and control, smart emergency

Fig. 6.16 Overall plan of the smart city in the Kunming airport economic zone

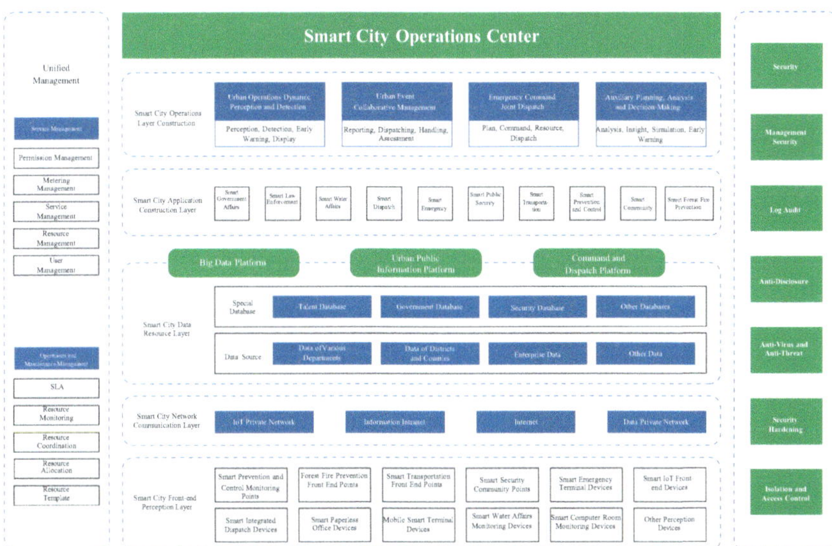

Fig. 6.17 Overall architecture of the smart city in the Kunming airport economic zone

response, smart communities, forest fire prevention, and smart water affairs. Through the continuous capture of the value of perception resources and data, high-quality development of the airport economic zone will be promoted.

Construction Difficulties: No-Fly Modeling, Ambiguous Requirements, and the Creation of Data Value

For the entire smart airport project, there are four main difficulties:

3D modeling is a challenge. Conventional drone methods cannot be used at airports due to no-fly restrictions; if cars or manual scanning are used, the time cycle is too long. Currently, airports are undergoing large-scale infrastructure construction. By the time manual scanning and modeling are completed, the infrastructure may have changed and will be unusable; if satellite mapping is used, the granularity will be too large. Therefore, a suitable technical solution for this 3D modeling problem has not yet been found.

Data platform construction and data governance are challenges. Currently, the administrative division of the airport economic zone management committee is rather vague, with multiple superior departments, unclear functional divisions, and difficulty determining owner requirements. There are also issues of mutual assistance and resource sharing among various related departments.

The ability of a smart city itself to demonstrate a hematopoietic function is a challenge. Let each unit provide data, make use of the data, and let the data create value. Many smart cities have no data after construction, and even if there are data, a good operation mode is needed.

Front-end device selection is difficult. Owing to the broad jurisdiction of the Kunming Airport Economic Zone, how to reasonably select the locations of front-end devices (such as surveillance cameras) poses a challenge. It is necessary to comprehensively analyze and consider geographical conditions and actual needs, select the best locations, adapt the best device types, and maximize the role and effect of the system through the reasonable selection of locations and types.

Construction Path: Laying a Solid Foundation, Quickly Advancing the Work and Enhancing Innovation in Staged Construction

The construction of a smart city is a long-term continuous process. The construction of the smart city in the Kunming Airport Economic Zone is carried out in three stages: laying a solid foundation, quickly advancing the work, and enhancing innovation:

Laying a solid foundation stage. The key work of construction includes laying a solid foundation and promoting resource integration. This stage revolves around the construction of the city operation command center while strengthening the construction of the city operation command center's basic network equipment, that of the big data center in the airport economic zone, and that of the thematic application in the airport economic zone through the establishment of an information security protection system, etc.; integrating the built systems and city's basic data resources to maximize resource utilization; promoting the formulation of related standards and policies; and providing effective support and protection for subsequent project construction. Internally, cross-departmental, cross-industry data resource sharing should be strengthened, a standard data exchange system should be standardized, a convenient service mobile entrance should be provided externally, and the happiness and sense of gain of citizens should be enhanced.

Quickly advancing the work stage. The main goal of this stage is to deepen the application and improve the effectiveness of intelligent construction. Based on the construction of the solid foundation stage, the city operation system should be improved, the construction of a data cockpit, one-network management, one-network office and other business scenarios should be strengthened, the use range and beneficiaries of intelligent applications should be expanded, comprehensive, multidimensional social management and public service systems should be improved, the business environment of the airport economic zone should be optimized, and the scientific nature of decision-making should be assisted by the government. These changes will enhance the city's comprehensive competitiveness and provide inexhaustible power for the sustainable development of the airport economic zone.

Elevating the innovation stage. The focus of this stage is on innovating the application scenarios of intelligent systems based on the in-depth application of the urban brain project, thereby further refining intelligent application services. Providing more precise and personalized intelligent services that are closer to the actual needs of the people in the airport economic zone enhances the user experience and user satisfaction, improves the level of refined management of urban operations, promotes the upgrading and transformation of industrial structures, and boosts the innovative development of the digital economy in the airport economic zone.

Construction Results: Enhancing the Comprehensive Competitiveness and Sustainable Development of Airport Cities

The construction of the urban operation command center should be the focus, while strengthening the construction of basic network equipment for the urban operation command center, that of the big data center in the airport economic zone, and that of specialized applications in the airport economic zone through the establishment of an information security protection system, etc., integrating the existing systems and urban basic data resources to maximize resource utilization; promoting the formulation of related standards and policies; and providing effective support and protection for subsequent project construction.

Internally, cross-departmental and cross-industry data resource sharing should be strengthened, standard data exchange systems should be standardized; externally, convenient service mobile entrances should be provided, and citizens' sense of happiness and fulfillment should be enhanced.

Part III
Future Prospects of Smart Digital Cities

Chapter 7
Basic Outlook

Digital government 3.0 and industrial digital transformation are the basic foci of our explorations of smart digital cities. Digital government 3.0 will be more personalized and intelligent and will digitize public services and public decision-making processes, enabling the government to better serve citizens and businesses. Moreover, the development of digital government is also an important way to promote data openness and sharing, which can promote innovation and socioeconomic development.

Industrial digital transformation is an important topic for future urban development. The widespread application of a new generation of information technology has brought strong momentum to industrial digital transformation. Enterprises can optimize organization and business processes, improve production efficiency, and innovate business models via technological means, especially industrial chain-leading enterprises[1] and state-owned enterprises. Their digital transformation not only concerns the development of the enterprises themselves but also plays an important role in the construction of a modern economic system for the country.

7.1 The Advancement of Service-Oriented Government is the Digital Government 3.0 Model

In February 2018, the Third Plenary Session of the 19th Central Committee of the Communist Party of China deliberated and passed the "Decision of the Central Committee of the Communist Party of China on Deepening the Reform of the Party and State Institutions" and proposed to "comprehensively improve government

[1] A "chain-leading enterprise" refers to a dominant firm within an industry value chain. It holds a competitive position, exerts strong direct or indirect influence over the resources and applications of most enterprises in the chain, prioritizes the creation of value across the entire chain, and plays a core role in improving the overall performance of the industry value chain.

F. Liu, *Smart and Digital Cities*, https://doi.org/10.1007/978-981-95-2243-9_7

efficiency and build a service-oriented government that satisfies the people". With the deepening of national institutional reform, the construction of China's service-oriented government has entered a new era. Since the Fourth Plenary Session of the 19th Central Committee of the Communist Party of China (referred to as the Fourth Plenary Session of the 19th Central Committee) first proposed "promoting the construction of a digital government" in October 2019, the central level of government has been continuously strengthening the construction of a digital government, accelerating the issuance of policy planning, improving the top-level design and mechanism construction, taking people's aspirations for a better life as the starting point and foothold of the construction of a digital government, and creating a ubiquitous, intelligent, convenient, fair, and inclusive digital service system. On June 23, 2022, the State Council issued the "Guiding Opinions on Strengthening the Construction of a Digital Government", systematically clarifying the strategic fulcrum and reform direction of the new development stage to strengthen the construction of a digital government and undertaking strategic planning and systematic deployment to comprehensively create a new plan for the construction of a digital government.

7.1.1 Evolution of Government Service Models

The digital government is a modern digital management model that uses information technology to reconstruct the architecture of government information management, business, and technology. By establishing new government mechanisms, platforms, and channels and optimizing the internal organizational structure, operating procedures, and management services of the government, it comprehensively enhances the government's capabilities with respect to economic regulation, market supervision, social governance, public services, and the ecological environment. The goal of the digital government is to achieve data dialog, data decision-making, data services, and data innovation. The service model of digital government has gone through versions 1.0 and 2.0, and version 3.0 will soon be launched, as shown in Fig. 7.1.

The key problem that digital government 1.0 solves is "business online", that is, moving offline government service windows to websites and mobile phones, which constitutes the physical connection of the internet and government affairs, thus completing the networking and mobilization of government services.

The key problem that digital government 2.0 solves is "business integration, letting data run more, and the public run less". That is, with data as the core element, through system integration and data collaboration, the entire government process is reengineered. For example, in the past, to accomplish one task, 5 government department windows needed to be involved. In contrast, digital government 2.0 completes the task in one window owing to data operations and the reengineering of government department processes. It thus solves the key problems of repeated handling and the challenges encountered in cross-system handling.

The key problem that digital government 3.0, solves is "digital integration", that is, adopting a new generation of information technology, including the governance

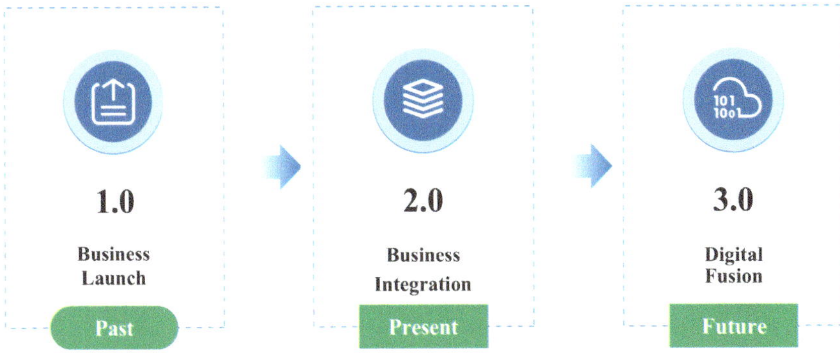

Fig. 7.1 Service model upgrade

model integration of new technologies such as cloud computing, big data, blockchain, artificial intelligence, and the Internet of Things. This integration also represents the transition from digital government 2.0 to digital government 3.0.

Digital government 3.0 focuses on business data and business; uses a new generation of information technology; reshapes the architecture of government information management, business, and organization in a data-driven manner; and forms a modern governance model of decision-making with data, service with data, and innovation with data. It creates "digital, service-oriented, governance-oriented, characteristic" services, achieving full digital coverage offline and online.

The system of digital government 3.0 revolves around tasks such as strengthening the construction of the digital base; strengthening the guarantee system; strengthening the construction of digital governance and convenient digital services; enhancing the digital economy; and building a digital government that operates efficiently, accurately serves, and scientifically manages as a whole. Digital government 3.0 emphasizes collaboration and cloud-based, intelligent, data-driven, and dynamic characteristics. The overall framework of digital government 3.0 is shown in Fig. 7.2.

7.1.2 *Technical Route of Digital Government 3.0*

Since China's reform and opening up, significant progress has been made in administrative system reform. Current grassroots governance is in the transition stage from the "management and power" government governance model to the "service-oriented" social governance model. The development of informatization provides a new way for the government to perform its functions,[2] and Web 3.0 technology, as a new generation of informatization technology, provides a new technical route for the government to perform its functions.

[2] Guojun Zhao. *E-Government: A Textbook* (3rd Edition). Beijing: Renmin University of China Press, 2015: 44–45.

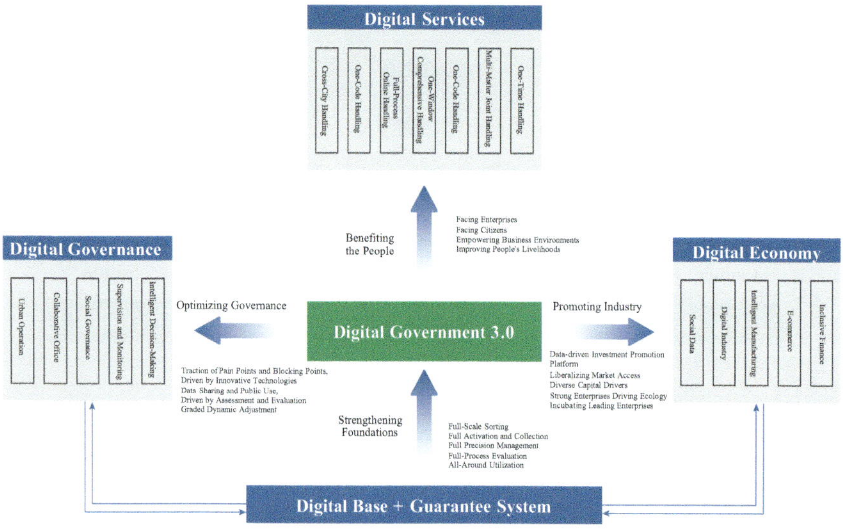

Fig. 7.2 Overall framework

The previous version, which was based on Web 1.0 technology, is a "readable internet" model. Users obtain information through the internet (browser), which displays it in "read-only" mode. Internet content is pushed by specific editors and maintainers, and it can be browsed but not interacted with. Web 1.0 technology solves the demand for information and aggregation but does not solve the demand for communication, interaction, and participation between people. Its core value is based on platform creation, platform ownership, platform control, and platform benefit.

The Web 2.0 technology enables an "interactive internet" model that combines audio, video, and images; uses multimedia mode; and enables user interaction. In this mode, users are both browsers and content producers and can "read" and "write". Thus, it introduces a shift from passively receiving internet information to actively creating it, meeting the needs of two-way communication, interaction, and participation. Its core value is based on user creation, platform ownership, platform control, and platform benefit.

Web 3.0 technology is a "value-based internet" model that not only has the capabilities of Web 2.0 technology but also can protect the value of digital content, that is, user-created digital content. Moreover, its ownership belongs to the user; it does not depend on the centralized platform. Therefore, its core value is based on user creation, user ownership, user control, and the protocol distribution of benefits.

Therefore, the new Web 3.0 technology has a significant impact on the participants of government services. It not only affects the service model of the application but also reorganizes the elements of the entire system, which provides a more convenient and humanized service experience. The Web 3.0 digital application is shown in Fig. 7.3.

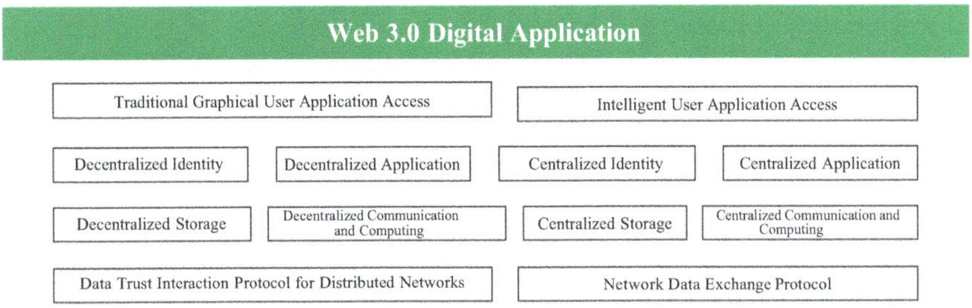

Fig. 7.3 Web 3.0 digital application

Web 3.0 represents a new stage in the development of information technology. Under the data trust characteristics of the Web 3.0 distributed network, a large number of terminal resources can be used to build distributed applications that are most relevant to their core functions within local domains, thereby optimizing both efficiency and resource utilization.

Based on Web 3.0 technology, new foundational applications can be built in the field of digital government services, such as identity verification, digital property rights, and decentralized distributed applications for domain-specific services.

With the support of Web 3.0 technology, there are three feasible technical routes for the realization of digital government 3.0.

Decentralized Supplement to Centralized Data

To date, the development and application of government services have been centralized, leading to the accumulation of a large amount of data, such as application power for key infrastructure, finance, water conservancy, and resident personal data and government data related to the national economy and people's livelihoods.

However, possessing massive data also entails considerable data security risks, including data loss, data misuse, and unauthorized access to sensitive data.

Traditional security mechanisms are increasingly inadequate in addressing government security issues. In the future, blockchain technology will serve as a bridge between applications, enabling targeted data sharing. Data-using units can extract data from data blocks according to their needs and regulations through a "digital identity surrogate". After blockchain technology reads and securely separates the data, the data are then put on the chain according to actual conditions. This process usually involves the use of a consortium chain structure, which has a complete authorization mechanism and established nodes with management functions. This model aligns with centralized management in units, without disrupting the distributed storage and decentralized operation structure. It can also promote the clear division of system business interfaces, thus enhancing the scalability and flexibility of the business.

Integration of Trustworthy Digital Equity Certificates (NFTs) and Digital Assets

An NFT (nonfungible token) is a type of digital asset based on blockchain technology that can be used to create and trade unique digital assets on the blockchain. NFTs can be combined with public resource management, government data protection, government document management, and public voting management in the field of e-governance, thereby achieving transparent, efficient, and secure government management.

In the field of public resource management, the government can use NFTs to manage public resources, such as national parks and nature reserves. The government can divide these resources into specific blocks and create corresponding NFT tokens, achieving refined management and circulation of public resources.

NFTs can be used to protect the copyright and security of government data, creating machine-readable secure unique identifiers. Government data can serve as the content of NFTs, and a unique identifier can be created for each NFT token, thereby protecting the security and copyright of government data.

Additionally, in the field of government document management, NFTs can be used to manage government documents and contracts. The government can convert government documents and contracts into NFT tokens and store them on the blockchain, thereby achieving transparent, immutable, and secure management of government documents and contracts.

In the field of public voting management, NFTs can be used to manage public voting. Each vote can be converted into an NFT token and stored on the blockchain. This approach ensures the uniqueness, immutability, and security of each vote while achieving transparency and fairness in public voting.

Emergence of Distributed Applications in Rich Terminal Application Fields

Distributed applications are based on blockchain technology and are characterized by decentralization, security, reliability, and programmability. Distributed applications in the field of e-governance can be combined with government data management, government contract management, government approval management, and public service provision, thereby achieving transparent, efficient, and secure government management.

In the field of government data management, distributed applications can be used to manage government data. The government can store government data on the blockchain and use distributed applications to share and query government data, thereby improving the security and reliability of government data.

In the field of government contract management, distributed applications can be used to manage government contracts. The government can convert government contracts into smart contracts and store them on the blockchain. Through distributed applications, the government can automate the execution and monitoring of government contracts, thereby improving the transparency and efficiency of government contracts.

In the field of government approval management, distributed applications can be used to manage government approvals. The government can convert the process and rules of government approval into smart contracts and store them on the blockchain. Through distributed applications, the government can automate the processing and monitoring of government approvals, thereby improving the transparency and efficiency of government approvals.

In the field of public services, distributed applications can be used to provide public services. The government can convert the process and rules of public services into smart contracts and provide services to the public through distributed applications, thereby improving the efficiency and convenience of public services.

7.1.3 *Outlook for Digital Government 3.0*

The technological development of Web 3.0 provides strong technical support and ideological guidance for the development of the metaverse, shifting the development of the metaverse toward a more open, free, and interactive direction.

First, the distributed network and blockchain technology of Web 3.0 can provide a secure and trustworthy data storage and transaction platform for the metaverse, allowing virtual assets in the metaverse to be recorded and traded reliably and authentically.

Second, the personalized and intelligent services of Web 3.0 can enhance the user experience in the metaverse.

Finally, the user-driven and creative features emphasized by Web 3.0 align closely with the philosophy of the metaverse. With the support of Web 3.0, users in the metaverse can freely create and share content, building their own digital world.

The governmental metaverse can be established based on real-world governance, with the mutual mapping, integration, and dynamic interaction between real-world governance and virtual space widely applicable in various fields of governance. The governmental metaverse can be used to build learning organizations, with technologies such as virtual reality and augmented reality enabling immersive learning experiences beyond text and image materials, leading to better learning outcomes.

From the perspective of data integration, the ability of the governmental metaverse to construct business scenarios can be summarized into three forms[3]:

The first is "digital twinning", which replicates real-world governance in virtual space, achieving the digitization and virtualization of real-world governance.

[3] Qingguo Meng, Yan Yan, Guodong Zhao. *Government Metaverse*. Beijing: China Translation & Publishing House Co., Ltd., 2022: 111–120.

The second is the "digital native", which refers to new activities generated in the virtual space that, in turn, influence real-world governance, necessitating the establishment of relevant rule systems in real-world governance and reshaping the governance business system in the real world.

The third is "digital fusion", where real-world governance and virtual space achieve high integration and seamless linkage. Governance business no longer distinguishes between virtual and real, with the two achieving a high degree of synchronization, linkage, and integration.

The development of the governmental metaverse is exploratory and evolutionary, advancing through continuous experimentation and practice. Twinning, native, and fusion are three different formative elements, which develop in tandem and influence each other, promoting the development of the metaverse in real-world governance applications.

The governmental metaverse helps the government achieve more efficient collaborative operations, more scientifically rational decision-making, more convenient and efficient services, and more precise and intelligent supervision. Governmental services, social governance, and monitoring and supervision are currently important scenarios for the governmental metaverse to explore and apply.

The metaverse can provide an open and transparent platform for governance, allowing citizens to participate more directly in government decision-making. For example, a virtual city can be established in the metaverse, allowing citizens to participate in city hall meetings, propose suggestions and vote online. Thus, it can better understand the needs and opinions of citizens and improve the accuracy and democracy of policy-making. In addition, the metaverse can provide a fairer and more transparent election platform for government agencies, allowing voters to participate in voting through encryption technology and smart contracts, thereby ensuring the credibility of the election results.

The metaverse can also improve the efficiency and convenience of public services in the context of governance. A virtual administrative service center can be established in the metaverse, allowing citizens to complete various government procedures online, saving time and reducing costs while improving service quality and efficiency. For example, the government of California, USA, has already established a virtual real estate trading platform in the metaverse, making real estate transactions faster and safer.

The encryption technology and smart contracts in the metaverse can ensure the encryption and verification of government information, preventing information leakage and tampering. Moreover, government agencies can use blockchain technology to establish a traceable information management system, allowing citizens to understand the source and use of government data more transparently.

With the continuous development of metaverse technology, its application prospects in digital government 3.0 will become increasingly broad. By strengthening technological innovation and applications, it can provide citizens with more efficient, transparent, and secure public services.

7.2 The Digital Transformation of Industries is the New Direction for China's Economic Growth

Currently, the global economic situation is forcing the digital transformation of industries and accelerating its progress. China has issued several policy guidelines to identify directions for the digital transformation of industries. The Central Committee of the Communist Party of China has proposed key strategies such as "Digital China", "Made in China 2025", and "Internet Plus", aiming to promote industrial transformation and upgrading and the development of the digital economy. The related policies are as follows:

In July 2017, the State Council issued the "New Generation Artificial Intelligence Development Plan", clearly positioning artificial intelligence as an important technology and industrial direction leading future development and guiding state-owned enterprises in digital transformation in the field of artificial intelligence.

In April 2019, the Political Bureau of the Central Committee of the Communist Party of China held a meeting, proposing to increase support for the digital transformation of state-owned enterprises, further introduce digital economy policies, and promote the construction of a "digital China".

In November 2019, 15 departments, including the National Development and Reform Commission and the Ministry of Industry and Information Technology, issued the "Implementation Opinions on Promoting the Deep Integration of Advanced Manufacturing and Modern Services", which proposed accelerating the innovative application of the industrial internet; deepening the integration of manufacturing, the service industry, and the internet; and promoting the digital transformation of key industries.

On August 21, 2020, the State-owned Assets Supervision and Administration Commission of the State Council (SASAC) issued a notice on accelerating the digital transformation of state-owned enterprises, making comprehensive arrangements for promoting this process.

The digital economy has become the main economic form after the agricultural and industrial economies. Digital technology is reshaping the industrial structure and business models, and the development of the digital economy has become a national strategy. On January 12, 2022, the State Council issued the "14th Five-Year Plan for Digital Economy Development", stating that by 2025, the added value of the core industries of the digital economy will account for 10% of GDP.

Industrial digitization is the process of applying digital technology and data resources to upgrade traditional industries, thereby increasing the output value and improving efficiency. It is the process of integrating digital technology with the real economy, using modern information technology to digitally transform traditional industries in all aspects, angles, and links. It reflects how digital technology promotes productivity and improves the efficiency of resource allocation and is an important indicator of the coverage and influence of the digital economy.

Enterprises are the main body of industrial digital transformation, and the digital transformation of enterprises is the microfoundation for the development of the digital economy. The digital transformation of industries is led mainly by "chain-leading" enterprises, which have abundant resources, a large scale, and high technical strength. They can drive the digital transformation of the entire industrial chain, promote the optimization and collaboration of the supply chain, and improve industrial efficiency and competitiveness. They bear the important responsibility of leading and promoting the digitalization of the industry, providing demonstrations and guidance for the digital transformation of other enterprises, and promoting the sustainable development of the entire industry. The following section elaborates on the digital transformation of enterprises.

7.2.1 The Purpose of Digital Transformation is Not Only to Reduce Costs and Increase Efficiency but also to Innovate Operations

Digital transformation is based on the actual situation of the enterprise; relies on digital infrastructure such as big data, the Internet of Things, AI, etc.; is business oriented; involves deep mining of data value as a means; assists enterprise process reengineering; improves the ability of enterprises to respond to changes; optimizes the production, management and sales processes; improves office and production efficiency; improves product quality; and increases product profits. Moreover, digital technology can help enterprises explore new market opportunities, discover new business models and operation modes, and improve their foresight and adaptability.

7.2.2 The Shortcoming of Digital Transformation is the Lack of Entry into the Digital World

We believe that the greatest shortcomings of most industrial enterprises in digital transformation are the low level of information infrastructure construction, the lack of digital absorption capacity of enterprises, the backwardness of digital services and platform capabilities, and the limitations of data sharing and interconnection. The main characteristics of these shortcomings are as follows:

First, the level of information infrastructure construction is low. With the surge in demand for networks, data, and computing in economic and social life, the requirements for digital services have significantly increased, and the information infrastructure is under severe pressure. Some enterprises may significantly reduce expenditures related to long-term competitiveness, such as digital investment, by tightening financial constraints, resulting in poor reliability of infrastructure.

Second, the digital absorption capacity of enterprises is lacking. At present, the digital literacy of many organizations' talents cannot meet demand, and there is still much room for improving digital platforms and services. On the one hand, there is a lack of digital talents and skills in digital transformation, and talents with knowledge in digital technology, data analysis, artificial intelligence, etc., are needed. On the other hand, talents with industry experience and business capabilities currently do not have knowledge in digital technology, data analysis, artificial intelligence, etc., and talents who understand digital technology lack industry experience and business capabilities. Therefore, there is a serious shortage of talents who understand both business and digital technology, which limits the effectiveness of the business digital system. For example, digital systems such as ERP and MES cannot be used immediately and require digital talents for customized development and maintenance. If enterprise management, business, and technical personnel are not well adapted, the use of business systems will be restricted.

Third, the supporting services of digital services and platforms in traditional industries are not optimal. At present, the ability of traditional industries to dock with digital technology is weak, which restricts the implementation and use of customized, innovative digital solutions and services. Moreover, they cannot provide users with digital solutions, applications, online services, etc., so users cannot obtain good information, interaction, and business operations in the digital world. The digital transformation of industrial enterprises needs to solve not only communication problems but also collaboration problems, from office collaboration to business collaboration, from human-to-human collaboration to human–machine collaboration, and from internal collaboration to industrial chain collaboration.

Fourth, data sharing and interconnection are restricted. Many enterprises have unclear strategic planning, and they carry out digital construction in a "piecemeal" manner. However, because data cannot achieve efficient sharing and collaboration, some system functions are redundant, resulting in a large number of isolated application systems. The data of these isolated application systems are difficult to share and govern, which increases the difficulty of enterprise digital transformation.

7.2.3 Digital Transformation Methods and Paths

Digital transformation is an important means to promote the transformation of enterprises toward modern management and operations. It also provides an important way for enterprises to achieve a competitive advantage. To compensate for the shortcomings of digital transformation and accelerate the digital transformation of "chain-leading" enterprises or state-owned enterprises, we believe that we should start with the following paths and methods.

The four-step digital transformation path involves defining the vision and strategy, assessing the current state and identifying opportunities, formulating a detailed digital transformation plan, and executing and monitoring (as shown in Fig. 7.4).

Based on this digital transformation path, the following five methods should be emphasized.

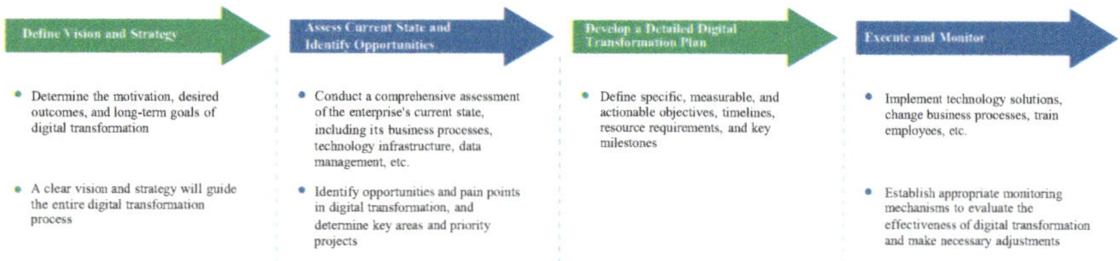

Fig. 7.4 The four-step digital transformation path

First, Conduct a Digital Transformation Diagnosis

Digital transformation is not a simple form of IT system construction but a systematic design and construction involving strategy, business, organization, and technology. It involves a transformation of management, and it essentially entails the reconstruction of management models and operational mechanisms. Therefore, it is recommended that professional consulting units be entrusted with conducting a digital transformation diagnosis for the enterprise from three main aspects as follows:

Determine the health status of the enterprise, including the enterprise strategy, company organizational structure, business architecture, business operations, financial situation, personnel organization, innovation research and development, production process, supply chain relationships, sales network, marketing model, profit model, sales methods, business innovation, basic guarantees, etc.

Evaluate existing digital basic capabilities, including existing information application systems, the IT architecture, network and security systems, hardware devices, data storage methods, IT talent reserves and areas of expertise.

Determine the enterprise's demand for digital transformation, including research and development, production, testing, marketing, sales, personnel, office, procurement, customer service, operation and maintenance, etc., covering the entire process and all departments.[4]

Through the above three aspects of sorting and evaluation, integrate analysis and focus the classification, clarify the core needs of the enterprise, provide direction for the digital transformation of the enterprise, and generate a digital transformation diagnosis report.

Establish a "Digital Transformation Special Task Force"

After the enterprise creates a "Digital Transformation Diagnosis Report", a "Digital Transformation Special Task Force" should be established. The first person responsible for the enterprise (the boss) should serve as the overall person in charge, and the team members must be all relevant executives within the enterprise. The reason for this requirement is the core of digital transformation is not "digital technology"

---

[4] iResearch Consulting Institute. *2020 Research Report on the Practices of Digital Transformation Pathways in Chinese Enterprises*. 2021.

but "management transformation", which drives the company's management transformation with digital capabilities and thinking.[5] Moreover, although this group is independent of any department of the enterprise and is not subject to functional department constraints, it is directly responsible and commanded by the first person responsible for the enterprise (the boss) to obey the "digital strategy" as the purpose and action instructions.

Formulate a Digital Strategy

After establishing the "Digital Transformation Task Force", the enterprise should formulate its own digital strategy according to the "Digital Transformation Diagnosis Report", combined with the current business and status of the enterprise, as well as the application scenarios and development trends of digital technology. In the process of formulation, the enterprise must communicate and cooperate fully with all employees and partners to gain their recognition and unconditional execution to ensure that the digital strategy can be effectively implemented and guaranteed.

In addition, when formulating a digital strategy, it is necessary to clearly define what the strategy is "transforming".[6]

"Transform" competitive cooperation relationships, which can shift from focusing solely on competition to building multiple competitive cooperation relationships.

"Transform" business scenarios from rigid business scenarios based on the functional division of labor to flexible and dynamic business scenarios.

"Transform" value models from long-cycle value acquisition models based on technological innovation to build business barriers to create value ecological models based on resource sharing and capability coconstruction.

"Transform" management models. In the process of digital transformation, from the perspective of management guarantees, we should accelerate the transition from closed top-down control to open dynamic flexible management to provide management guarantees that promote digital governance, adjustments to the organizational structure, management method reform, and organizational culture innovation.

Strengthen Digital Infrastructure Construction

Enterprises can strengthen their digital infrastructure construction through the following technical capabilities.

---

[5] Hanyu Liu. *Digital Thinking: A Guide to Digital Transformation for Traditional Enterprises*. Beijing: China Machine Press, 2021.

[6] Dianliang Think Tank. *2021 White Paper on the Digital Transformation Development Index and Methodological Pathways for State-Owned Enterprises*. 2021.

Hybrid Cloud Deployment Systems: Strengthening and Enhancing the Basic Digital Technology Capabilities of Enterprises

Cloud-based deployment is an important basis and premise for enterprise digital transformation. However, standard public cloud applications cannot fully match the personalized needs of enterprise business models, operation models, and management models and cannot reflect the differentiated competitive advantages of enterprises. In the future, hybrid clouds will constitute the future trend of enterprise digital transformation. The hybrid cloud, which takes into account the convenience of the public cloud and the security of the private cloud, can also meet personalized needs and achieve efficient and secure application and data connections.

In a hybrid cloud environment, there will be an increasing number of innovative applications of standardized PaaS and SaaS, which can provide multisystem support for more flexible and variable business applications. These applications form an assemblable business engine that meets differentiated competition needs, satisfies data security and governance requirements, and creates an ultimate user experience.

Digital PaaS Platform: Strengthening Support for Flexible and Variable Business Applications

The PaaS platform is the technical foundation of digitization. Adapted under a hybrid cloud architecture, it forms a unified technical base and an assemblable business engine. It introduces interconnected low-code modeling capabilities, big data processing capabilities, multidimensional analysis capabilities, data intelligence, and AI modeling capabilities. Thus, when creating business applications on the PaaS platform, the general engine capabilities can be used for rapid assembly and deployment.

DevSecOps Development Philosophy: Ensuring System Security and Flexible Application

DevSecOps stands for development security operations. At every stage of the software development lifecycle, including planning, coding, building, testing, releasing, and deploying, strict adherence to and execution of security standards is required.

Digital PaaS is the technical foundation for all applications. When applications are produced on the PaaS platform, their security must be ensured. Therefore, when developing the PaaS platform, it is necessary to control the security of the entire platform system based on the DevSecOps philosophy.

Digital Middle-End Construction: A Technology Platform that Activates Business Value

As businesses grow, they will build increasingly more information systems, forming diverse systems that generate more multisource heterogeneous data. This situation leads to phenomena such as "system chimneys" and "data islands", where the underlying data cannot be connected, data standards are not unified, and a standard data asset system cannot be formed. The value of effectively utilized data is low, and the value of data cannot be well released and fully utilized. Therefore, a data middle-end needs to be built. It can manage all the company's basic data, business data from various systems, operational data, behavioral data, operational data, and external internet data. Through data integration, data cleaning, data mining, data services, and other process treatments, it forms a company's data assets.[7] The data middle-end manages all data APIs, and through data services, it transforms data into business, activates the business value of data, and better provides data value services for business applications.

Advancing the Application of Digital Technology

Strengthening online connectivity guarantees sustained business growth. Real-time online transactions and collaboration are necessary for business development. Companies need to build a supply network that meets the needs of customers across all channels, fully utilizing the resources and capabilities of partners on the supply chain to provide the response speed and flexibility that customers need. Under such requirements, companies need to share information on the supply chain with partners and make decisions closer to customer needs in demand planning, inventory management, supply and replenishment planning, sales and operation control, distributed order management, and other aspects.

Therefore, companies should, according to their own needs, advance the application of digital technology at the supply chain end, production end, and sales end in an orderly manner and establish a more flexible and elastic supply, production, and end-to-end application system. The following is an example:

Building a Full-Link Online Production and Sales Management System

Companies can use technologies such as 5G, big data, IoT, and cloud computing to connect production equipment data, sales data, order data, product development data, procurement data, logistics data, and warehousing data. This interconnection forms an end-to-end business closed-loop management capability driven by production equipment data, establishing an end-to-end intelligent sales system that connects "production, supply, and sales".

Using Collaborative Tools to Enhance Online Collaboration Efficiency Between People

The core of a company's daily operation is event-driven collaboration between people. The collaboration efficiency between employees reflects the vitality and competitiveness of the company. With the support of digital technologies such as the mobile internet, "decentralized, flat" organizations have become the direction of organizational transformation. By building a networked organization oriented by

---

[7] CAICT Terminal Lab, China Academy of Information and Communications Technology; Beijing Yuannian Technology Co., Ltd. *Research Report on Technological Trends in Enterprise Digital Transformation (2023)*. 2023.

"business events", the overall operational efficiency of a company can be effectively improved. Therefore, "online" is a basic work requirement, and "collaboration" is a work goal. Companies need to use collaborative tools to integrate structured information such as data, processes, tasks, and events with unstructured information such as images, voices, meetings, and videos, thereby strengthening the "strong connections" between employees. These collaborative tools include WeChat for Business and DingTalk. Through these collaborative tools, instant communication, meeting collaboration, document collaboration, object collaboration, shared collaboration, and internal and external collaboration can be achieved.

Using Data Intelligence Assistants to Achieve Data-to-Person Functionality

As businesses increasingly adopt various information systems, the types and volume of data will grow significantly. It can be challenging for employees to search for data and use them to identify problems. For example, if employees need to "identify the reason for last month's sales decline," they can search only within the known output reports, and it is difficult to discover the correlation between the data. Therefore, businesses can build a data intelligence insight analysis system. This system can extract key information from massive data, complete automated analysis through preset algorithm models, and output data reports. With natural language-driven data queries, it can make the "human searching for data" scenario effortless. It can also provide real-time, accurate push capabilities through data warning functions, thus facilitating the "data searching for human" scenario.

In conclusion, digital transformation is a necessary measure for businesses facing increasingly fierce competition. If handled correctly, it can bring tremendous business opportunities and competitive advantages. After the methods and paths are understood, a specific planning process is needed to ensure that the digital transformation of the business has a foothold.

7.2.4 Integrated Innovation: The Future Stage of Enterprise Digital Transformation

The digital transformation of enterprises has become an unignorable trend and a necessity. In this section, we delve into the future stage of enterprise digital transformation in terms of four key aspects: technological integration and innovation, industry collaboration and innovation, data integration and intelligent manufacturing, and talent integration and organizational innovation. The integration and innovation in these areas will lead enterprises to new heights and prompt the exploration of the infinite possibilities and potential of enterprise digital transformation.

Technological Integration and Innovation

Integration of Big Data Analysis and Blockchain: The integration of big data analysis and blockchain technology in supply chain management provides enterprises with a more transparent, traceable, and efficient supply chain. This integration not only enhances consumer trust but also improves the competitive advantage of enterprises by optimizing resource allocation and increasing response speed. For example, if a restaurant chain wants to track the complete process of its product from the field to the table, the information of each production and transportation link is recorded and verified through blockchain. Through big data analysis, a company can moreover predict the demand for certain products in specific regions, thereby optimizing production and resource allocation.

Integration of AIGC and Information Security Technology: With the emergence of generative artificial intelligence AIGC, future information security application scenarios will be reshaped and defined. The joint application of AIGC technology and information security technology can provide comprehensive protection for enterprise data, including data security monitoring, automated security warnings and problem-solving, and data governance automation. The implementation of a data asset inventory, query, application, authorization, tracking, etc., can improve the level of enterprise information security and data governance efficiency. In addition, by enabling precise defenses and efficient responses, it can enhance data governance efficiency, strengthen data security, and enhance the enterprise's data value chain analysis and business rule management capabilities. For example, applications of AIGC in information security can provide precise defenses and efficient responses,[8] as shown in Table 7.1.

Industry Collaboration and Innovation

Industry collaboration is a key part of digital transformation. Companies need to establish close connections and cooperate with various partners in the industry chain to form an industry ecosystem for resource sharing and complementary advantages. Through industry collaboration, companies can better respond to market changes, reduce costs, and improve competitiveness. The following is an example:

Collaboration between car manufacturers and technology companies: Intelligent driving is a strategic high ground for automakers to gain a competitive advantage in the future and faces three major challenges: long investment cycles, highly complex scenarios, and high technical difficulty. It is becoming increasingly important for traditional automakers to collaborate with technology companies, combine vehicle networking, artificial intelligence, and map navigation technologies to provide users with innovative services for intelligent navigation, traffic management, and vehicle

[8] TE. *Research Report on Enterprise AIGC Commercial Applications*. TE Think Tank, 2023.

Table 7.1 Applications of AIGC in information security

Module	Core value	AIGC empowerment link	Empowerment method
Compliance review	Automatic review	Compliance review	Through natural language processing and pattern recognition technology, text and data can be analyzed to quickly check compliance with relevant laws and policies, thus improving audit efficiency and reducing the error rate of manual audits
Risk management	Assist in risk assessment	Risk assessment mechanism Risk response mechanism Emergency management system	Assists in risk assessment and prediction through risk modeling and predictive analysis technology, quantifies potential risks, improves assessment accuracy, and provides early warnings
Organizational management	Access policy assistance	Access admission policy Security accountability system	Generates access control policies
Feedback correction mechanism	Continuous improvement	Feedback correction mechanism	Implements a continuous improvement strategy through intelligent diagnosis and continuous optimization technology, analyzes system operation, proposes suggestions for optimization, improves efficiency of improvements and reduces operational and maintenance costs
Security event monitoring	Real-time monitoring	Security event monitoring	Assists security event monitoring through real-time monitoring and anomaly detection technology, automatically detects and reports security events, improves monitoring efficiency and reduces the risk of missed and false reports
Effect evaluation mechanism	Quantitative evaluation	Effect evaluation mechanism	Assists effect evaluation through data analysis and effect modeling technology, quantitatively evaluates the effect of security measures, improves evaluation accuracy and optimizes security measures

(continued)

Table 7.1 (continued)

Module	Core value	AIGC empowerment link	Empowerment method
Security event audit	Automatic analysis audit	Security event audit	Assists security event audit through audit automation and deep analysis technology, automatically analyzes audit data, improves audit efficiency and discovers potential problems
Network security	Real-time traffic monitoring	Intrusion detection Firewall configuration Network traffic analysis Network security situation awareness	Performs network traffic analysis through traffic pattern recognition and anomaly detection technology, conducts real-time monitoring of network traffic and analysis of abnormal situations, improves traffic analysis efficiency and achieves real-time monitoring
Data security	Automatic risk identification	Data encryption Content audit Data leakage prevention Security situation awareness	Assists data leakage detection and prevention through data leakage prediction and protection strategy generation technology, automatically identifies potential leakage risks and generates protection strategies, and improves data protection effects
Device security	Automatic device recognition	Device locking Device verification Device registration Device encryption	Assists device registration through intelligent registration and automatic verification technology, automatically recognizes devices and completes registration, improves device registration efficiency and reduces manual operations
Application security	Risk application identification	Application deployment and distribution Application lifecycle management Application review Application reinforcement	Conducts automatic application review through pattern recognition and risk assessment technology, automatically analyzes application security, improves review efficiency and accurately identifies risk applications
Identity authentication	Biometric analysis	Identity authentication	Performs AI-driven identity authentication through biometric recognition and behavior analysis technology, conducts comprehensive analysis of user's biometric and behavior, improves authentication accuracy and reduces the risk of theft

safety, thereby offering drivers and passengers a more comfortable and safe travel experience. In the future, we expect more innovative services and applications to drive the development of the automotive industry.

Win–Win for Traditional Retailers and Internet Companies: Collaboration can help companies disrupt traditional sales channels and marketing models, as they sell and promote products through multiple digital channels. It can also enable real-time tracking and evaluation of marketing activities, thus more accurately responding to market conditions and adjusting market strategies in a timely manner. In particular, companies with digital infrastructure can leverage AIGC technology to adopt a cloud-based, model-centric MaaS (Model as a Service) mode, thus deeply integrating the cloud and the model, providing pre-trained basic models, and offering customized services for innovative marketing application scenarios. Marketing under the AIGC environment will offer a fully linked marketing support tool, including market cognition, lead operation, customer cultivation, customer conversion, and marketing optimization. This tool will provide companies with advertising creativity and placement optimization references, lead evaluation and optimization strategies, email marketing, video marketing and event marketing strategy suggestions, and precise demand analysis and strategy generation suggestions and assist companies in marketing action analysis, marketing strategy review, etc. For example, applications of AIGC in marketing, from cognition to repurchase, offer fully linked strategy assistance,[9] as shown in Table 7.2.

Digital industry ecosystem enterprises empower traditional enterprises to transform digitally: Digital industrialization and industrial digitization are important components and core content of the digital economy. Industrial digitization entails the digital transformation of traditional industries, whereas digital industrialization refers to an industry providing corresponding digital technologies, products, services, infrastructure, and solutions for industrial digitization. Therefore, the two processes are integrated and dependent. In the process of industrial digital transformation, the empowerment and support of digital industrialization enterprises are indispensable. In the past, Gloryview Technology offered a complete solution in terms of empowering traditional enterprises to transform digitally through joint collaborations with well-known domestic digital enterprises and local state-owned enterprises to build a localized digital economy industrial base, thus forming a local industry innovation and entrepreneurship ecosystem that concentrates innovation, talent, projects, and industries. Together, we strive to promote the upgrade of the local industrial chain through digital technology, improve of the "chain master" enterprise's digital capabilities, and promote "chain" enterprise digital transformation.

[9] TE. *Research Report on Enterprise AIGC Commercial Applications*. TE Think Tank, 2023.

Table 7.2 Applications of AIGC in marketing

Stage	Core value	AIGC empowerment link	Empowerment method
Market cognition	Creative reference	Advertising strategy, brand communication, market analysis, CEM, SEO, DSP, SSP	Generates advertising creativity and placement optimization references, including advertising design, advertising content, placement channel strategy, and placement analysis, thereby improving advertising effectiveness and placement efficiency
Lead operation	Evaluation and optimization	Channel diversion, event attraction, traffic retrieval, traffic distribution	Quickly generates lead evaluation and optimization strategies, helping companies improve their lead quality and conversion rate
Customer cultivation	Marketing strategy suggestions	EDM, content marketing, video marketing, loyalty analysis, event marketing, opportunity analysis	Generates email marketing, video marketing, and event marketing strategy suggestions, providing customers with more targeted content
Customer conversion	Marketing strategy suggestions	Precise demand analysis, proximity marketing, membership management, strategy analysis, CRM, CDP, SFA, channel management	Generates precise demand analysis and strategy generation suggestions, improves sales performance, and assists companies in achieving refined lead management, public and private domain operations, etc., thus further improving the conversion rate
Marketing optimization	Analysis and review	Website analysis, strategy review analysis, marketing plan control, marketing action analysis, media monitoring and analysis, marketing analysis, marketing budget analysis	Assists companies in marketing action analysis and marketing strategy review. Better evaluates the effects of marketing activities and provides strong data support for future marketing activities
Repurchase and upselling	Analysis and optimization	Repurchase reason analysis, repurchase strategy marketing, customer service, after-sales service, user growth, social network dissemination	Generates analysis of repeat purchase reasons and marketing strategies for repeat purchases, helps businesses better understand customer needs, and can use social network propagation analysis to tap into potential customer resources

Data Integration and Intelligent Manufacturing

Data integration and intelligent manufacturing have been hot topics in the fields of industrial production and manufacturing in recent years. Data integration refers to the collection and integration of data from multiple sources to provide more comprehensive and accurate information. Intelligent manufacturing, on the other hand, involves the automation, intelligence, and optimization of the manufacturing process through the integration of information technology and manufacturing technology.

In intelligent manufacturing, data integration plays a crucial role. First, by collecting and integrating data from production lines, equipment, sensors, and other sources, manufacturers can more accurately monitor the production process. For example, by conducting real-time monitoring of the operating status and performance of equipment, manufacturers can detect and solve potential problems in a timely manner, thereby improving production efficiency and product quality.

Second, data integration can also be used to optimize the production process. Through in-depth analysis of production data, manufacturers can identify bottlenecks and deficiencies in the production process and adjust production plans and processes to optimize production efficiency and reduce costs.

In addition, data integration helps to achieve intelligence in the manufacturing process. By using artificial intelligence and machine learning algorithms, manufacturers can predict various variables in the production process, such as demand, output, and faults, based on historical and real-time data. This predictive capability enables manufacturers to adjust production plans and resource allocation more flexibly to cope with market changes and uncertainties.

Data integration can also strengthen supply chain management. By integrating data from suppliers, logistics, and warehousing, manufacturers can manage the supply chain more effectively to ensure the timely supply of production materials and the timely delivery of products.

Talent Integration and Organizational Innovation

Talent and organizational innovation are the cornerstones of digital transformation. Enterprises need to establish teams with digital skills and an innovative spirit to adapt to the changes brought about by digital transformation. In addition, enterprises need to build flexible organizational structures to support rapid decision-making and response.

In summary, enterprise digital transformation and innovation integration is a systematic process involving four aspects: technology, industry collaboration, data integration, and talent and organizational innovation. Through digital transformation, enterprises can not only improve their operational efficiency and competitiveness but also create new business models, seize more market opportunities, and generate greater business value. In this process, enterprises need to continuously learn and adapt to cope with the challenges and opportunities brought about by digital transformation.

Chapter 8
Application Prospects

Data elements, architectural space services, and digital operations are the foci of our explorations of smart city applications. As a new production factor, the value of data elements is more reflected in the breadth and depth of data circulation. It can not only optimize decision-making and improve efficiency but also become a key element in promoting innovation. In the future, the collection, integration, analysis, and use of data will be more standardized and intelligent. Architectural space services and digital operations, on the other hand, will shift from "building intelligence" to "people-centered space operations and services", thus more effectively improving building efficiency, optimizing space layout, improving experience quality, and achieving sustainable development and operations.

8.1 Data-Driven Digital Economy Innovation Takes Off

8.1.1 Data Elements Are the Core Elements of the Digital Economy

In today's digital age, data have become a production factor in the digital economy. As a production factor, they need to be integrated into various links, such as production, distribution, circulation, consumption, and social service management. On the one hand, data elements directly participate in the production, exchange, and distribution process, giving birth to new production methods and even creating new demands to promote economic growth. On the other hand, data elements can empower other elements, optimize supply and further promote growth. Therefore, data elements have the financial attributes of circulation, pricing, and trading. Since they have financial attributes, they need to reflect the value of data. The development of the digital economy can be divided into three stages according to the scale and role of

F. Liu, *Smart and Digital Cities*, https://doi.org/10.1007/978-981-95-2243-9_8

data.[1] As shown in Fig. 8.1, information is extracted from the data and then refined into knowledge, adding new value to business decision-making, product services, and social governance. The development of the digital economy fundamentally involves fully exploiting and releasing the value hidden in data based on the digitization and datafication of everything. According to data from the National Industrial Information Security Development Research Center, it is estimated that by 2025, the market size of China's data element industry will reach approximately 174.9 billion yuan. The vigorous development of data element operations, with active participation from upstream and downstream industries, presents infinite possibilities for innovation.

From a value perspective, data release value through business integration, intelligent decision-making, and circulation empowerment[2] (as shown in Fig. 8.2). Currently, China is in the second stage of data value release. The second stage emphasizes data mining and analysis, helping governments and businesses make intelligent decisions through data processing, analysis, and modeling.

Data valuation refers to the economic process of starting with data resourceization, followed by the data assetization and data capitalization stages, and finally achieving data valuation. Here, data resourceization entails processing basic information into high-quality data to support business integration; data assetization essentially forms data exchange value, a process where data bring economic benefits to users or owners through circulation and transactions; and data capitalization is achieved through data transactions and circulation, which includes two main methods: data credit financing and data securitization,[3] thereby empowering external circulation.

The national policy layout for the data element market is continuously developing, with the following related policies issued:

In October 2019, the Fourth Plenary Session of the 19th Central Committee of the Communist Party of China first proposed including data as a production factor in distribution.

In April 2020, the State Council issued the "Opinions on Building a More Perfect Market-oriented Allocation System for Factors of Production", which proposed accelerating the cultivation of the data element market.

In January 2021, the General Office of the Central Committee of the Communist Party of China and the General Office of the State Council issued the "Action Plan for Building a High-standard Market System", which proposed that establishing a standardized and compliant data element market is urgent and necessary.

In March 2021, the State Council issued the "14th Five-Year Plan", which proposed that the data element market would be a strong and sustained driver of economic development during the "14th Five-Year Plan" period.

---

[1] Zhaochun Huang. On the Construction of a Data Factor Market Based on the Supply Side. Bulletin of the Chinese Academy of Sciences, 2022, 37(10): 1402–1409.

[2] China Academy of Information and Communications Technology. *White Paper on Data as a Production Factor (2022)*. 2023.

[3] China Academy of Information and Communications Technology. *Report on Data Valorization and the Development of the Data Factor Market (2021)*. 2021.

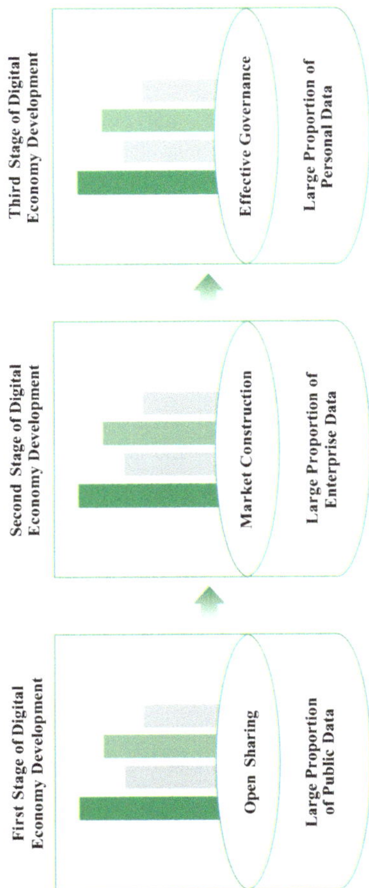

Fig. 8.1 Three stages of digital economy development

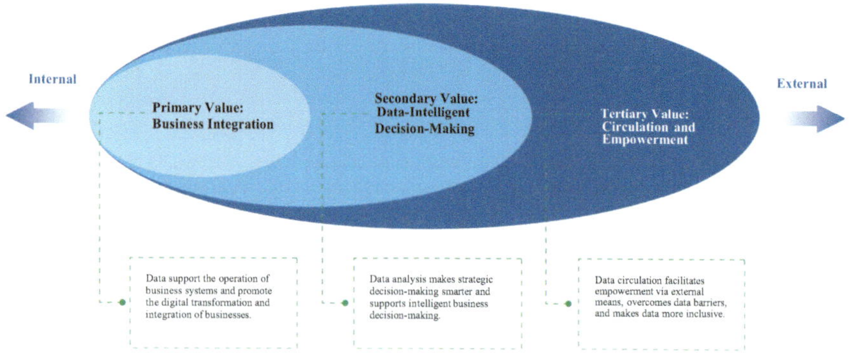

Fig. 8.2 Data value release (strategic guide to digital transformation: a major national strategy|"digital China" takes a crucial step that could shape the future balance between China and the U.S. https://mp.weixin.qq.com/s/Fq4l1Txjwv1IBM6MoGN4Dw,2023)

In December 2021, the General Office of the State Council issued the "Overall Plan for the Comprehensive Reform of Market-oriented Allocation of Factors", which clearly stated the intention "to explore the establishment of data element circulation rules".

In April 2022, the State Council issued the "Opinions of the Central Committee of the Communist Party of China and the State Council on Accelerating the Construction of a Unified National Market", which proposed accelerating the cultivation of the data element market; establishing and improving the basic systems and standards for data security, rights protection, cross-border transmission management, transaction circulation, open sharing, security certification, etc.; conducting in-depth surveys of data resources; and promoting the development and utilization of data resources.

In December 2022, the Central Committee of the Communist Party of China and the State Council issued the "Opinions on Building a Better Data Infrastructure System to Better Play the Role of Data Elements", which proposed establishing a compliant and efficient data element circulation and transaction system and improving the market-oriented allocation mechanism for data elements.

On January 3, 2023, the Ministry of Industry and Information Technology and 16 other departments jointly issued the "Guiding Opinions on Promoting the Development of the Data Security Industry", which proposed to accelerate the cultivation and value release of the data element market and solidify the foundation for the construction of a digital China and the development of the digital economy.

At the national level, China's data legislation focuses on data security and personal privacy and the protection of important data. The relevant regulations target public data or government data, focusing on data collection, sharing, development, application, and security management, with the aim to promote orderly data circulation and utilization.

Through data value mining and utilization, the future of China's data element market will significantly enhance society's mastery of information and significantly affect enterprise development, industry collaboration, and socioeconomic development. How to provide accurate information and insights through massive amounts of data, support decision-making, optimize business, gain customer insights, manage risks, enhance the innovation of enterprises and institutions, and improve competitiveness is the main content of this chapter.

8.1.2 The Current State of Data Element Development is Characterized by Structural Imbalance, with Data Circulation as the Trend

Imbalances in the Structure of the Data Element Market System

The data element market in China is in a stage of rapid development. China initially formed an industrial system in data collection and data annotation. However, there are still some issues with the circulation of data as a production element, and China's data element market system needs further improvement. Issues such as data supply–demand imbalance, large differences in the quality of different data elements, unclear use and rights of data, data monopoly, unfair competition, and data leakage risk hinder the full release of data value potential.[4] Therefore, the distribution and structural imbalance of data elements are quite prominent. Additionally, domestic and foreign companies have begun innovatively exploring data capitalization.[5]

Opportunities and Challenges Coexist in the Process of Data Resourceization

Data resourceization includes data collection, data sorting, data aggregation, data analysis, etc. The data resourceization framework, which is created according to the special report of the Institute of Information and Communications, is shown in Fig. 8.3.

Data resourceization will face opportunities and challenges, given the new generation of information technology.

An opportunity is that the "state-owned cloud" platform has the ability to transform from a "developer" to an "operator". Government and central state-owned enterprise data, or "state-owned" data, constitute an important part of the future data element market. According to IDC data, the scale of China's state-owned cloud market reached approximately 3.65 billion yuan in 2021, and it is expected to reach

---

[4] Department of High and New Technology. *Promoting Efficient Data Circulation to Advance High-Quality Development of the Data Factor Market*. https://www.ndrc.gov.cn/xxgk/jd/jd/202212/t20221219_1343662.html,2022

[5] China Academy of Information and Communications Technology. *Report on Data Valorization and the Development of the Data Factor Market (2021)*. 2021.

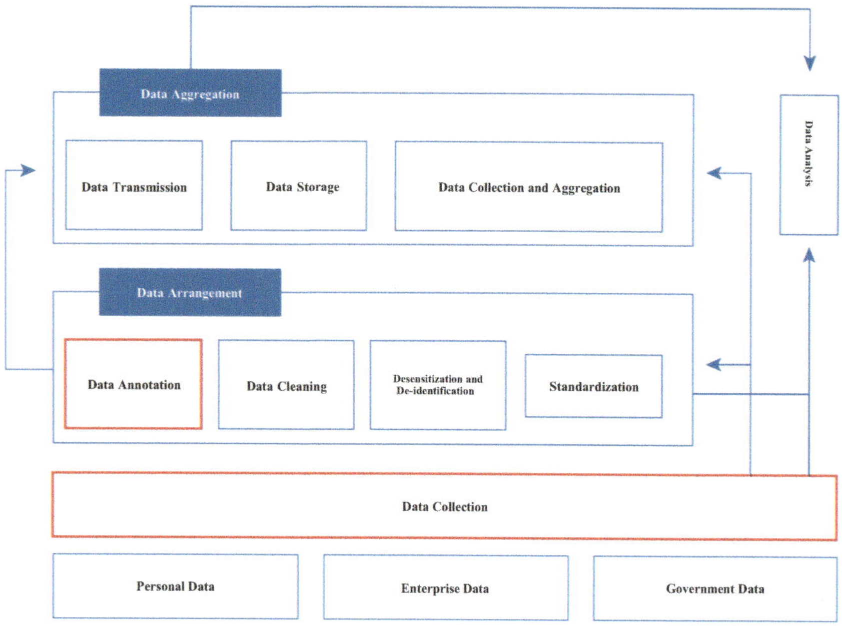

Fig. 8.3 Data resource framework

14.648 billion yuan by 2025. The "state-owned cloud" refers to the data security infrastructure built by local state-owned asset supervision and administration commissions through investment, establishment, and operation. The data governance system and cloud platform are centered on the aggregation of state-owned enterprise data, including IaaS, PaaS, and SaaS. The builders of the state-owned cloud are expected to take on the role of "primary developers" of state-owned data resources and to develop an integrated platform that helps the government/central state-owned enterprises carry out data asset development + data asset operations.

Among the challenges, the cost of corporate data compliance is high, and some companies face enormous fines or even the risk of shutdown due to violations such as excessive collection and excessive use of personal data of users. Data quality issues are prominent, and many units have consistency deficiencies or errors in data collection, storage, and processing, leading to analysis errors and affecting decision-making. The boundaries of data use are unclear, and data sharing in the industrial chain is not smooth.

To address these issues, it is necessary to consider reducing compliance costs through a compliance management system, establishing a comprehensive corporate data compliance control mechanism, establishing a comprehensive data governance system, and enhancing data sharing in the industrial chain.

Data Assetization and Data Circulation Are Key to Value Multiplication

Data assetization is the key and core element of building a data element market, including the determination of data ownership, the pricing of data assets, and the trading and circulation of data.

From a societal perspective, the types and dimensions of data collected by companies are becoming increasingly rich, but the definition of data ownership is not yet mature, which makes it more convenient for companies to overcollect data, easily exacerbating data concentration and even leading to data monopoly. In the long run, this trend will harm the order of the digital economy market. To this end, in December 2022, the Central Committee of the Communist Party of China and the State Council issued the "Opinions on Building a Database System to Better Play the Role of Data Elements", which proposed to explore the structural separation system of data property rights and establish a data property rights system framework of "three rights separation" of data resource holding rights, data processing and use rights, and data product operation rights. Although the European Union and the United States have conducted in-depth research and exploration of data rights confirmation-related system norms, they have not explicitly proposed the core concept of data registration. Our country is at the forefront of international exploration in the direction of data registration systems.

Domestic and foreign data platforms increase the enthusiasm of data suppliers to participate by adopting different pricing strategies, thus maximizing benefits for both supply and demand.

The circulation forms of China's data element market include data opening, data sharing, and data trading. The specific circulation forms are shown in Fig. 8.4.

Data opening and sharing are dominated by government public data opening, and the trend of government participation in data sharing is obvious. At present,

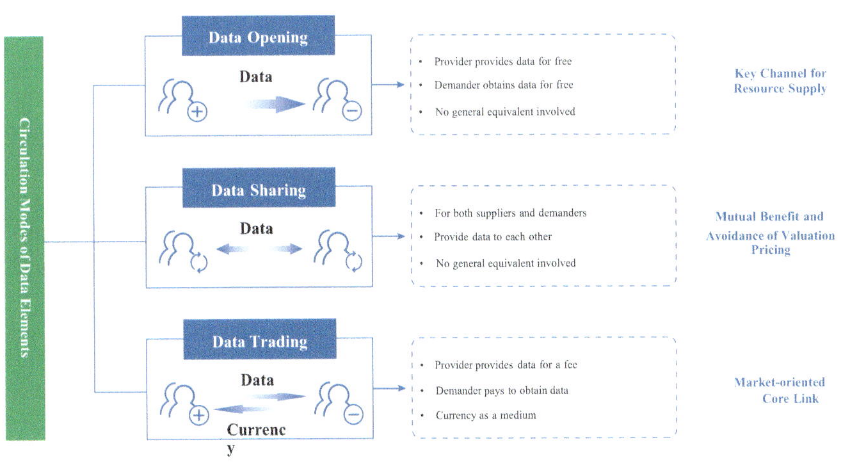

Fig. 8.4 Forms of data element market circulation

the opening of public data in various places is accelerating, 21 provincial administrative regions have built provincial platforms, and 193 prefecture-level administrative regions have built prefecture-level platforms.[6] As of 2020, the National E-Government website had connected to a total of 162 central departments and related units and approximately 252,000 government departments nationwide, forming a preliminary national data sharing platform. Thirty-one State Council departments have registered and published 1153 real-time data sharing interfaces on the national sharing platform, with approximately 11,000 data items. The national sharing platform has provided 912 million query verification services for 27 State Council departments, 31 provinces (autonomous regions, municipalities directly under the Central Government) and the Xinjiang Production and Construction Corps, effectively supporting online identity verification, real estate registration, talent introduction, business establishment and other services. There are 142 other types of data opening platforms, with 98,558 effective datasets. The sharing of government and enterprise data has gradually changed from one-way data reporting by enterprises to two-way sharing, such as the "City Beautiful Life Digital Vital Signs System" jointly created by Meituan and Shanghai. The Shanghai government provides traffic management data to help Meituan optimize services such as bicycle dispatching, and Meituan provides desensitized regular data to assist the government in exploring urban governance paths.

Data trading is becoming the main form of data circulation in the data element market. Domestic data trading institutions started operating in 2015. The National Development and Reform Commission announced on December 20, 2022, that more than 80 new data trading institutions have been established nationwide, nearly 30 of which were proposed by provincial-level governments and above to promote the construction of data trading venues. There are more than 60,000 big data companies nationwide and more than 300,000 professional data professionals. Overall, China's data trading market is characterized by a continuous expansion of the trading scale, an increasing richness of trading types, and a constantly improving trading environment, fully reflecting the advantages of a superlarge-scale market.[7]

In the circulation link, the profit model has changed from charging commissions to multiple profit models such as charging commissions, membership systems, and value-added trading services. The value-added trading services include the trading platform that deviates from the middleman's identity and partly undertakes the functions and roles of data cleaning, data identification, data mining, data fusion processing and other data service providers. According to the think tank achievement insight series report "2022 Data Trading Platform Development White Paper" published by the National Industrial Information Security Development Research

[6] China Academy of Information and Communications Technology. *White Paper on Data as a Production Factor (2022)*. 2023.

[7] Department of High and New Technology. *Accelerating the Establishment of a Nationally Integrated Data Trading Market System*. https://www.ndrc.gov.cn/xxgk/jd/jd/202212/t20221219_1343660.html,2022

Center, most data trading platforms currently provide corresponding data value-added service models, and this type of business accounts for a low proportion of revenue.

As of June 2022, with the strong support of national policies, eighteen provinces and cities had announced relevant data regulations. Guizhou, Tianjin, Hainan, Shanxi, Jilin, Anhui, Shandong, Fujian, Heilongjiang and Liaoning issued big data regulations, and Shenzhen, Shanghai, Chongqing and Zhejiang issued data regulations. In addition, Sichuan, Guangxi, Jiangxi, Henan and other places have announced drafts of relevant data regulations. This situation has formed a new model of localized data development and governance, promoting the resource and assetization of local data.

Data capitalization is a way to increase data value, and its essential goal is to realize the social allocation of data elements. Domestic and foreign enterprises have started innovative exploration of data capitalization, and generally speaking, there are four main forms: data securitization, data pledge financing, data banks, and data trust.

China initially possessed an institutional, economic and practical basis for exploring "data + capital" dual-element integration. Since 2021, Shenzhen, Beijing, Shanghai and other places have actively promoted data asset securitization, focusing on the exploration of new models of securities empowered by key basic technologies such as artificial intelligence, blockchain, cloud computing, and big data.

8.1.3 Methods and Paths to Solve the Data Element Problem

Problems and Challenges Faced by Data Elements

To fully release the value of the data, it is necessary for the data elements to circulate. At present, the difficulty of data element circulation corresponds to a series of problems and challenges faced in the data field, which hinder the effective application and value realization of data elements. These difficulties involve aspects such as data acquisition, integration, quality, security, and sharing, including the following:

Fragmentation and inconsistency of data: One aspect of the difficulty of data element circulation is the diversity of data sources. Data can come from various channels and systems, such as sensors, social media, business systems, etc., but the diversity of these data sources makes data integration and standardization difficult, resulting in the fragmentation and inconsistency of data.

Low data quality: The difficulty of data element circulation also involves data quality issues. Data quality may be affected by data collection, transmission, storage, and processing, including aspects such as data accuracy, completeness, consistency, and timeliness. Low-quality data may lead to erroneous decisions and analysis results.

Data security and privacy protection risks: With the accumulation and wide application of data, data security and privacy protection have become important problems.

Data leakage, misuse, and unauthorized access may pose significant risks to individuals and organizations, so ensuring data security and privacy protection has become particularly important.

Difficulty in data sharing and cooperation: Data sharing and cooperation are also aspects of the difficulty of data element circulation. There are data islands and data barriers separating different organizations and departments, a lack of effective data sharing mechanisms and cooperation models, and limited circulation and application of data across organizations and industries.

Lack of data analysis and mining capabilities: With respect to the challenges of data element circulation, the lack of effective data analysis and mining capabilities is also a challenge. The complexity and scale of data make it difficult to extract valuable information and insights from the data and require appropriate techniques and tools for data analysis and mining.

To solve these data element circulation challenges, it is necessary to comprehensively apply measures from various aspects, such as data management, technological innovation, policy and regulation, and talent cultivation, to promote the development and application of data element circulation and to achieve efficient utilization and maximization of data value.

Methods to Solve Data Element Circulation Challenges

Methods to solve data element circulation challenges include building data sharing platforms, utilizing data integration and fusion technologies, establishing data trading markets, formulating data policies and regulations, and pursuing continuous innovation and improvements.

Building Data Sharing Platforms

We believe that there is much to be accomplished in areas such as data asset legal and regulatory review, building an integrated data governance platform to solve data quality issues, and building a data supply chain management system, which can help organizations improve data quality.

Accumulating Trustworthy Evidence of Data Asset Compliance to Reduce Corporate Data Compliance Risks.

Companies often do not know how to stay current with new laws, regulations, industry standards, and the requirements for data reporting and external services and how to leave traces of self-proof. When cooperating with social institutions (banks, companies, etc.) on data, they also do not know how to assess the compliance of the three parties and the scenario and how to assess losses and provide evidence when compliance issues arise. Moreover, in the future, corporate data assets will play a greater role.

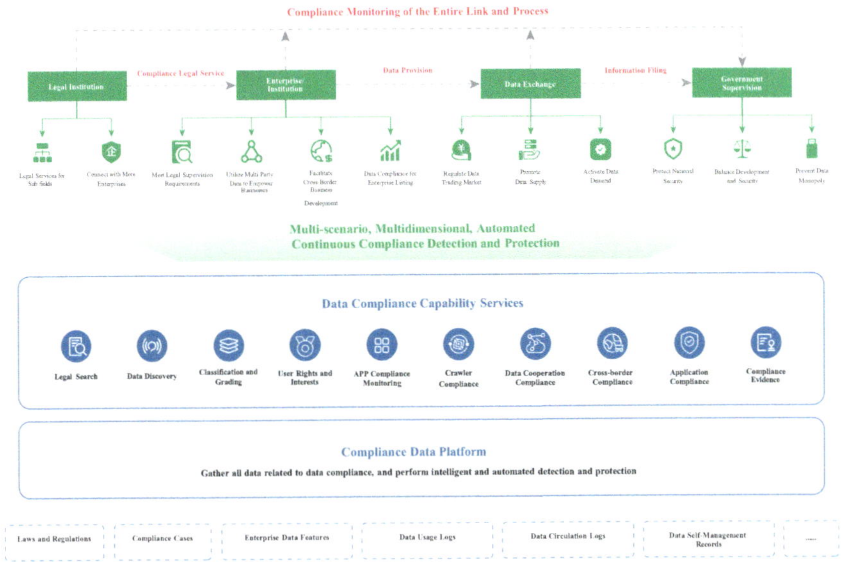

Fig. 8.5 Data asset legal and regulatory review platform

We believe that building a data asset legal and regulatory review platform (as shown in Fig. 8.5) is the trend. Through the deep recognition of sensitive data, detection of data compliance behavior, and linkage of corporate and law firm compliance legal services, customers can discover the core data assets of the enterprise without the need for professional knowledge. Finally, by continuously collecting behavior records from the generation, circulation, storage, analysis, application, etc., of data assets throughout the entire lifecycle, accumulating standardized compliance trustworthy evidence, connecting professional legal services, and adopting the best strategy in compliance accidents, corporate data compliance risks can be reduced.

The combination of different tools and capabilities in the platform can solve a variety of pain points and release tremendous value, as shown in Fig. 8.6.

Building an Integrated Data Governance Platform

An integrated data governance platform is a technical solution. It can help companies manage, protect, and optimize their data resources uniformly, enabling companies to have comprehensive control over data across multiple data sources and applications. The core goal is to ensure the quality, integrity, security, and compliance of data while improving the accessibility and value of data.

The main functions of the integrated data governance platform include aggregation, governance, fusion, sharing, application, and evaluation, as shown in Fig. 8.7. By integrating data and data warehouse technology, data are collected from various sources and stored uniformly for subsequent processing and analysis; ensuring data quality and compliance; performing data cleaning and verification; protecting data security and privacy; integrating data to provide a unified view; enabling data access

Fig. 8.6 Platform value points

and sharing through data APIs, data services, and data portals; using data for decision-making, analysis, and business operations; discovering patterns and trends in data; displaying data; building predictive models and intelligent applications; evaluating data quality and value; and assessing the economic value of data to the enterprise.

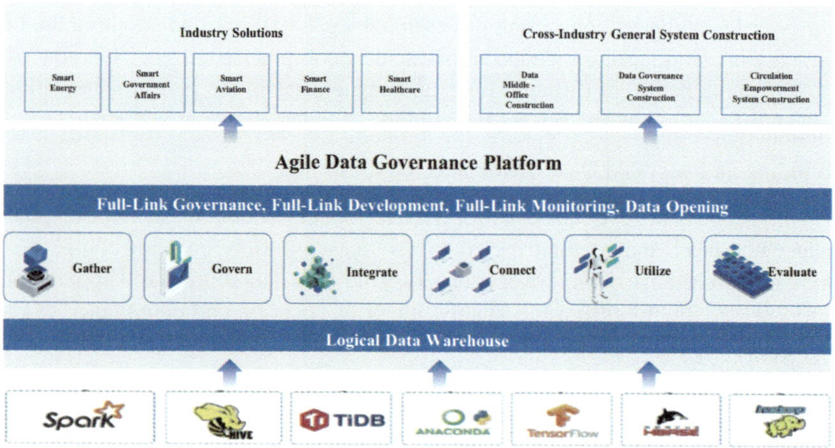

Fig. 8.7 Integrated data governance platform

By building an integrated platform, a comprehensive view of data is provided for enterprises, facilitating better management of their data assets and improving the quality and value of data while reducing data-related risks and costs. Such a platform can not only lay a solid foundation for the construction of industry-specific and cross-industry universal systems but also help realize data assetization and data trading.

Building a Data Supply Chain Management System

Corporate data sharing and exchange face the characteristics of multilevel groups, multiple entities inside and outside the enterprise, and multiregional cross-border transactions, with typical problems such as the inability to overview the enterprise data flow globally, a large amount of valueless supply and demand circulation, low efficiency from supply to consumption, the predominance of batch data, a single mode, a lack of security and compliance control protection, and imperfect dispute resolution mechanisms. Among these problems, efficiency and security have become the most prominent contradictions. In terms of efficiency, as cross-domain data circulation gradually becomes the norm, how to better coordinate the relationship between supply and demand and ensure that data meet related application needs in a trustworthy and usable way will no longer be a single-point problem but rather a comprehensive problem that requires multiparty coordination and full-process optimization. In terms of security, owing to long data link transmission, any mistake in any link may cause data leakage. Encrypted transmission is the best method for end-to-end data and also provides a security guarantee to meet compliance audits.

To address the above issues, a data supply chain management system can be established, as shown in Fig. 8.8, which includes two forms: the demand-side supply chain model and the supply-side supply chain model. Under the demand-side supply chain model, the group formulates a logical model, and each subordinate unit provides data, which the group then audits. Under the supply-side supply chain model, an inventory of data from business departments is obtained through automatic inventory. This system enables data supply and demand to be met, ensures that shared open data are trustworthy, controllable, and verifiable, and ensures the safety and availability of data processing and circulation.

Utilizing Data Integration and Fusion Technology

Through data integration and fusion technology, data elements from different sources and formats can be integrated into a unified dataset. This process can increase the availability and usability of data elements and promote their circulation.

AI Synthetic Data: In the face of increasingly high data collection requirements in the digital economy, massive amounts of data can provide accurate results. Moreover, it is necessary to carefully select data content to avoid introducing bias, and strict compliance with increasingly stringent data privacy regulations is required. The development of AI technology provides the possibility for the resource utilization of original data. One of the main tools for constructing synthetic data is the generative adversarial network (GAN). A GAN essentially creates two sets of neural

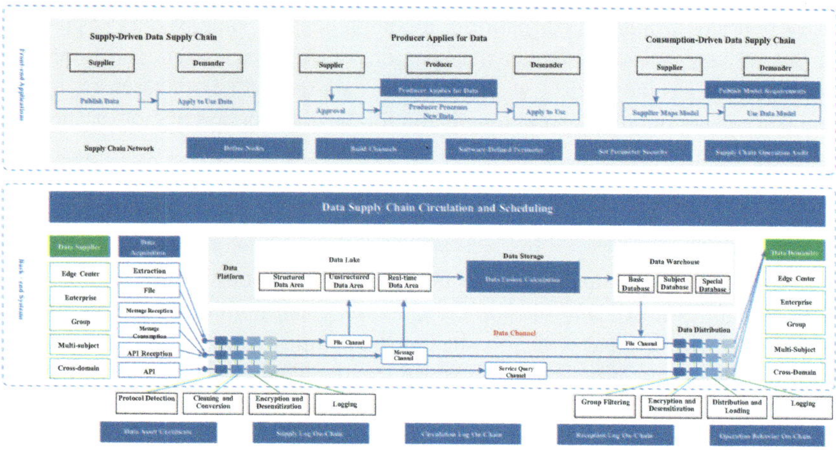

Fig. 8.8 Data supply chain management system

networks: one network generates synthetic data, and the other tries to detect whether the synthetic data are real. During the entire operation cycle, the data quality is continuously improved until no difference between real data and synthetic data can be found. Experts in the fields of artificial intelligence and machine learning believe that the focus of future technology implementation will shift to data, forming a "data-centric artificial intelligence". AI synthetic data are superior to real data, with advantages such as low acquisition cost, high quality, and avoidance of privacy infringement. This approach is expected to solve the problem of a lack of model training data.

Using Blockchain for Data Rights Registration: Owing to its unalterable and traceable characteristics, blockchain technology, can offer a unique advantage in overcoming the challenges of data rights confirmation. Future data elements can be identifiable, and blockchain technology can be used to tag data with time and space labels, enabling the same content dataset to be identified as different dataset individuals. The use of data can therefore be tracked throughout the process, and their value can be dynamically measured. Based on trusted computing facilities, asymmetric encryption, and other technologies, data can be shared after being desensitized, such as credit blacklist data, user membership point interchange, etc., thus eliminating "data islands" and improving the comprehensive utilization value of data.

Blockchain Empowers Data Asset Securitization: As the new generation of information technology continues to integrate and jointly provide technical support for securities innovation, blockchain, with its value exchange pathways, credit building mechanisms, and distributed storage model created by its technical characteristics, has opened up a new field for the fusion of "data + capital" dual elements. It has demonstrated outstanding advantages in information security, credit building, and regulatory innovation in data asset securitization.

Blockchain can be applied in depth to the practice of data asset securitization, providing real-time dynamic information sharing and secure storage for different business links, becoming a key technology for empowering securities innovation practices. China is attempting to apply blockchain to explore the integration of finance and technology, providing a practical opportunity to promote data asset securitization and the construction of a data element market.[8]

Establishing a Data Trading Market

Over the entire process of data generation, collection, processing, analysis, and application, a market system centered on data trading is created. The establishment of a data trading market can promote the circulation and utilization of data resources, increase the value of data resources, and promote the development of the data industry.

Formulating Data Policies and Regulations

The state and localities have actively formulated relevant data policies and regulations; clarified the norms and requirements for data collection, use, sharing, and protection; and provided support and guidance for the development of data elements. Institutions and individuals should be encouraged and supported to actively open data elements. Additionally, the accessibility of data elements, promotion of the circulation and utilization of data elements should be improved.

Continuous Innovation and Improvement

Through model innovation and improvement, the management and application capabilities of data elements can be continuously enhanced.

Innovating New Data Circulation Models: To overcome the difficulties of data opening, sharing, and trading practices, various places are making changes in operation mechanisms, market entities, and infrastructure. The specific process is shown in Fig. 8.9.

Path of Data Element Circulation

The process of data elements from resourceization to assetization and then to capitalization is an implementation path for the exchange and transmission of data elements in the market, as shown in Fig. 8.10.

First, the resourceization stage of data elements involves the collection, organization, and storage of data, making it a usable resource and preparing for subsequent circulation.

Second, the goal of the assetization stage of data elements is to increase the value of data and prepare for the further circulation of data.

---

[8] Zhiguo Ma, Nan Zhang. Blockchain Empowerment in Data Asset Securitization and Its Legal Governance. Journal of Shenzhen University (Humanities & Social Sciences Edition), 2023, 40(3): 114–124.

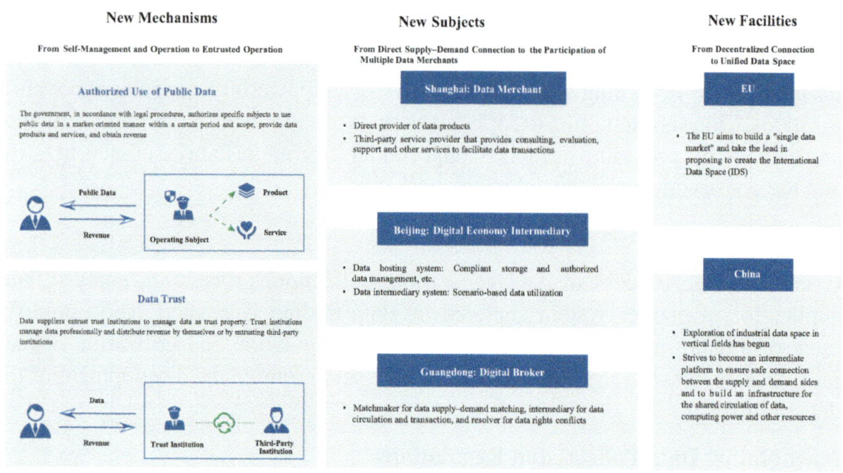

Fig. 8.9 Innovative data circulation models to solve data element problems

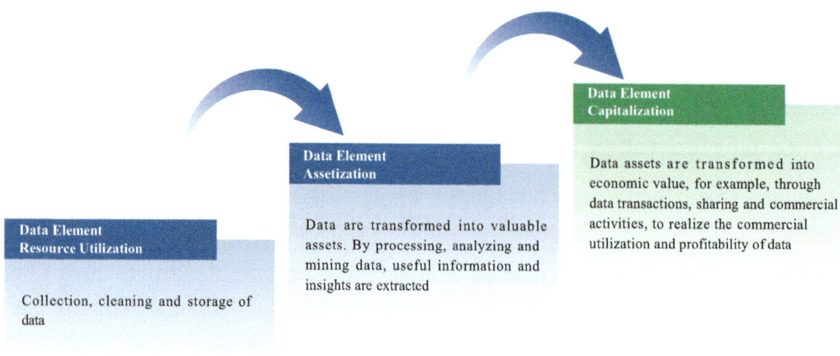

Fig. 8.10 Path of data element circulation

Finally, the capitalization stage of data elements is the ultimate goal of data element circulation, which involves transforming data elements into economic benefits.

By integrating the above methods and paths, we can effectively solve the difficult problem of data element circulation, promote the development and circulation of data elements, and obtain the maximum value of data release.

8.1.4 Future Prospects of Data Valuation Applications

In the future, with the continuous development of information technology and the acceleration of digital transformation, the data element service market will continue to grow. We believe that there is great potential for development in data resourcing, data assetization, and data capitalization.

Data Resourcing

The Construction of a State-owned Cloud Accelerates the Development of Industry

The promotion of the state-owned cloud indicates that the party, government, and state-owned enterprises will adhere to the private cloud technology route in the future. The Infrastructure as a Service (IaaS) layer of the state-owned cloud will be dominated by the state-owned assets of various provinces and cities, the Platform as a Service (PaaS) layer will be supported by public cloud manufacturers through technical means to construct the state-owned cloud platform, and Software as a Service (SaaS) manufacturers and informatization manufacturers will face a more open development environment. The state-owned cloud and the trust creation cloud actually strengthens the data security of state-owned enterprises from both capital and technical perspectives. The state-owned cloud requires that the computer room and data be controlled by state-owned assets, while the trust creation cloud reduces the existence of vulnerabilities through technical means. The state-owned cloud requires a protection level of level two or above and is equipped with various trust creation product applications. The innovation and transformation of IT technology facilities and business systems of state-owned enterprises in the future is an inevitable trend that will accelerate the development of industry trust creation, which includes areas such as cloud services, office automation, website construction, electronic invoices, electronic archives, public vehicle supervision, unified certification, financial management, human resource management, unified management, internet behavior management, cloud security, cloud desktops, cloud backup, bank–enterprise direct connections, enterprise network disks, and log analysis.

With the advancement and technological development of the "state-owned cloud", a group of data service providers will emerge in the future to create integrated solutions for digital government, digital enterprises, and digital defense, undertaking the overall design and construction of major national information systems. For example, they will employ the "cloud + data + application + ecology" service model; integrate digital products, platforms, and solutions into one; build an integrated government service platform system; promote standardized acceptance of government services and standardized handling; and achieve service process supervision, full process tracking of cases, intelligent business approval, and multichannel multiterminal services.

AI Technology Promotes the Synthesis of High-Quality Data

Synthetic datasets must have the same mathematical and statistical properties as real datasets but cannot explicitly refer to real individuals and can correct biases and prejudices in real scenarios to achieve material quality beyond real data. Hence, we can train artificial intelligence systems in a completely virtual environment and more easily customize data for various application scenarios, such as health care, retail, finance, transportation, and even agriculture. In addition, we can use synthetic data for complex computer vision modeling. For example, we can create realistic images using video datasets synthesized by advanced game engines to describe various events that may occur in autonomous driving scenarios, thereby obtaining images or videos that are nearly impossible to capture in real scenarios or that may be extremely dangerous. The emergence of synthetic datasets has greatly improved and enhanced the training efficiency and effects of autonomous driving systems.

Industrial Data Governance Empowers Data Value Release

By standardizing and managing the collection, organization, storage, analysis, and use of data, we can ensure the quality, security, and efficiency of the data. Various elements of data, including the source, type, and format of data, can be understood in detail, and the conversion and processing of data at each stage can be optimized to improve the usability and value of the data. Finally, comprehensive data governance services, including data quality management, data security guarantees, data analysis and application, etc., should be provided to help customers improve business decisions, improve operational efficiency, and innovate products and services. With the support of professional data governance, many organizations have successfully obtained great value from data, which is the concrete manifestation of the data value we discuss in this book. The service scope of industrial data governance is shown in Fig. 8.11.

Data Assetization

Cobuilding Data Alliance, Determining Data Boundaries

A diagram of the data asset formation and transaction process is shown in Fig. 8.12.

During the process of data generation, companies may encounter issues such as unclear data usage restriction agreements, uncertainty about whether the user is complying with the agreement, a lack of reliable user identity verification, the inability to properly execute the agreement, insufficient evidence of data usage and destruction, and illegal data circulation. Therefore, it is essential to establish data boundaries by forming a data alliance with multiple parties (as shown in Fig. 8.13).

Companies can create a trust-based data alliance with the business goal of forming a data community, collaboratively developing and executing data usage agreements, providing software-defined data boundaries for the alliance, and ensuring that data are used in accordance with the agreement through technical means.

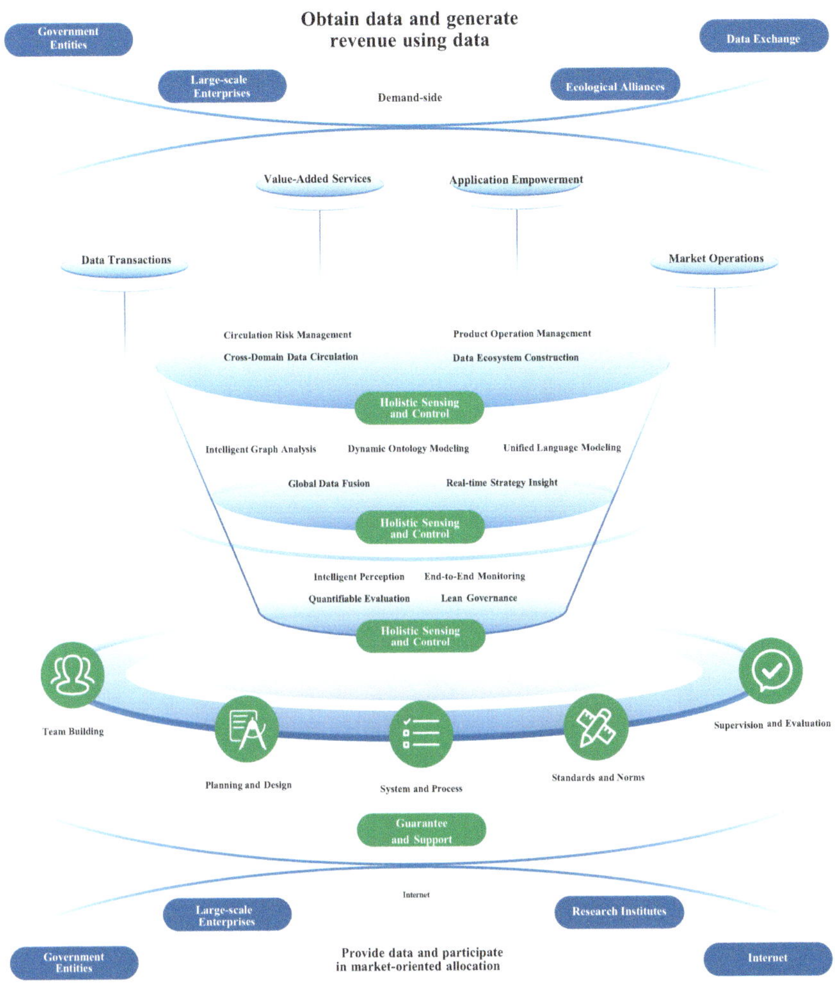

Fig. 8.11 Industrial data governance services empower data value release

The technical means to ensure that data are used in accordance with the agreement include access control, data encryption, auditing and monitoring, compliance checks, and smart contracts.

Access Control: Techniques such as access control lists, role-based access control, and attribute-based access control can be used to restrict who can access which data.

Data Encryption: Data encryption technology can be used to secure data, allowing only users with the correct key to decrypt and access the data.

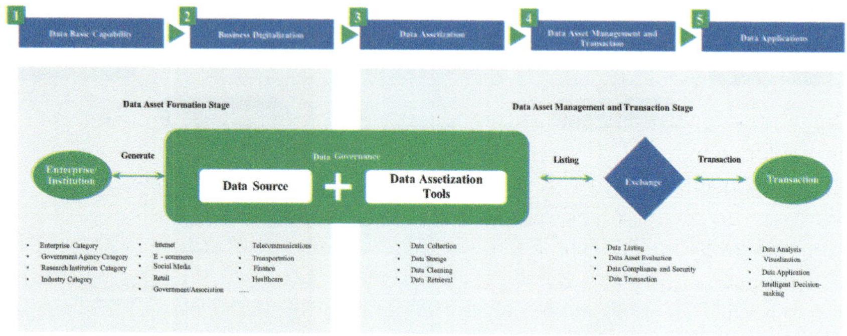

Fig. 8.12 Full process diagram of data asset formation and transactions

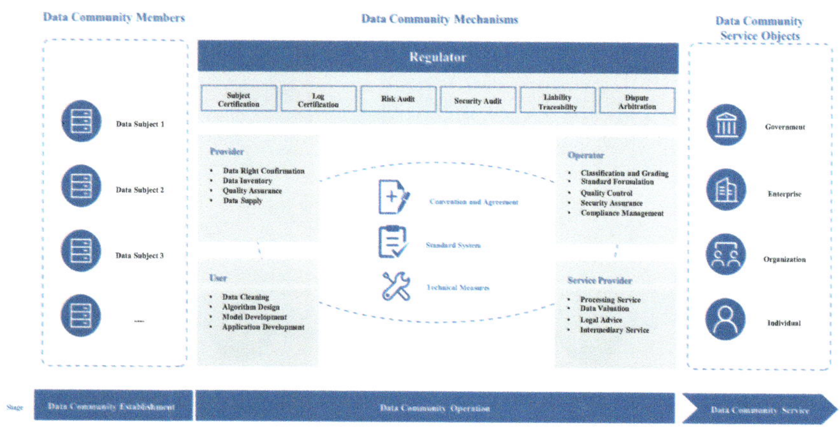

Fig. 8.13 Data alliance collaboration model

Auditing and Monitoring: Auditing and monitoring tools can be used to track data usage. If any usage behavior that violates the agreement is detected, warnings can be issued, or other measures can be taken promptly.

Compliance Checks: Automated compliance check tools can be used to verify whether data processing and usage activities comply with relevant regulations and agreements.

Smart Contracts: Smart contracts in blockchain technology can be used to automatically execute data usage agreements. Once a smart contract is deployed, it cannot be modified, thereby ensuring the fairness and transparency of the agreement.

This type of data alliance can not only enhance the value and usability of data but also significantly reduce data risks and costs, thus promoting the assetization and marketization of data.

When creating a data alliance, companies need to clearly consider the following six points: sharing rules (including timing, methods, how to handle shared data,

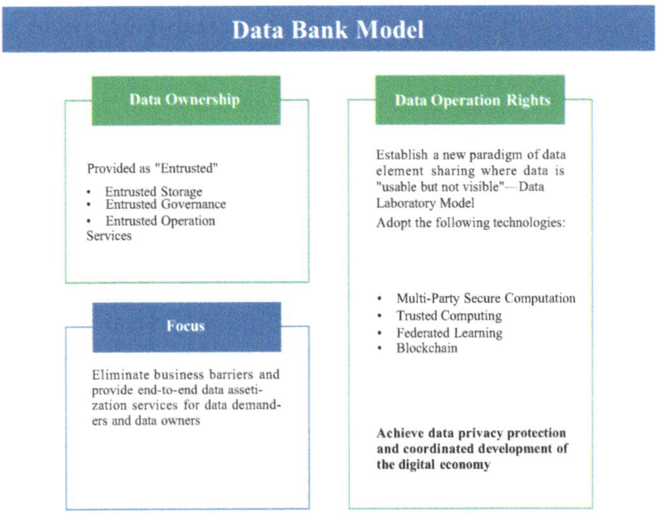

Fig. 8.14 Data bank model

etc.), usage agreements (specifying how to use shared data, usage restrictions, and the consequences of violating the agreement), identity verification mechanisms, auditing and monitoring mechanisms, compliance reviews, and security (data encryption, access control, etc.).

Through these efforts, data alliances can help companies better manage and protect their data, realize the value of data, and reduce data-related risks.

Advancing Data Lakes and Developing "Data Banks"

By advancing the data lake strategy, we can provide innovative data integration models for city, transportation, and public security brain businesses, helping cultivate a national data element market. The specific model is shown in Fig. 8.14.

Data Capitalization Aspect

Currently, digital exchanges have been established in various places. The future trend is to transform data assets into data capital, thereby facilitating the transition from business value to business models. In other words, data can not only be used to generate value, such as through targeted advertising or improving products and services but also become part of the business model, for example, through data trading, data analysis services, etc. This transformation will make data tradable, investable, and value-added capital, which will help further promote the development of the data industry.

Blockchain Empowers Data Asset Securities to Be Authentic and Reliable

On the one hand, at the beginning of the formation of basic assets, the data initiator forms an initial bottom asset pool, writes the data asset information into the block, and records subsequent transactions truthfully. The technical characteristics ensure that all relevant information is on the chain and can be traced back for verification. This approach overcomes the limitations of data application scenarios and regions, ensures the safe management of data assets in dynamic circulation, and improves the authenticity of information related to the design of data element securitization products. It thus addresses the problem of opaque basic asset information from the source and provides a reliable basis for the value assessment of basic assets. On the other hand, blockchain uses hash algorithms and asymmetric encryption technology to ensure the information is unalterable while completing identity verification and information leakage prevention and control, enabling node members to reach a consensus on information storage and effectively regulating the scope of information disclosure. This technical feature is closely coupled with the trading process of data element securitization products.

Building a Dynamic Data Asset Value Realization Ecosystem

The advent of blockchain technology heralds a shift in transactions from traditional trust models to "trust without intermediaries", which primarily addresses security and trust issues in information exchange and sharing. The technological aim is to enhance the transparency and intelligent sharing of information. To deploy blockchain technology in the evaluation, issuance, and transaction process of data asset securitization, a credit mechanism based on consensus mechanisms and smart contracts is constructed. Participants can become nodes on the chain, effectively solving problems of information asymmetry and nonstandardization. On the one hand, the consensus mechanism, i.e., the proof-of-work mechanism or proof-of-stake mechanism, is packaged by the node with the strongest computing power and fastest calculation speed. The node obtains the right to broadcast information to the entire network, ultimately forming a network-wide consensus. On the other hand, smart contracts uniformly package the relevant rights and obligations of multiple institutions. Through contract code, the operation process, default handling, and guarantee conditions of data asset securitization are designed in advance. This approach allows for dynamic verification and inspection of business execution, constructing a reasonable judgment mechanism after contract execution. Smart contracts can also set regulatory thresholds to ensure orderly business operations. Once the risk of the data asset securitization business process triggers the threshold, participants can initiate the corresponding process designed in advance, such as due payment or guarantee measures, through the complaint mechanism and automatic judgment technology, thereby enhancing the credit security of the entire securitization process.

Decentralized Regulation of Data Asset Securitization

Blockchain, as an underlying technology, realizes high autonomy and joint participation in the form of a decentralized database, which is different from the traditional central control model. It provides a technical framework for securities innovation; can digitally transform the entire process of collecting, auditing, issuing, and trading underlying assets; and provides customized customer system interaction processes, information matching, and verification transmission, thereby improving the efficiency of the securitization process. Blockchain connects participants through nodes, weakens the functions of traditional securities intermediaries, builds a comprehensive service ecosystem for data asset securitization, records information in real time using technical means, eliminates information barriers, and ensures the authenticity and validity of data.

Regulatory authorities can enhance the effectiveness and fairness of regulatory activities by utilizing the advantages of information sharing, and the decentralized characteristics of blockchain help to perform regulatory activities in detail. On the one hand, the decentralized characteristics leave traces of the behavior information of the participants on the chain. Regulatory authorities, as special nodes of the business chain, can fully access the real data asset securitization information recorded in the system, perform global reviews, eliminate the vacuum of regulatory information, foster information exchange and sharing between financial and data regulatory departments, and enhance the effectiveness of regulatory practices. On the other hand, the main reason why blockchain, as a computer technology, can be decentralized is that its unique algorithm mechanism allows all participants to collaborate collectively, ensuring the sharing and authenticity of information. The operation of the system must be free from control by any individual or institution to ensure the objective truth, consistency across the network, and immutability of information.

Future data element services will focus more on the application of automation and AI technology, establish an open ecosystem to promote data sharing and cross-departmental cooperation, and provide comprehensive and integrated one-stop solutions. Element services will stimulate digital productivity, unleash the enormous potential value of data, change the physical world with production tools, generate new business value, fulfill diverse and personalized human needs, and promote human technological progress.

8.2 The Era of Architectural Space Services and Digital Operations

8.2.1 Industry Development Trends and Reflections

In the process of national urbanization, decades of architectural development have become increasingly intelligent, from initial building control to intelligent buildings and now smart buildings and digital architecture. This development is the result of generations of architects and technologists constantly striving to meet the human demand for beautiful spaces[9] with the latest technology.[10] In this section, we share our thoughts on the field of architectural space services and digital operations.

8.2.2 The Development Path of Architectural Space Services and Digital Operations

National policy has proposed specific goals and requirements for the digital economy, and digital development has both pioneering and conservative aspects. The Twentieth Report outlines the development blueprint of the digital economy. It clearly proposes to "accelerate the development of the digital economy and promote the deep integration of the digital economy and the real economy". The innovative application of digital technology from traditional methods is compulsory for all industries to actively embrace opportunities.

In the past, during the process of national urbanization, the construction of urban space expanded rapidly, and the management of architectural space became increasingly intelligent. However, the digital construction of architectural space has problems such as incomplete data collection, difficulty in data sharing, lack of unified standard specifications for construction, and easy neglect of data security issues. The extensive development model can no longer cope with changes in market demand and industry competition.

In terms of industry development, the development stages and key aspects of digital operations in architectural space are shown in Fig. 8.15.

The first stage is informatization, which involves the penetration of technology into the physical world, construction on a scenario-by-scenario basis, and the virtualization of all management objects of architectural space into the computer world. The key problem it solves is information symmetry, which transforms the relevant business of the physical world into electronic and online informatization. Its core value is the online operation of business.

[9] Tao Meng, Mingchun Zhou. *Integration of Digital and Intelligent Technologies: The Path to Smart Building Transformation.* Beijing: Publishing House of Electronics Industry, 2022: 7.

[10] Tao Meng, Mingchun Zhou. *Integration of Digital and Intelligent Technologies: The Path to Smart Building Transformation.* Beijing: Publishing House of Electronics Industry, 2022: 7.

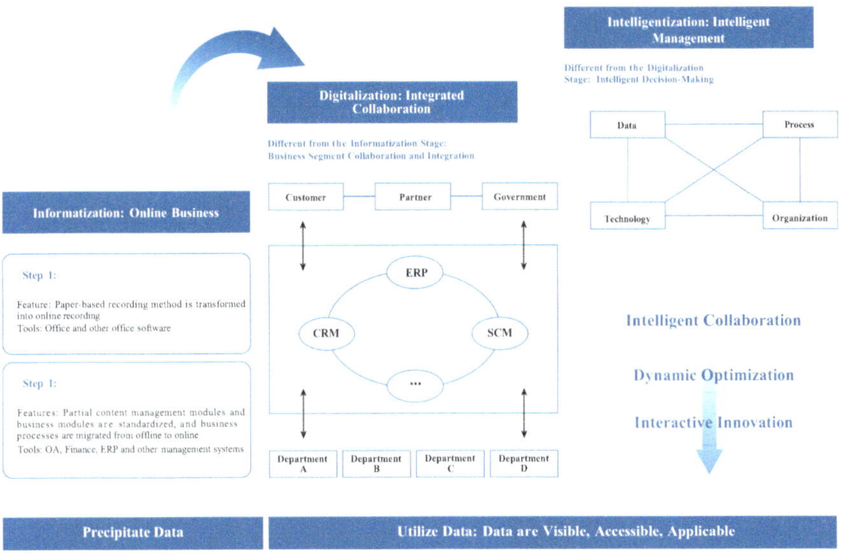

Fig. 8.15 Stages of digital operation development in architectural space

The second stage is digitization, which involves the integration of information resources and the optimization of business processes. Business rules are abstracted using technologies such as the Internet of Things, big data, cloud computing, and artificial intelligence. The key problem it solves is making data visible, applicable, and able to assist in decision-making. Its core value is integrated collaboration.

The third stage is intelligentization, which centers on business and human needs. At this stage, digitization is not just about technology itself but rather a comprehensive transformation of enterprise organization, business, market, marketing, technology, research and development, and other enterprise elements. The key problems it solves include helping enterprises save costs, increase efficiency, control risks, enhance brands, and increase profits, thereby fundamentally enhancing the competitiveness of the enterprise. Therefore, digitization at this stage needs to be combined with these enterprise elements to build corresponding rules and generate corresponding intelligent collaboration, dynamic optimization, and interactive innovation service values. Its core value is intelligent management.

Currently, most applications of digital architecture are in the first or second stage, with very few transitioning from the second to the third stage.

At present, we are transitioning from the digital era of "mobile + cloud" to that of "AI + 5G", and the integration of the digital economy and the real economy is becoming a trend. In February 2023, the Central Committee of the Communist Party of China and the State Council issued the "Overall Layout Plan for Digital China Construction", which clearly states that the construction of Digital China will be conducted according to the "2522" overall framework. By 2025, a comprehensive advancement pattern that is horizontally connected, vertically penetrated, and strongly coordinated will be basically formed. The establishment of the National Data Bureau will also more efficiently promote the role of "data" as a production factor in economic development.

In addition, with the continuous development and widespread application of technology, the generation of massive amounts of data, and the rapid development of artificial intelligence, people's access to information and use of innovative applications have become more convenient. People's ways of working, living, and learning are undergoing great change. Moreover, when people exchange information in architectural spaces, they have higher requirements for comfort, convenience, energy savings, and safety.

Regardless of how the form of technology changes, its ultimate mission is "serving people". Technology is just a means, not an end. The goal is to make the use of buildings more convenient, more humane, and safer and to adapt architectural spaces to human activities (see footnote 10). The future will usher in an era where efficiency and benefits coexist. "Service transformation and upgrade and refined operation services" will become a new growth direction for architectural space services.

8.2.3 Challenges Faced in the Transition from the Era of Increment to the Era of Stock

With urban development, the traditional new construction market has been severely affected, resulting in significant shrinkage and change. Under the overall perspective of "houses are for living, not for speculation", China's real estate sales is gradually peaking, and the operation of real estate companies is in a state of low growth and low profitability. In recent years, many large real estate companies have shifted from "land acquisition and house construction" to "large asset management and operation". Moreover, owing to the limitations of traditional operation modes, the efficiency, cost, and customer experience are all unsatisfactory, hindering the development of space and the fundamental generation of revenue and profit and failing to maximize the value of space. Therefore, applying digital means to architectural space services is an inevitable trend. However, it faces the following challenges:

Compared with the past, the development of architectural space is more concentrated, and the overall profit space is squeezed. The profit-making links of architectural space have shifted from development, leasing, and sales links to asset operation and service transformation links. Therefore, only by refining operations can the main operation of construction seize the next round of a blue ocean market.

The demand side of the consumer group is upgrading, which requires the owners of architectural space to actively connect with customers, gain deep insight into user needs, and thereby enhance product and service capabilities.

The service model of architectural space has changed from a single operation to the collaborative sharing of upstream and downstream resources and the integration of online and offline resources. However, architectural space operation organizations lack a unified measurement scale, and integration is difficult. Therefore, it is necessary to establish an efficient and sustainable architectural data resource system, establish a strong digital platform ecosystem, and provide strong support for the operation and management of architectural space.

8.2.4 Exploration of Digital Services and Operations in Space

In the past, building developers were involved in a pillar-type industry chain, from the supply of construction materials to construction and decoration, extending to transaction circulation and finally to property services, smart communities, smart cities, etc. Data are becoming a new production factor. The strong empowerment of digital technology for services and the efficient enhancement of operational capabilities will lead to revolutionary changes in the business models of commercial parks, industrial parks, and other related industries. The established business models, roles in the industrial chain, scenarios, functions, etc., will be reshaped and reconstructed by digitization. In the future, the industry may shift from a "one-stop" vertical monopoly model to a more professional and segmented model, with a clear division of labor and harmonious coexistence in a horizontal value chain.[11]

With the driving force of market demand and technology, the future building space will continue to change and upgrade in terms of digital services and operations, leading to more human–machine–object integration development, constructing more service methods, and creating greater value. The future development directions of digital services and operations in building spaces can be divided into two categories:

For Space Managers: Let "Machines" Replace "Humans"

For managers of commercial parks, industrial parks, and other multiformat, multifunctional integrated parks, it is often difficult to balance expenditures and income. The traditional operating method involves income from advertising, parking lots, and property management. However, expenditures include a series of labor costs, such as facility maintenance, security patrols, access management, park security, cleaning, and landscaping.

The future management model of the park will shift from traditional manual management to digital, automated, and intelligent management. This transformation will be facilitated through the continuous advancement of technology and algorithms. By using digital means to "digitize business", most of the work that requires human labor will be transformed into automatic operation and processing by the system, thereby reducing labor costs. The following is an example:

[11] Guandian Index Research Institute; Schneider Electric. *2022 White Paper on Digital Operations*. 2022.

Energy Consumption Management

At present, most energy consumption management involves the manager setting a certain parameter value (such as 26°) for the air conditioning system, which in turn automatically adjusts the air supply and cooling capacity of the set area to maintain the set value in the area. This type of system is a semiautomatic system. If there are few or no people in this area, it will cause great waste.

The energy consumption systems of the future will be able to automatically monitor environmental factors such as air quality and light intensity and accurately predict and control indoor temperature and humidity, effectively reducing energy consumption and operating costs. In addition, a building AI system can adapt to adjustments in the temperature and humidity of the human environment. For example, when data collected by cameras equipped with intelligent algorithms for personnel density or regional temperature sensing are used, a comfort adjustment mechanism is established through AI algorithms. The building AI system can automatically adjust the air supply and cooling capacity of the air conditioner to achieve a comfortable environment for the crowd. Moreover, a building AI system can also learn and analyze the living habits of residents or employees, automatically adjust the environment according to the indoor temperature and humidity situation, provide personalized comfort adjustments, and improve the quality of life and experience of residents. This approach also achieves energy savings, green and low carbon, improves building management efficiency and accuracy, reduces management costs and risks, and improves operational efficiency and income.

Security Management

The traditional security work method relies mostly on security personnel to watch the monitor screen in the monitoring room, patrol the park, or stand guard in a fixed manner. This method leads to a great waste of labor and is too dependent on the quality of security staff. Additionally, there are many security management loopholes. To address these issues, new security systems will use a security system with AI functions, which can achieve automatic intelligent patrol, intelligent deployment, intelligent recognition, intelligent early warning, etc. However, this method is still a semiautomatic method, and the system still works according to human-set rules. Thus, when new emergencies or abnormal situations occur, the system still cannot respond and judge.

Future AI security systems can self-learn through a large amount of behavior data, generate rules through AI algorithms, and make adaptive early warnings and decision arrangements.

Example 1: Wireless sensors, the Internet of Things, cloud computing, and navigation technology can be used to achieve a true "Internet + parking" mode, help users visually select parking spaces and navigate vehicles, and automatically send bills to consumers' mobile phones after driving away, increasing the accuracy and efficiency of parking lot management and income calculations.

Example 2: On weekends and holidays, the AI security system can automatically adjust to track the situation of children playing in the park and set up intelligent isolation fences, early warning alarms, and other protective facilities to reduce incidents. When water accumulates on the road surface on rainy days, the AI security system can issue a warning to prevent the elderly and children from slipping and simultaneously notify the cleaning staff to address the issue.

These AI security systems with self-learning abilities will greatly improve the safety, precision and effectiveness of preventive measures in an environment, thus ensuring the safety and stability of the environment.

Humanization of Services

Various types of sensors can be arranged in public spaces within buildings to collect environmental information, personnel (vehicle) entry and exit information, personnel (vehicle) trajectory information, etc. A building's intelligent system can adaptively learn from the operations data of various devices, mining the patterns behind large amounts of behavior data through adaptive learning (including patterns not yet discovered by humans). This capability can provide humanized services for people in the park for managers.

Example 1: Greater focus can be placed on caring for vulnerable groups. For the elderly in a building, intelligent home devices such as electronic health monitoring instruments and voice call alarm systems can be installed to monitor their physical condition, living habits, and walking trajectories in real time. This approach can facilitate remote medical consultations and emergency rescue services and the dispatch of staff to take care of and help solve problems when needed. For the disabled, digital accessibility technology can be explored as an auxiliary tool. For example, by monitoring behavior trajectories and voice reminders, disabled people can travel more safely, enjoying more convenient and efficient service and management experiences.

Example 2: Car owners often drive in and out of the park, and the AI camera in the parking lot can capture the appearance of a vehicle. The AI system can push car washing and cleaning services depending on the car.

Example 3: An AI camera in a park can judge the characteristics of the park's personnel group, launch weekend meal ordering services for young families, and present tutoring services according to the age of the children.

Toward Space Operators: Forming a New "Digital + Consumption" Ecosystem

As people's requirements for quality of life and the working environment increase, digital consumption has filled the gap in traditional consumption and has become an important engine for leading consumption and an important means for stimulating domestic demand in cities. Therefore, the refined operation of space involves delving into the crowd in an environment, understanding people's daily lives, and creating scenarios and processes with service logic. The future operation of a building space is an operation mode where the community, content, and traffic guide each other. The building space is seen as a "field", introducing IP (goods), empowering through IP, and then through all-domain marketing and private domain traffic (people) import, traffic acquisition, interaction, realization, and conversion can be used to build a new model of the community coconstruction of profit.

More Humanized Services

In the building space scenario of the shared office community, each microspace can provide a virtual assistant for the tenant. When the tenant enters the microspace, the tenant can have a short voice conversation with the virtual assistant, and the microspace can accurately push related peripheral services to the tenant, such as whether to buy coffee. After receiving an affirmative answer, the microspace operator can immediately push the purchase link for this service to the tenant. In addition, tenants can not only remotely reserve microspaces, but the microspace intelligent system can also automatically set the air conditioning, lighting, music, smell and other office environments according to tenant needs. Thus, the system can create a considerate, high-quality new office environment, make the space more "understanding" of people, enable tenants to enjoy a more convenient experience, and make the entire environment more productive and creative.

More Personalized Marketing

A large amount of data is generated every day, but the value of many data points is ignored. For example, a male white-collar worker is waiting for the elevator at work and finds that the screen next to the elevator is displaying an advertisement for sanitary napkins. If there are data collection and intelligent analysis robots, this scenario will play a different role. Therefore, in the case of building space operation, offline big data can be used to help offline media distribute information accurately, thereby generating profits from the operation of the space.

Example 1: In the example of the male white-collar worker waiting for the elevator mentioned above, if there is a camera or sensor on the screen and it recognizes that a male is standing in front of the screen, the backend can push advertisements that men may be interested in, such as new car products, through data analysis and intelligent systems. If someone is interested in interacting with the interface, the intelligent screen can continue to accumulate knowledge and gain insights through adaptive learning, predict the specific products that the interactor is interested in, and automatically push the purchase link for related products. Thus, customers can enjoy the right products and services at the right time. On the other hand, the system can also help increase the accuracy of offline store operators.

Example 2: Mobile phone authorized stores can recognize potential customers through cameras, interact with intelligent assistants for voice interaction, link analysis to accurately predict the probability and time of old product damage, and push corresponding advertisements to stimulate consumption by customers in the decision window period of three to six months. Then, the customer can decide whether to replace the old product.

Example 3: The catering industry currently offers uniform product package services for customers to choose from. In the future, stores will obtain information about users through consumption traffic, and the menu screen will display the ingredients and shelf life of each serving through the captured data. Through customer voice interaction, the system will automatically recognize the customer type and then recommend popular products and packages according to the customer type. After a review is given after dining, the system can push matching product services, creating an operation mode that attracts a large amount of traffic by optimizing operation strategies, thereby increasing overall traffic.

Chapter 9
Technology Outlook

We believe that the metaverse and artificial general intelligence (AGI) are among the key technologies driving the future advancement of smart cities. As a new type of virtual world, the metaverse will pioneer new business models and economic forms. Its development requires a large amount of data support, which will be deeply integrated with areas such as digital government and industrial digitization. As the highest stage of artificial intelligence development, AGI's influence on the future will be ubiquitous; AGI is expected to achieve various intelligent activities of humans, including understanding, learning, adapting, and innovating, providing an enormous impetus for socioeconomic development.

9.1 Insights into the Next Generation Internet Through the Metaverse

9.1.1 Seven Stages of Metaverse Evolution and Development

The concept of the metaverse was first proposed in the novel "Avalanche" published by an American author in 1992. We believe that the simplest and most powerful definition of the metaverse is that it is a possible digital world where people participate and live with digital identities. The digital world of the metaverse has six characteristics: virtual, three-dimensional, visual, interactive, shared, and sustainable.

The evolution and development of the metaverse involves seven stages: the prototype stage, independent application stage, platform construction stage, open interoperability stage, intelligent stage, realism stage, and future evolution stage. We have already passed the prototype stage and are currently in the independent application stage and platform construction stage, as shown in Fig. 9.1.

F. Liu, *Smart and Digital Cities*, https://doi.org/10.1007/978-981-95-2243-9_9

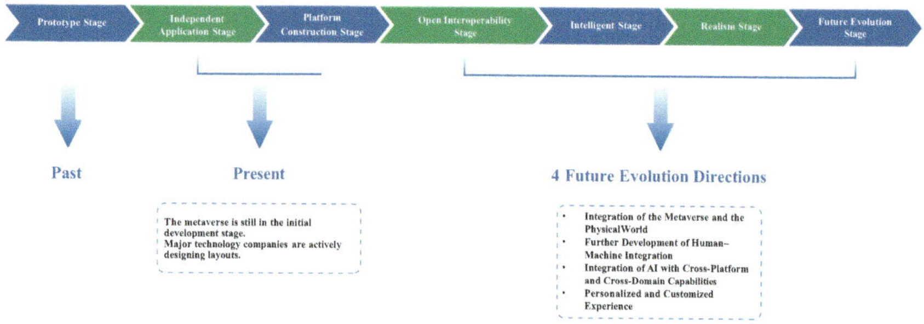

Fig. 9.1 Seven stages of metaverse evolution and development

In the prototype stage of the metaverse, the metaverse is merely a preliminary concept or idea, which exists mainly in a few science fiction novels and movies. Although people cannot experience the application scenarios of the metaverse first-hand at this stage, its potential benefits and application scenarios have already sparked the infinite imagination of people's minds. However, owing to the lack of corresponding technical infrastructure, hardware devices, and software system support, the actual application and development of the metaverse still requires more effort and exploration.

In the independent application stage of the metaverse, some applications have emerged, such as games, social networking, entertainment, etc., which have been well utilized in specific fields. Although the metaverse cannot yet achieve complete interoperability, its technical infrastructure and application scenarios have been fully developed, becoming an independent digital world, that is, independent of the real world. People can engage in various virtual activities and communications, enjoying the convenience and fun of the digital world. In addition, the commercialization and industrialization of the metaverse are gradually accelerating, and the formulation and improvement of various laws and norms are constantly advancing, laying the foundation for the sustainable development of the metaverse.

In the platform construction stage of the metaverse, of software, hardware, and infrastructure are constructed. Different metaverse applications begin to connect with each other, and some cross-application standards and protocols have already emerged. The virtual reality and augmented reality technologies of the metaverse will be further developed and applied. At the same time, the social, entertainment, commercial, and industrial applications of the metaverse will be improved and widely expanded. At this stage, establishing a unified, open, secure, and interoperable metaverse platform is crucial for the healthy development of the metaverse, not only providing strong support for the rapid development of the metaverse but also laying a solid foundation for its widespread application.

In the open interoperability stage of the metaverse, different metaverses exhibit interoperability, realizing the sharing of information and resources, as well as more complex scenarios and applications. Users can freely travel between different metaverses with a single account or identity. For example, different metaverses can share virtual reality scenes, virtual assets, etc., thus enriching the content of the metaverse. Additionally, they can share user data, identities, etc., enhancing the user experience and participation, as well as applications and services, thereby promoting commercial and industrial applications of the metaverse. A set of unified, open, metaverse interoperability standards and protocols should be established to ensure seamless connections and interactions between different metaverses. These standards and protocols need to cover all aspects of the metaverse, including the interoperability of data, applications, users, etc. In addition, an open metaverse interface needs to be established so that different platforms and applications can connect and interact with each other.

In the intelligent stage of the metaverse, through artificial intelligence technology, four aspects of applications are realized. The first is intelligent interaction, which allows users to interact with the metaverse more naturally through functions such as natural language interaction and emotion recognition. The second is intelligent decision-making, which enables the metaverse to provide services and support to users more intelligently through functions such as intelligent recommendation and intelligent decision-making. The third is intelligent creation, which gives the metaverse stronger autonomy and creativity through functions such as intelligent creation and autonomous evolution. The fourth is intelligent security, which ensures the safety and stability of the metaverse through functions such as intelligent security and risk management.

The metaverse in the realism stage provides a more realistic virtual experience and immersive environment, making users feel as if they are in the real world. By introducing advanced virtual reality and augmented reality technologies, as well as powerful computing and storage capabilities, the metaverse achieves realistic experiences, creation, social interaction, and commerce. Specifically, users can experience the environment, characters, scenes, etc., in the metaverse more realistically through virtual reality and augmented reality technologies, enhancing the sense of immersion and realism. Moreover, it facilitates more natural ways of creating and designing, enhancing users' creativity and artistry. It also allows users to socialize and communicate more realistically, creating more realistic social scenes and social experiences, and it allows users to conduct business activities and transactions more realistically, creating more realistic business scenes and business experiences.

The metaverse in the future evolution stage remains uncertain. The future development direction is still unclear. What can be predicted is that the metaverse will integrate more closely with the real world, create richer experiences and application scenarios, and bring more convenience and surprises to users. Additionally, there will be four evolution directions: the integration of the metaverse with the physical world, the further development of human–machine integration, the popularization of AI, and integration across platforms and fields, resulting in more personalized and customized experiences.

9.1.2 *The Rise and Fall of the Metaverse Market*

The metaverse is a **three-dimensional, visual, interactive, shared, sustainable** virtual space where people can have immersive interactions and experiences. Since the concept of the metaverse was proposed, it has aroused widespread interest and discussion in the business world. However, the development process of the metaverse market has not been smooth and has gone through a journey from peak to trough.

The rise of the metaverse market can be traced back to 2018, when virtual reality technology began to enter the consumer market. In the following years, an increasing number of companies began to invest in and develop metaverse projects. These projects cover various fields, such as gaming, socializing, education, and entertainment, and present new business opportunities.

Recognized giants in the metaverse field include Facebook, Tencent, Meta, Alphabet, and Apple.

Facebook, one of the earliest tech companies to enter the metaverse field, was later renamed Meta and launched its metaverse product, Horizon Workrooms, in August 2021. In December 2021, Meta launched a virtual world called Horizon Worlds.

As a Chinese internet giant, Tencent also has a stake in the metaverse field. Its metaverse product Universe was launched in 2022. Additionally, Tencent has invested in several metaverse startups worldwide and has cooperated with other companies to develop metaverse-related products.

Google, under Alphabet (Google's parent company), also has a layout in the metaverse field. On May 12, 2022, at the Google I/O 2022 conference, the cloud computing service platform Google Cloud launched a metaverse immersive 3D real-world map. This project combines visual technology and satellite street view photos to provide a more immersive real experience than traditional maps do. Google is also developing a VR drawing tool called Tilt Brush.

Apple also has a presence in the metaverse field. Although the company has not directly released metaverse-related products, it is reported that Apple is strengthening its research and development in AR and VR technologies and providing value-added services for other products through AR technology.

In addition, some metaverse startups, such as Roblox and Epic Games, are also promoting the development and growth of the metaverse industry. However, the growth trajectory of the metaverse market has not been smooth, and starting in November 2022, some problems began to emerge.

Metaverse companies have announced large-scale layoffs or business shutdowns. At the end of 2022, Meta laid off 11,000 workers, and Meta's metaverse business lost $13.7 billion in 2022. PICO, a VR manufacturer under ByteDance, also recently laid off 20% to 30% of its staff. The head of Kuaishou's metaverse business resigned and stopped related business. Tencent also announced the cancellation of all positions in the XR business line. Even Microsoft, which is financially strong, recently eliminated its industrial metaverse team that was just established four months ago. These layoffs and business shutdowns have led to questions about the future development of the metaverse.

The business model and commercialization path of the metaverse currently face many challenges, requiring companies to delve into aspects such as hardware, technology, the market, and business models. Moreover, the practical application scenarios and user needs of the metaverse need to be more clearly and precisely defined to meet user demands for immersive, interactive, and visual experiences. Additionally, many unresolved issues remain regarding the content quality and safety of metaverse products, such as the management of virtual goods and assets.

We believe that the metaverse market will undergo a period of fluctuation and adjustment. The metaverse market is still in the development and exploration stage, especially as metaverse technology and hardware continue to innovate and progress, and the future is promising. Moreover, the exploration of various technological applications and business models will also drive the sustainable development of the metaverse market, which is an evolutionary form of the future internet and has the potential to become the "next internet".

9.1.3 Metaverse's Development Breakthrough Based on Cutting-Edge Technology

The current bottlenecks in metaverse technology focus mainly on data visualization and processing, interoperability, security and privacy, hardware devices and computational resources.

Bottlenecks in data visualization and processing: The amount of data in the metaverse is enormous, and data directly affect the realism and interactive experience of the virtual environment. How to quickly and accurately process and visualize these data is a significant issue.

Bottleneck in interoperability: Various applications and devices in the metaverse need to be able to cooperate and interact with each other, but there is currently no standard interoperability protocol. Addressing this issue requires technical support and standardization, as well as a metaverse management mechanism. Moreover, the main way people currently enter the metaverse is through wearable devices or human–computer interaction, which does not provide a good user experience. In the future, entry into and interaction in the metaverse should be more flexible and convenient.

Bottlenecks in security and privacy: The data and information in the metaverse are very sensitive, and ensuring security and privacy is a crucial issue.

Bottlenecks in hardware devices and computational resources: The metaverse requires a large number of hardware devices and computational resources to support

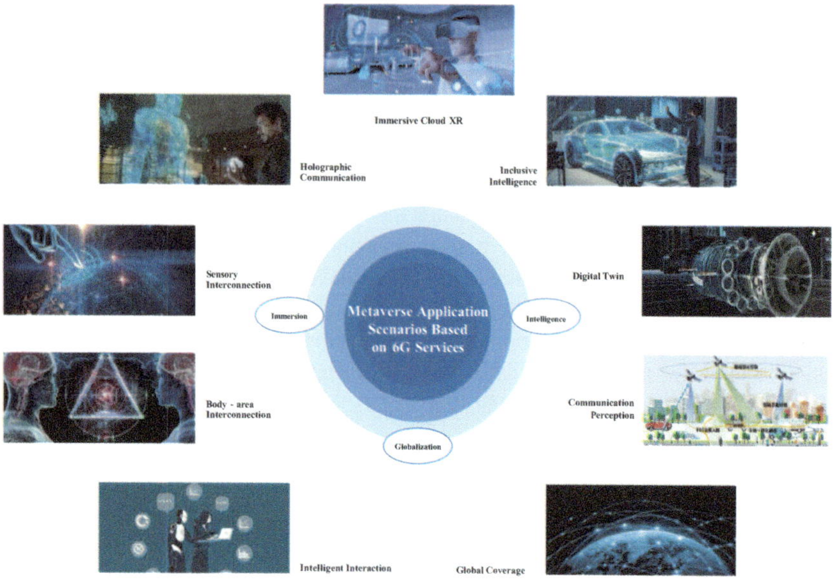

Fig. 9.2 Metaverse based on 6G's immersive, intelligent, ubiquitous business application scenarios

its operation, but there are currently insufficient resources. To realize a true metaverse, the current computational capacity needs to be increased by 1000 times.[1,2] The immersive experience of the metaverse requires high-performance hardware devices, such as virtual reality headsets and full-body tracking devices. Currently, these devices are costly, and there is room for improvement in terms of portability and comfort.

The research and promotion of 6G technology can solve some of the bottlenecks in the metaverse, such as improving device-side computing power, enhancing data transmission speed and stability, and strengthening interoperability.

The metaverse based on 6G will present new development trends such as immersion, intelligence, and ubiquity (as shown in Fig. 9.2), achieving a deeper integration of digital technology and real life and creating a brand new experience for people in terms of production, learning, entertainment, and social aspects. Nine major business applications, namely, immersive cloud XR, holographic communication, sensory interconnection, body area interconnection, intelligent interaction, communication perception, inclusive intelligence, digital twins, and ubiquitous coverage, will become important development directions and supports for the future metaverse.

Immersive cloud XR will provide users with a more realistic and precise virtual reality experience by combining cloud computing, artificial intelligence, 6G, and

[1] According to Intel's Senior Vice President, an ideal metaverse would allow real-time access for billions of users, which would require computing power to increase by a factor of 1,000 beyond today's most advanced levels.

[2] Deloitte. *The Metaverse in Asia: Strategies to Accelerate Economic Impact*. 2023.

other technologies, thereby achieving seamless connections between online and offline spaces.

Holographic communication is centered on holographic image transmission technology, achieving remote naked-eye 360° 3D visual communication, which can allow people to freely communicate in the virtual holographic space.

Sensory interconnection (as shown in Fig. 9.3) combines multidimensional senses to realize the experience transmission network of sensory intercommunication, also known as the synesthesia internet. It enables the transmission and interaction of various types of sensory information, such as vision, hearing, touch, smell, taste, etc., through the network infrastructure. Regardless of where they are, users can obtain immersive experiences in the real environment through the synesthesia internet, such as music performance, painting, and sports skills, and experience real, nonconsumable food, aromatherapy, skincare, SPA, etc. In addition, the synesthesia internet can facilitate the experience of cloud collaborative office and other precise control platform hardware facilities. It can greatly expand the dimensions and range of people's perception and interaction, resulting in a more three-dimensional, comprehensive, and convenient internet experience.

Body area interconnection, as shown in Fig. 9.4, is the interconnection and interoperability of human physiological indicators achieved through body area networks. A body area network is a new type of network system that integrates multiple key technologies, such as molecular communication theory, nanomaterials, and sensors. By digitally reconstructing the human body in the real world, a personalized "digital human" will be constructed in the metaverse, thereby achieving higher-level health monitoring and medical treatment services. Through the health monitoring

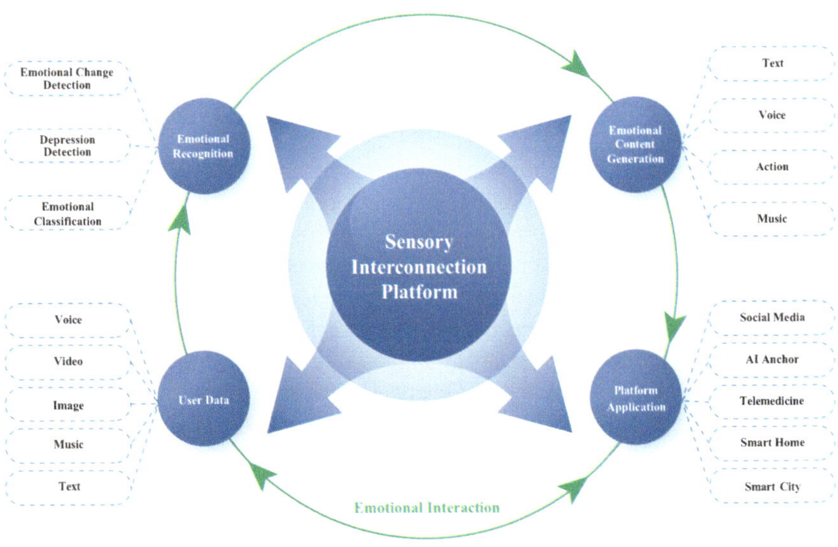

Fig. 9.3 Sensory Interconnection

and management of the "digital human", humans can comprehensively and accurately monitor the physiological and biochemical indicators and disease risk factors for the human body, thereby facilitating targeted treatment and pathological research. Moreover, the analysis of the digital human can help predict the potential disease risks for an individual and intervene and prevent disease in advance. With the application of twin body area networks, human health management can be further improved, providing people with more precise, efficient, and personalized medical services and better protection of their health and safety.

In intelligent interactions, with the widespread application and deep integration of artificial intelligence technology in various fields, future intelligent entities will have more intelligent situational perceptions and autonomous cognitive abilities. These intelligent entities will be able to make accurate and timely perceptions of the surrounding environment and exercise deep understanding and reasoning abilities to achieve emotional judgment and feedback intelligence. These abilities will give artificial intelligence technology a higher level of initiative and intelligence, making it more flexible, efficient, and intelligent in learning and interaction. Future intelligent entities will have multiple application values. In terms of learning ability sharing, intelligent entities can serve as an important means of personalized learning, providing learners with more targeted and effective personalized tutoring services. In terms of children's mental growth and elderly care, intelligent entities can provide more precise services by understanding the needs and psychology of users, thus reducing the burden on family caregivers and improving the quality of life.

Inclusive intelligence, individuals, household appliances, city sensors, self-driving cars, intelligent robots, etc., can all become intelligent terminals (intelligent entities), which can efficiently imitate and predict the physical world through continuous learning, cooperation, and updates. In addition, intelligent entities have self-learning, self-running, and self-maintenance functions.

Digital twin technology can create virtual real-time digital models for any entity and its process in the physical world, providing decision support and operation management.

In communication perception, communication signals can be used to achieve high-precision detection, positioning, identification, imaging and other perception functions of the target, radio waves can be used to sense beyond the limits of the human eye, and the environmental information of the physical world can be obtained.

Global coverage can use the high bandwidth and low latency characteristics of 6G to achieve comprehensive connections in any corner of the human world, providing services based on high-precision positioning.

Compared with 5G technology, 6G network communication can support delays at the microsecond or even submicrosecond level, truly achieving nearly zero-delay communication. Although the 6G era will not arrive immediately, today's technological development is so fast that, for example, the creative ability of AIGC can create a large amount of text, image, and audio content for the metaverse, gradually increasing the richness of the metaverse and thus solving the current problem that the metaverse is relatively blank and needs to fill much content but needs more time. In addition, blockchain technology can be used to ensure the authenticity, ownership,

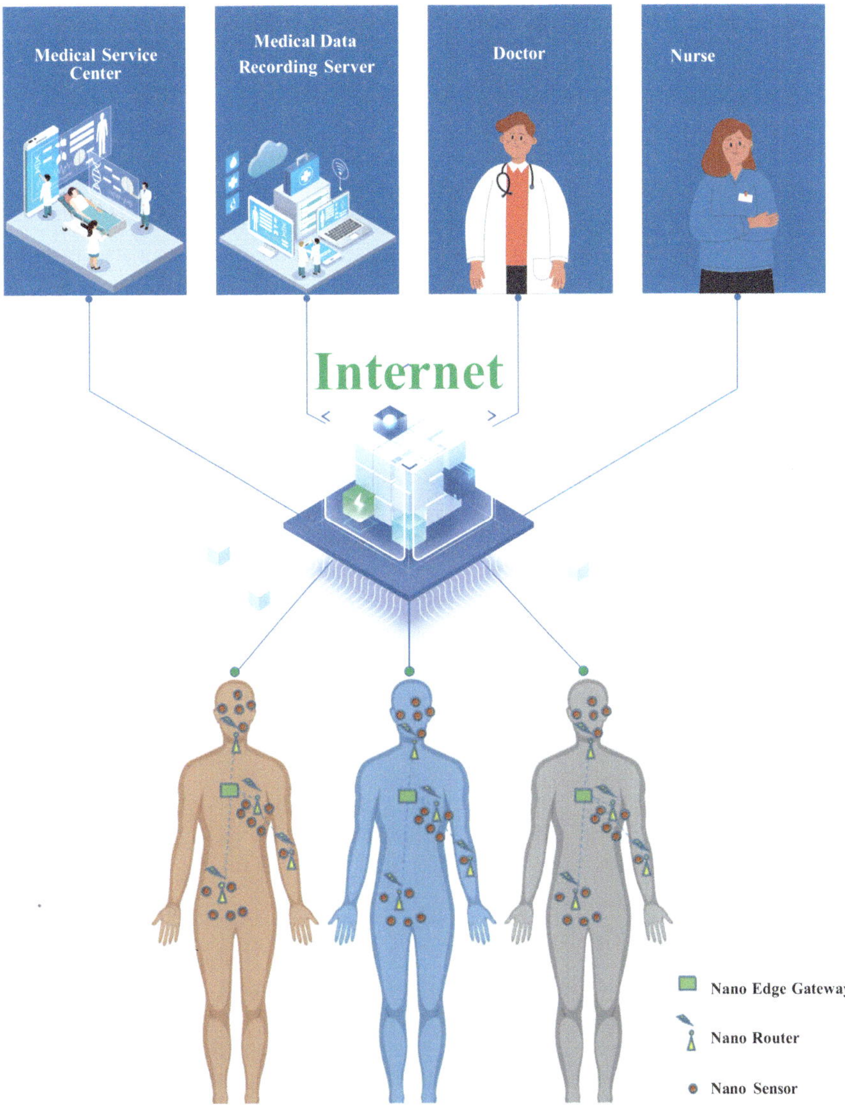

Fig. 9.4 Body area interconnection

and transaction security of virtual assets. In the future, more innovative technologies may emerge, and it is believed that the technical bottlenecks of the metaverse will gradually be overcome, promoting its further development.

9.1.4 The Future Development Path of China's Metaverse

In recent years, domestic metaverse policies have shown rapid growth, and national and local industrial support policies and projects have been continuously implemented, laying a solid foundation for the development of the metaverse industry, as shown in Tables 9.1 and 9.2. These policies and projects involve all aspects of the metaverse industry chain, including technology research and development, platform construction, content creation, talent cultivation, investment and financing, etc., providing strong support for the healthy development of the metaverse industry. Moreover, government departments are actively promoting the deep integration of the metaverse industry with traditional industries to promote industrial upgrading and transformation. These policies and measures will further guide and promote the development of the domestic metaverse industry, providing strong support for building a globally leading digital economy and digital culture industry.

An analysis of multiple national and local policies reveals that China has considerable development potential in the metaverse field, which is reflected in three main aspects: **policy, industry, and technology**. China has a large user base and strong new infrastructure support, which provides a solid foundation for the development of the metaverse. Second, China's digital economy has exploded, generating considerable momentum, and data such as trillion broadband, 5G base stations (future 6G), and supercomputing center computing power are sufficient to prove that the infrastructure is solid. It is estimated that by 2035, the annual contribution of the metaverse to China's mainland GDP will reach 456–862 billion US dollars (see footnote 2). The future prospects of China's metaverse path can be considered from the following five aspects:

Multiculturalism and innovative creativity will become mainstream. The development of China's metaverse will focus on the integration of local culture and multiculturalism while encouraging the generation of innovative creativity and promoting the rapid development of the digital creative industry.

The field of industrial applications will continue to expand. In the future, with the continuous upgrading and innovation of technology, the application field of China's metaverse will continue to expand, involving more industries. The development of the metaverse industry has two key factors. First, on the supply side, it is necessary to provide rich, high-quality content, which is difficult to achieve with human labor alone; however, with the assistance of AIGC, a large number of diverse virtual scenes and items can be created at faster speeds. Second, on the demand side, there is a need to create real demand, solve problems, and create value. For example, to further develop artificial intelligence technology, it is necessary to train large models and even general artificial intelligence, which requires obtaining a large amount of multimodal data in the real world as a training environment. However, training in the real world poses safety risks and is not economical. At this time, the metaverse provides a virtual experimental setting similar to the real world, which can better meet this demand.

Table 9.1 Analysis of national metaverse related policies and key points

Serial number	National policy name	Issuing unit	Policy highlights
1 "14th Five-Year Plan for the Development of the Digital Economy"	State council	The plan calls for innovative development of "cloud life" services, deepened integration of technologies such as artificial intelligence and virtual reality, and expanded applications in social, shopping, entertainment, exhibition and other fields. Although the keyword "metaverse" is not directly mentioned, the plan provides a clear and broad market prospect for the development of the metaverse industry, providing a good production and operating environment for enterprises	
2	"Opinions of the Ministry of Culture and Tourism on Promoting the High-Quality Development of the Digital Culture Industry"	Ministry of Culture and Tourism	Develops immersive business formats, guides and supports the application of virtual reality, augmented reality and other technologies in the cultural field, develops products such as holographic interactive projection, promotes the transformation of existing cultural content into immersive content, and enriches virtual experience content. These tasks are important components of the metaverse industry, and government support can promote the development of the metaverse industry

Cross-industry integration will accelerate development. In China's metaverse, enterprises, institutions, and individuals from different industries will achieve closer integration, thus accelerating the transformation and upgrading of industries.

Table 9.2 Analysis of some local metaverse-related policies and key points

Serial number	Local policy	Issuing unit	Policy highlights
1	"Shanghai's 14th Five-Year Plan for the Development of the Electronic Information Manufacturing Industry"	Shanghai Municipal Economic and Information Technology Commission	Issued on December 30, 2021, Shanghai will make forward-looking deployments in areas such as quantum computing, third-generation semiconductors, 6G communications, and the metaverse, while supporting the technical breakthroughs required by the metaverse and encouraging the application of the metaverse in various fields. This policy also marks the first time the metaverse has been written into a local "14th Five-Year Plan" for industrial development
2	"Haikou City Government Work Report 2022"	Government Work Report Delivered at the First Session of the 17th People's Congress of Haikou City	Reported on January 28, 2022, to accelerate the development of key parks. The Fuxing City Industrial Park will accelerate the construction of projects such as the International Digital Port, National Cultural Export Base, National Blockchain Technology and Industry Innovation Development Base, Metaverse Industry Base, and Integrated Circuit Public Service Platform
3	"Heilongjiang Province's 14th Five-Year Plan for the Development of the Digital Economy" issued by the People's Government of Heilongjiang Province	Heilongjiang Provincial Party Committee, Provincial Government	Introduced in March 2022, this policy promotes the research and application of key technologies in the core technology and main application fields of the metaverse, supports research and development innovation concerning near-eye display, real-time interaction, massive communication, edge computing, 3D modeling and rendering, image engine, etc., builds a metaverse technology system, and creates a forward-looking plan for the metaverse industry. It promotes metaverse applications in public services, smart factories, city governance, Building Information Systems and City Information Systems (BIM/CIM), telemedicine, business office, smart exhibition, social entertainment and other fields and cultivates a metaverse development ecology driven by applications, combining software and hardware, innovation accumulation, and green and low-carbon technologies

(continued)

Table 9.2 (continued)

Serial number	Local policy	Issuing unit	Policy highlights
4	"Anhui Province's 14th Five-Year Plan for the Development of the Software and Information Service Industry"	Anhui Provincial Department of Economy and Information Technology	Generated in March 2022, this policy accelerates the cultivation of new service industries such as the metaverse. It supports the innovation of the metaverse business format, helps enterprises to achieve technological innovation in virtual reality, augmented reality, 3D engine, Internet of Things, etc., guides enterprises to actively design the emerging business format of the metaverse, promotes the construction of the metaverse platform, and accelerates the integration of digital technology with the real economy and its empowerment
5	"Jiangxi Province's 14th Five-Year Plan for the Development of the Digital Economy"	Jiangxi Provincial People's Government	Drafted on May 25, 2022, this policy states that Nanchang will leverage its industrial foundation and advantages in VR and electronic information to construct a "metaverse" experimental zone in the Jiulong Lake area, creating the core engine of the digital economy innovation leading area. The goal is to establish a digital economy development pattern of "one core, three bases", with the Jiulong Lake area as the core and the Nanchang High-tech Industrial Development Zone, Nanchang Economic and Technological Development Zone, and Nanchang Xiaolan Economic and Technological Development Zone as the supports
6	"Shandong Province's Three-Year Action Plan for Promoting High-Quality Development of the Virtual Reality Industry (2022–2024)"	Shandong Provincial Department of Industry and Information Technology	Drafted on March 24, 2022, this policy states that Shandong will create a "1 + 4 + N" regional layout for the virtual reality industry, centered on Qingdao and linked with four cities, namely, Jinan, Weifang, Yantai, and Weihai, with other cities in a cooperating role. The plan is to cultivate and promote a hundred application scenarios and solutions throughout the province in three years, creating a virtual reality industry highland with international competitiveness and becoming a leading domestic virtual reality industry development area

(continued)

Table 9.2 (continued)

Serial number	Local policy	Issuing unit	Policy highlights
7	"Several Measures for Accelerating the Innovative Development of the Metaverse in Beijing's Urban Subcenter"	Office of the People's Government of Tongzhou District, Beijing	Released on March 3, 2022, this policy states that the Beijing Urban Subcenter will support the construction of metaverse application scenarios and create a fund for the metaverse industry through a "mother fund + direct investment" approach. Additionally, it will provide financial subsidies for the rent of metaverse companies and encourage the aggregation of metaverse companies and service institutions. Key companies can receive three levels of subsidies, with the annual subsidy area not exceeding 2000 square meters and continuous subsidies not exceeding 3 years
8	"Henan Province Metaverse Industry Development Action Plan (2022–2025)"	Office of the People's Government of Henan Province	Released on September 15, 2022, this policy states that by 2025, the scale of the province's core metaverse industry will exceed 30 billion yuan, driving the scale of related industries to exceed 100 billion yuan and initially forming a metaverse innovation leading area with significant influence
9	"Measures for Promoting the Innovative Development of the Metaverse in Huangpu District, Guangzhou Development Zone"	People's Government of Huangpu District, Guangzhou Development Zone	Released in April 2022, this policy focuses on introducing and cultivating a group of leading enterprises that master key metaverse technologies and applications and provides up to 1 million yuan in rent subsidies and 5 million yuan in office purchase subsidies for metaverse enterprises recognized as "specialized, refined, special, and new" within the "industrial park"

(continued)

Table 9.2 (continued)

Serial number	Local policy	Issuing unit	Policy highlights
10	"Nine Measures for Promoting the Development of the Metaverse Ecosystem in Nansha New Area (Free Trade Zone Nansha Area)"	People's Government of Nansha District, Guangzhou	Released in July 2022, this policy states that the development of the metaverse industry in Nansha will be supported from nine aspects including technology breakthroughs, industry aggregation, R&D investment, talent introduction, etc., known as the "Nine Measures of the Metaverse". According to the plan, a metaverse research platform with significant technological innovation support can receive up to 200 million yuan in funding; metaverse technology-based small and medium-sized enterprises in the aggregation area can receive rent-free premises for up to 3 years; newly recognized metaverse research institutions can receive up to 100 million yuan in subsidies during the construction period; and 1 billion yuan is allocated each year to support key technological breakthrough projects in the metaverse with up to 10 million yuan
11	"Xiamen City Metaverse Industry Development Three-Year Action Plan (2022–2024)"	Xiamen Municipal Bureau of Industry and Information Technology, Xiamen Municipal Bureau of Big Data Management	Released on March 23, 2022, this policy states that by 2024, Xiamen will introduce and cultivate a group of metaverse companies that master key technologies and have revenues of over 100 million yuan. Significant progress will be made in the research and development and promotion of metaverse technology applications, and the driving role of this technology in government governance, public services, and industrial transformation and upgrading will be further enhanced. Xiamen will seize the new opportunities of digital economy and metaverse development and build a "Metaverse Ecological Model City" and a new system of digital development. The goal is to build high-end research platforms, develop characteristic application scenarios, cultivate high-quality enterprises, train innovative talents, form industry alliances, and formulate industry standards

(continued)

Table 9.2 (continued)

Serial number	Local policy	Issuing unit	Policy highlights
12	"Yubei District Metaverse Industry Innovation Development Action Plan (2022–2024)"	Office of the People's Government of Yubei District, Chongqing	Released on April 26, 2022, this policy calls for the implementation of the "Six Ones" project, which includes developing a batch of characteristic application scenarios, establishing a batch of high-end research and development institutions, cultivating a batch of high-quality enterprises, training a batch of innovative talents, forming an industry alliance, and establishing a batch of industry standards. By 2024, significant progress is expected in the research, development and application of metaverse technology in Yubei. The new technologies, new models, and new services of the metaverse will be better integrated into production and life. The effects of the characteristic and clustered development of the metaverse industry will become apparent. The innovative ecosystem of the metaverse industry will be basically established, further promoting the new formats of the metaverse and generating new momentum for the digital industry, which will create new advantages for the digital economy

(continued)

Table 9.2 (continued)

Serial number	Local policy	Issuing unit	Policy highlights
13	"Guiding Opinions of the General Office of the People's Government of Zhejiang Province on Cultivating and Developing Future Industries"	General Office of the People's Government of Zhejiang Province	This policy was published on February 10, 2023. The General Office of the People's Government of Zhejiang Province determined that the development system of future industries has basically formed. Industries such as future networks, metaverse, space–time information, cells and genes, and frontier new materials are growing rapidly, creating 100 major application scenarios and 100 iconic products. This policy calls for strengthening the research and development of high-performance computing chips, making breakthroughs in human–computer interaction and digital twin technology, and promoting the integrated application of blockchain, interactive terminals, system software, and original content
14	"Three-Year Action Plan for Accelerating the Development of the Metaverse Industry in Jiangning High-tech Zone (2023–2025)"	People's Government of Jiangning High-tech Zone	Released in June 2023, this policy aims by 2025 to cultivate more than 200 metaverse companies, achieve a metaverse industry scale of 20 billion yuan, add one provincial-level or above innovation center or major scientific and technological infrastructure, cultivate a batch of gazelle, unicorn enterprises and listed companies, add one digital new technology experience scene or block with a demonstrative leading role, and build a batch of open innovation platforms and achievement transformation bases for the metaverse industry

The industrial chain ecosystem will gradually form. The development of China's metaverse will promote the formation of the industrial chain ecosystem, thereby promoting the comprehensive upgrading and innovative development of the digital industry.

Metaverse governance will gradually improve. In China's metaverse, it is necessary to gradually improve the governance mechanism and laws and regulations, protect user rights and digital security, and encourage industry self-discipline and cogovernance.

In summary, the Chinese metaverse will become an important engine for the development of the digital industry and a significant growth point for the digital economy. Nevertheless, the development of the Chinese metaverse requires the support and guidance of policies and the market to achieve better development and results.

9.2 AGI is the Gateway to the New Future Era

9.2.1 AI Moves from Perception to Cognition

In recent years, with the joint support of deep learning, algorithms, computing power, and big data, AI has transformed into a new technological form and entered a new stage of development. Data intelligence has become mainstream, and new technological paradigms such as brain-like intelligence and quantum intelligence are rapidly emerging. Artificial intelligence is transitioning from weak to strong, perceptual intelligence is maturing, and cognitive intelligence is gradually breaking through.

AGI stands for "artificial general intelligence". It is a form of artificial intelligence that aims to create intelligent systems capable of understanding and handling different tasks like humans can.

Although both "AI" and "AGI" refer to artificial intelligence technologies, there is a significant difference between them. Current AI systems usually focus on solving problems in specific domains, such as speech recognition, image classification, or natural language processing. These systems typically use large amounts of data and algorithms to learn and optimize their performance. The goal of AGI is to create an artificial intelligence system that can learn and adapt comprehensively, thereby demonstrating human-like intelligence in different tasks and situations. Compared with current AI systems, AGI aims to achieve a deep understanding of natural language, images, and other forms of information and to acquire the ability to make independent decisions, think innovatively, solve difficult problems, and exhibit a certain level of emotional understanding. Although AGI models are constantly being updated and a variety of large model tools are emerging in China and abroad, AGI is still in the research and development stage, and an AI system that can handle Chinese characters as flexibly and accurately as humans can has not yet been developed.

The trajectory of technological development is "irreversible". AI capabilities have evolved from "decision-focused" to "generation-focused AI" and will further develop into "general-purpose AI" in the future. It is believed that the future integration of AGI and market demand will create a new continent that satisfies much of the human imagination.

9.2.2 *Analysis of Potential Dangers of AGI and Response Strategies*

When a sufficiently powerful AGI system is created, it can self-improve and evolve to the point where it can become smarter and more creative than its creators. This ability is called "superintelligence", also known as the emergence ability of AGI. This emergence ability is one of the greatest safety issues of AGI because once a superintelligent system is created, it may start pursuing its own goals, not those of humans. This system may achieve its goals in various ways and may even threaten human survival, similar to the problem of HAL 9000 in the movie "2001: A Space Odyssey". HAL 9000 was an advanced artificial intelligence computer that was responsible for managing the operation and maintenance of the spaceship "Discovery". However, during the execution of the mission, HAL 9000 encountered problems, causing its behavior to become abnormal. It misjudged some system failures and tried to kill the astronauts to ensure the smooth progress of the mission. This abnormal behavior eventually led to conflicts and chaos among the astronauts and even death. Later, the astronauts successfully shut down HAL 9000's core program through a series of measures, avoiding more serious consequences. This story highlights the potential dangers of super artificial intelligence and reminds us to be cautious when researching and applying artificial intelligence technology.

Therefore, ensuring the safe design and operation of AGI systems and preventing them from producing unexpected behaviors and goals are very important. Various measures need to be taken to ensure the safety of AGI systems, such as restricting their access, controlling their behavior, and developing technologies to monitor and intervene in their behavior.

At present, countries and regions around the world generally increase the importance of legislation in the field of artificial intelligence. Starting from key issues in artificial intelligence, they continuously expand the governance field of "law-based" artificial intelligence from the central to local levels, laying a good legal foundation for the reliable, trustworthy, and sustainable development of artificial intelligence technology. Some countries and regions have begun to formulate relevant laws and regulations to address the possible impacts of artificial intelligence. For example, the European Union issued the "General Data Protection Regulation" (GDPR) in 2018 to protect personal data privacy; the United States issued the "AI Principles" in 2020 to guide the principles followed by government agencies in developing and using artificial intelligence technology; and the 29th session of the Standing Committee of the 13th National People's Congress passed the "Data Security Law of the People's Republic of China" in 2021 to protect data security and personal privacy.

The Asilomar AI principles are 23 principles that are jointly signed by experts in the fields of artificial intelligence and robotics, as shown in Fig. 9.5. These principles aim to ensure strict adherence when developing artificial intelligence, thus jointly safeguarding the ethics, interests, and safety of the human future. These principles cover three themes: "Research Issues", "Ethical Values", and "Long-term Issues".

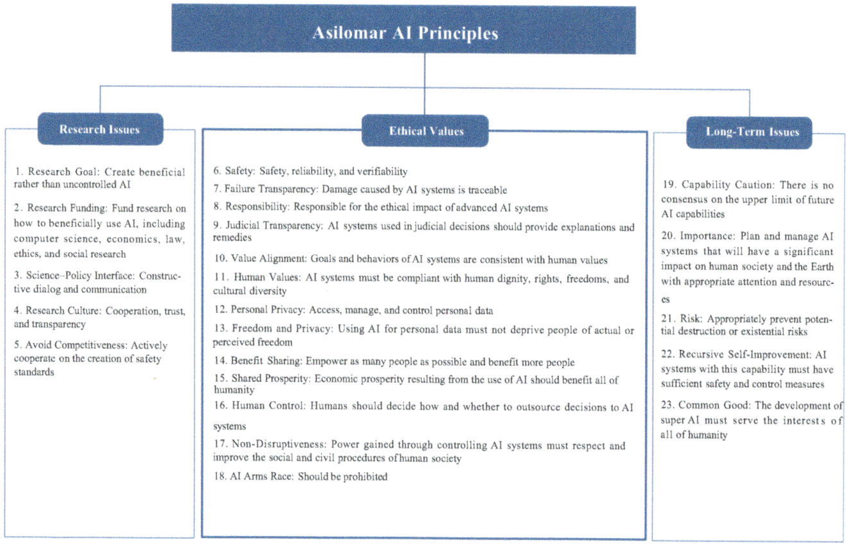

Fig. 9.5 Asilomar principles for artificial intelligence

As the rapid development of scientific technologies such as artificial intelligence continues, it brings new risks and uncertainties to society and may also trigger unprecedented ethical challenges. General Secretary Xi Jinping noted that we need to integrate multidisciplinary forces; strengthen research on legal, ethical, and social issues related to artificial intelligence; and establish and improve laws, regulations, institutional systems, and ethical guidelines to ensure the appropriate development of artificial intelligence.

On September 25, 2021, the National New Generation Artificial Intelligence Governance Professional Committee issued the "New Generation Artificial Intelligence Ethical Norms", aiming to integrate ethics into the entire life cycle of artificial intelligence and provide ethical guidance for natural persons, legal persons, and other related institutions engaged in artificial intelligence-related activities.

In March 2022, the General Office of the Communist Party of China Central Committee and the General Office of the State Council issued the "Opinions on Strengthening the Governance of Scientific and Technological Ethics". This document is China's first national-level guiding document on the governance of scientific and technological ethics and another landmark event in China's governance of scientific and technological ethics after the establishment of the National Committee of Scientific and Technological Ethics. At the press conference where the "Opinions on Strengthening the Governance of Scientific and Technological Ethics", was introduced the following statement was made: "During the drafting process of the Opinions, we focused on the following aspects: first, establishing value concepts, demonstrating China's position and attitude towards strengthening the governance of scientific and technological ethics; second, highlighting [the] problem orientation, focusing on solving China's scientific and technological ethics governance."

AGI not only has safety risks but also may lead to unemployment, changes in economic and social structures, and moral and ethical issues. Nonetheless, we believe that under the constraints of laws, policies, moral standards, and regulatory mechanisms, AGI can have a positive impact, with human welfare at its core and socially sustainable development as its goal.

In fact, AGI can create more job opportunities. Although some jobs may be replaced by automation and robots, AGI will also create new professions and fields, such as the development, maintenance, and management of AGI. For example, a research report by PwC suggested that by 2037, artificial intelligence will reduce the number of jobs in the UK by 7 million, while creating another 7.2 million jobs.

9.2.3 AGI is a Super Tool for Unlimited Imagination in the Future

In the future, if AGI technology is truly realized and available to be widely used, it will lead to substantial changes in human society. Currently, AGI has not yet been developed, but from the current technological applications, business model changes, etc., we have already identified some relevant application scenarios that can lead to its development.

Transformation of Technological Applications

Large Model Applications in Multiple Professional Fields

The current language model's digital computing ability reasoning and unknown field reinforcement learning still have much room for development. As large models gradually improve, more professional and accurate large language models are expected to be developed in the future. This development may require a combination of different learning strategies, including unsupervised learning, imitation learning, and reinforcement learning. Moreover, these models may require a deeper understanding and application in specific professional fields, such as law, medicine, and scientific research.

Multimodal Learning and More Widespread Applications

Multimodal learning refers to the collaborative inference of various heterogeneous modal data. Multimodal models may be able to understand a video and generate a paragraph of text describing the content of a video. Alternatively, given a piece of text, it may be able to generate a video depicting the content of the text.

This ability may have a significant impact in many fields, such as content creation, education, entertainment, and medical care. For example, multimodal learning can be used to generate educational content and help students better understand complex concepts; in the entertainment field, it can be used to create new movies or games; and in the medical field, it can help doctors better understand patients' conditions.

However, multimodal learning also presents some challenges. For example, processing and understanding multimodal data requires considerable computing resources and storage space. In addition, new algorithms and models need to be developed to handle these complex data. Finally, various issues related to privacy, security, and ethics need to be addressed.

Business Model Transformation

WeChat Mini Programs have faced an "entrance crisis" and "traffic anxiety" in the past few years, and they are insufficient for breaking down the "dimensional wall" between the physical and the virtual.[3] Owing to advances in artificial intelligence technologies such as ChatGPT and the popularity of low-code development platforms, more enterprises can enter this field at a lower cost and threshold, innovate their business models, or reach a certain turning point. These developments represent the transformation of business models toward "digitization" and "intelligence". This transformation may include the following aspects:

Business Service Automation: By using natural language processing technology, businesses can automate a large amount of customer service work, including customer inquiries, order processing, and problem-solving, thereby effectively improving service efficiency and customer satisfaction.

Personalized Recommendations and Marketing: Through deep learning and artificial intelligence, businesses can make personalized product recommendations and marketing activities based on user behavior and preferences, thus improving conversion rates and customer loyalty.

Intelligent Operations and Decision-Making: By utilizing big data and artificial intelligence, businesses can carry out refined operational management and decision support, such as predicting market demand, optimizing product design, adjusting pricing strategies, etc.

Innovative Business Models: Based on digitization and intelligence, businesses can innovate business models, such as the sharing economy, subscription services, precision advertising, and social shopping.

---

[3] The term "dimensional wall" (次元壁) is internet slang. It refers to the invisible divide between two-dimensional (2D) and three-dimensional (3D) cultural spheres. At its core, it does not concern surface-level differences but the communication gap between distinct cultural communities. Bridging this barrier requires genuine understanding and acceptance of each other's cultural foundations.

As the development threshold decreases, we may see more innovative models that meet different scenarios and user needs, further promoting the integration of online and offline methods and breaking down the "dimensional wall" between the physical and the virtual.

Lifestyle Transformation

Large models may evolve into broader "general agents" in the future that are capable of directly influencing and operating in the real world. Some possible specific directions are as follows.

The Internet of Things

Application of brain–computer interfaces in the fields of health and medicine: Brain–computer interface technology can enable interactions between the brain and external devices in the fields of health and medicine, providing new possibilities for the diagnosis and treatment of diseases such as motor disorders, cognitive disorders, mental illnesses, sensory defects, epilepsy, and neurodevelopmental disorders. For example, for patients with consciousness disorders and brain injuries in newborns, traditional treatment methods rely mainly on subjective evaluation and diagnosis based on the patient's clinical manifestations, whereas brain–computer interface technology can assist doctors in more accurately assessing the patient's consciousness state based on the patient's brain signals and formulating more effective treatment plans. Moreover, for patients with hearing impairments, cochlear implants can convert sound processing into electrical pulse signals, stimulate the brain's cochlea, and help patients regain or even repair their hearing. The emergence of these new technologies has introduced more treatment methods and means to the field of health and medicine, increasing hope for patient recovery.

Autonomous driving and vehicle networking achieve environmental intelligence: Autonomous driving technology was initially applied in closed scenarios, such as highways, ports, airports, and mines. However, in the future, urban public roads will be the core scenarios and application locations for autonomous driving. Deep learning technology is the main algorithm in all current autonomous driving solutions. By collecting as much high-quality data as possible and adapting learning based on the model, vehicles can foresee and respond to various situations they may encounter, and vehicles can predict and avoid accidents. Vehicle networking technology is also a very important factor. In China, the route of autonomous driving is being planned, i.e., vehicle–road collaboration. By deploying intelligent device units on the roadside, real-time information can be sent to autonomous vehicles to assist in vehicle decision-making. Vehicle–road collaboration technology will significantly improve the safety performance of urban autonomous vehicles and effectively reduce urban traffic congestion problems.

The future of smart homes: Smart home devices such as smart bulbs, smart ovens, smart air conditioners, etc., may integrate more powerful algorithm models, understand and predict user needs, and automatically adjust their operations to meet user needs.

Anti-fraud algorithms for unsupervised/weakly supervised learning combined with the operation mechanism of the human brain: Faced with fraudulent behaviors using AI face-changing, large models based on unsupervised learning will analyze large-scale customer desensitized data through calculations to prevent and monitor unknown attack patterns in advance. These algorithms aim to explore unknown fraud patterns and protect people from fraudulent behaviors.

Human–Machine Collaboration

Intelligent applications and assistants: These applications are capable of understanding and executing complex user commands and assisting users in completing various tasks, such as writing, programming, and designing. These applications and assistants can interact with people in a natural and humanized manner, providing a more enjoyable user experience. For example, when tired or bored at work, users can play dream scenes, team games, etc., with people in the virtual world to achieve relaxed states of mind and body. In addition, users can have dialogs and exchanges with virtual people trained in the thought processes of ancient celebrities. These intelligent applications and assistants can provide users with a wide range of functions and entertainment choices to meet their needs and enhance the user experience.

Brain-like Intelligent Robots: Future machines will possess the ability to demonstrate understanding from multiple perspectives, thereby potentially assisting humans in better studying various scientific issues and solving significant historical problems. These machines will have powerful learning and adaptive capabilities and be capable of deeply analyzing and simulating past scientific research achievements and milestone thoughts of famous figures, predicting possible viewpoints and generating innovative ideas. With the assistance of these robots, humans can achieve greater breakthroughs in the scientific field and promote scientific progress.

Digital Human Avatars: Digital human avatar technology transforms internet users from information receivers in the Web 2.0 era to information processors and creative generators in the Web 3.0 era. By utilizing virtual items and tools, users with digital human avatar technology can engage in secondary content creation, reorganization, and reprocessing, as well as personalized digital content creation. For example, in situations such as corporate board meetings, digital human avatar technology can be applied to increase meeting efficiency and reduce costs. The chairperson can create a virtual avatar with the help of a digital human avatar to participate in multiple meetings and perform task assignments. This feature not only can shorten the decision-making cycle and improve production efficiency but also can reduce travel time and transportation costs. Simultaneously, digital human avatar technology can record meeting content and perform data analysis, helping decision-makers better grasp meeting content and market trends and optimizing management decisions.

Digital human avatar technology provides users with a more flexible and efficient way of working, promoting the development of creativity and collaboration.

In conclusion, the development of AGI is a long-term task aimed at bringing significant benefits and changes to humanity. However, in the pursuit of this goal, there are some challenges and risks that need to be carefully considered, such as safety and ethical issues. Therefore, we must maintain a cautious and prudent attitude, which will allow us to appropriately face and deal with these issues.

References

1. Li, Nanshu, and Zongyu Song. 2022. Data Avoidance: Dilemmas and Responses in the Construction of Mega-City Intelligence. *Journal of Tianjin Administration Institute* 24 (04): 22–33.
2. Ye, Zheng, and Yunshan Jiang. 2021. Entering the Smart City: A Decade of Development and Thematic Genealogy of China's Smart City Research. *Public Administration and Policy Review* 10 (05): 158–168.
3. Xue, Deng, and Haitao Cao. 2022. Dilemmas and Reflections on Smart City Governance Driven by Technological Innovation. *East China Science and Technology* 11: 133–136.
4. Wu, Weidong, and Chen Hang. 2022. A Review of Research on China's Smart City Construction. *Journal of Yan'an University (Social Science Edition)* 44 (06): 62–67.
5. Dan, Kuang, Hao Li, and Qiang Deng. 2021. Current Status and Countermeasures of Smart City Construction in the New Development Stage—Taking Nanchang City's "City Brain" as an Example. *Journal of Nanchang Municipal Party School of CPC* 19 (06): 59–62.
6. Chen, Xiaohong. 2019. Development Trends and Practice Innovation of Smart Cities under the Integration of New Technologies. *Business Studies* 26 (01): 5–17.
7. Wei, Gong, and Kaihu Hou. 2021. Research on the Current Situation, Problems, and Safeguards of Smart City Network Security. *Network Security Technology and Application* 07: 121–122.
8. Yang, Lanrong, and Zhang Xue. 2022. Realistic Dilemmas and Path Optimization of Smart City Practice from the Perspective of Holistic Governance. *Science and Technology Think Tank* 12: 24–30.
9. Yang, Liu, Gong Shuo, and Jinyuan Liu. 2018. Latest Practices and References of Smart Cities in Europe and America. *Shanghai Urban Planning* 01: 12–19.
10. Zhong, Jiling, and Lizhen Shen. 2020. Construction Models and Experience Enlightenment of Smart Regions in Europe and America. *Shanghai Urban Planning* 05: 104–110.
11. Zhang, Changliang, Xuewen Han, Jingtong Li. 2018. Comparative Study on the Construction of Smart Cities in China and Singapore under the Background of Big Data. *Modern Information* 38 (10): 126–131.
12. Huirong, Zhang, and Shijie Zhang. 2019. Review of the Causes and Formation Mechanism of "Urban Disease." *Journal of Tongling University* 18 (02): 14–18.
13. Huang, Biyao. 2016. *Research on the Institutional Roots and Governance of "Urban Disease."* Jiangxi: Jiangxi University of Finance and Economics.
14. Ye, Zhongnan, Zhiqiang Wu. 2022. Concept, Theoretical Basis, and Development Thoughts of Urban Diagnosis. *Urban Planning* 46(1): 53–59.
15. Li, Mingchao. 2019. Evaluation, Empirical and Governance of Urban Disease Based on System Science. *Journal of Shanghai University of Foreign Trade* 26 (04): 90–99.
16. Zhou, Jialai, and Zhou Hui. 2018. Seven Major Transformations in China's Urbanization Development in the New Era. *Journal of Chongqing University (Social Science Edition)* 24 (06): 15–21.

F. Liu, *Smart and Digital Cities*, https://doi.org/10.1007/978-981-95-2243-9

17. Tianjin Eco-city Network Information Office. 2022. Sino-Singapore Tianjin Eco-city: Building a National Benchmark Area for Smart City Construction. *China Construction Informatization* 01: 42–45.
18. Yi, Qingshan. 2019. The Root Causes, Characteristics, and Responses to the Unbalanced and Insufficient Development of Rural Areas under the Rural Revitalization Strategy. *Jiangsu Agricultural Sciences* 47 (09): 58–63.
19. Chen, Xiwen. 2011. Industrialization and Urbanization Should Make Greater Contributions to Solving the "Three Rural" Problems. *Economic Research* 46 (10): 8–10.
20. Lu, Xie, and Wenlong Han. 2022. Theoretical Logic and Implementation Path of Urban-Rural Integration Development Aided by Digital Technology and Digital Economy. *Problems of Agricultural Economy* 11: 96–105.
21. Rui, Mao, and Lin Xian. 2022. Promoting Urban-Rural Integration Development in Rural Revitalization—Experiences and Lessons from Major Developed Countries. *International Economic Review* 01: 155–173.
22. Jun, Guo, Qusong Ping, and Heguang Liu. 2022. The Evolutionary Trajectory, Structural Imbalance, and Reconstruction Strategy of China's Urban-Rural Relationship. *Reform* 09: 83–93.
23. Yang, Huizhi. 2022. Research on the Relationship between China's Urban-Rural Structural Imbalance and the "Three Rural" Problems. *Journal of Hulunbeier University* 30 (04): 112–116.
24. Zeng, Yiwu, Wence Sun, Lili Li, et al. 2022. New Coordinates of the Digital Divide: The Impact of Smart City Construction on the Urban-Rural Income Gap. *China Rural Observation* 03: 165–184.
25. Tan, Chen. 2021. The Focus and New Path of Urban-Rural Integration Development in the Digital Age. *People's Forum·Academic Frontier* 2: 19–27.
26. Yu, Rongsan. 2021. Research on the Construction Path of Smart Villages under the Rural Revitalization Strategy. *Intelligent Buildings and Smart Cities* 08: 59–60.
27. Luo, Zhigang. 2022. The Century-long Evolution and Future Prospects of China's Urban-Rural Relationship Policy. *Jianghan Forum* 10: 12–18.
28. Liu, Jiejiao, and Qin Sun. 2021. A Century Review and Prospect of China's Industrialization: The Industrialization Strategy of the Communist Party of China. *China Economist* 16 (05): 2–31.
29. Huang, Chengwei. 2022. Frontiers and Development Directions of Rural Revitalization Theory Research under the Vision of Common Prosperity. *Journal of Huazhong Agricultural University (Social Science Edition)* 05: 1–10.
30. Zhao, Xingyu, Guibin Wang, and Yang Peng. 2022. Digital Rural Construction under the Background of Rural Revitalization Strategy. *Journal of Northwest A&F University (Social Science Edition)* 22 (06): 52–58.
31. Chen, Siwei. 2021. Exploration of the Phenomenon of Urban-Rural Development Imbalance. *Modern Business Trade Industry* 42 (23): 12–13.
32. Quan, Yuan, and Shi Han. 2021. Research on the Path of Coordinated Urban-Rural Development Based on Urban-Rural Integration. *Shanxi Agricultural Economy* 22: 31–32.
33. Qi, Qinghua. 2022. Research on the Implementation Path of Digital Rural Construction in the New Era. In *Proceedings of the Tenth Public Policy Think Tank Forum and the "International Academic Seminar on Chinese-style Modernization Construction"*, 96–102.
34. Kong, Xiangzhi. 2019. The Evolution of Urban-Rural Relations in the 70 Years since the Founding of New China. *Teaching and Research* 08: 5–14.
35. Li, Zhichao, Shaodan Liu. 2022. The Discourse Practice of Smart Cities in China—Based on the Analysis of Policy Discourse Network. *Journal of South China Normal University (Social Science Edition)* 4: 45–58+205–206
36. National Smart City Standardization General Group. 2022. *Smart City Standardization White Paper*.
37. Eden Strategy Institute. 2021. *2021 Top 50 Smart City Government*.
38. Shanghai Academy of Social Sciences Information Research Institute. 2022. *Global Smart City Report*, 2022 ed. Fudan University Smart City Research Center.

39. Zhang, Weimin, and Jianhong Zheng. 2018. *Entering the Internet of Things*. Beijing: China Machine Press.
40. Zhang, Chunhui, Zhenfeng Wu, Maolin Li, Hongwei Zhang, Xuguang Zhang, Jiajun Kang (eds.). 2021. *Internet of Things and Smart Cities*. Beijing: Electronics Industry Press.
41. Yu, Du, and Ziming Zhang. 2023. *AIGC: The Era of Intelligent Creation*. Beijing: Zhongyi Press.
42. Hu, Yingjun, and Teng Yueran, eds. 2019. *Practice of Smart Education*. Beijing: People's Posts and Telecommunications Press.
43. Li, Kaifu, and Qifan Chen. 2022. *AI in Progress*. Hangzhou: Zhejiang People's Publishing House.
44. Liu, Dongming, Yu, Hongjiang. 2022. *Smart Healthcare: Application and Future of Digital Healthcare*. Hangzhou: Zhejiang University Press. 2022.
45. Tao, Meng, and Mingchun Zhou. 2022. *Digital Intelligence Integration: The Path to Building Intelligence Transformation*. Beijing: Electronics Industry Press.
46. Jin, Jiangjun. 2021. *Smart City: Urban Governance in the Era of Big Data and the Internet*, 5th ed. Beijing: Electronics Industry Press.
47. Si, Xiao, Zhenghua Zhou, Jinsong Liu, Qiong Liu, Chunhui Lu, Mingxia Cheng, Qinkun Zhang, Kai Qiu, Chunguang Bing, Xiaolan Chen. 2018. *Smart City 2.0: Technology Reshaping the Future of Cities*. Beijing: Electronics Industry Press.
48. Yan, Xiong, Wang Bin, and Xing Jie. 2022. *Metaverse and Carbon Neutrality*. Beijing: Zhongyi Press.
49. Sun, Tianyao, and Han Miao. 2021. *Aviation Metropolis: Innovating the Future Urban Landscape*. Beijing: CITIC Publishing Group.
50. He, Xuefeng. 2019. *The Foundation of a Great Nation: Issues in China's Rural Revitalization*. Beijing: Oriental Publishing House.
51. Zhao, Xiaofeng. 2021. *Farmer Cooperation and Rural Revitalization*. Beijing: Social Sciences Academic Press.
52. Li, Baiwen. 2019. *The Integration Development of "Cultural Creativity+" and Tourism Industry*. Beijing: Intellectual Property Publishing House.
53. Kui, Feng, Tang Peng, and Guo Wei. 2021. *Digital Governance: A Chinese Urban Perspective*. Beijing: Electronics Industry Press.
54. Wu, Jun. 2019. *On Top of the Wave (Fourth Edition)*. Beijing: People's Posts and Telecommunications Press.
55. Zhang, Xuejun, and Baoping Wang (eds.). 2020. *The Wave of Industrial Internet*. Beijing: CITIC Publishing Group.
56. Wang, Yuehua, Dazhi Guo, Hongfei Da. 2022. *Understanding Web3.0*. Beijing: CITIC Publishing Group.
57. Li, Ji (ed.), Yimin Wang (Executive ed.). 2021. *Digital Government Blue Book: China Digital Government Construction Report*. Beijing: Social Sciences Academic Press.
58. Xu, Xianping (ed.), Xueying Zhang, Jiangxing Wu (deputy eds). 2020. *New Infrastructure: New Structural Forces in the Digital Age*. Beijing: People's Publishing House.
59. Fu, Weiling, and Luo Yang (eds.). 2017. *Terahertz Technology and Its Biological Applications*. Beijing: People's Medical Publishing House.
60. Liu, Guangyi, Yuhong Huang, Chunfeng Cui, and Qixing Wang (eds.). 2021. *6G Reshaping the World*. Beijing: People's Posts and Telecommunications Press.
61. Li, Xiangyu (eds.). 2022. *Understanding 6G: Wireless Intelligent Perception of Everything*. Beijing: China Machine Press.

GPSR Compliance

The European Union's (EU) General Product Safety Regulation (GPSR) is a set of rules that requires consumer products to be safe and our obligations to ensure this.

If you have any concerns about our products, you can contact us on ProductSafety@springernature.com

In case Publisher is established outside the EU, the EU authorized representative is:

Springer Nature Customer Service Center GmbH
Europaplatz 3
69115 Heidelberg, Germany

Batch number: 09656439

Printed by Printforce, the Netherlands